VOLUME TWO HUNDRED AND SEVENTEEN

ADVANCES IN IMAGING AND ELECTRON PHYSICS

VOLUME TWO HUNDRED AND SEVENTEEN

ADVANCES IN IMAGING AND ELECTRON PHYSICS

Quantitative Atomic-Resolution Electron Microscopy

ANNICK DE BACKER

JARMO FATERMANS

ARNOLD J. DEN DEKKER

SANDRA VAN AERT

Edited by

MARTIN HŸTCH and PETER W. HAWKES

CEMES-CNRS

Toulouse, France

ELSEVIER

ACADEMIC PRESS

An imprint of Elsevier

Cover photo credit:
"Atom counting" by Annick De Backer

Academic Press is an imprint of Elsevier
125 London Wall, London EC2Y 5AS, United Kingdom
525 B Street, Suite 1650, San Diego, CA 92101, United States
50 Hampshire Street, 5th Floor, Cambridge, MA 02139, United States
The Boulevard, Langford Lane, Kidlington, Oxford OX5 1GB, United Kingdom

ISBN: 978-0-12-824607-8
ISSN: 1076-5670

For information on all Academic Press publications
visit our website at https://www.elsevier.com/books-and-journals

Publisher: Zoe Kruze
Acquisitions Editor: Jason Mitchell
Developmental Editor: Jhon Michael Peñano
Production Project Manager: James Selvam
Designer: Alan Studholme

Typeset by VTeX

Contents

Contributors

Annick De Backer
Electron Microscopy for Materials Science (EMAT), University of Antwerp, Antwerp, Belgium
NANOlab Center of Excellence, University of Antwerp, Antwerp, Belgium

Arnold J. den Dekker
imec-Vision Lab, University of Antwerp, Wilrijk, Belgium

Jarmo Fatermans
Electron Microscopy for Materials Science (EMAT), University of Antwerp, Antwerp, Belgium
NANOlab Center of Excellence, University of Antwerp, Antwerp, Belgium
imec-Vision Lab, University of Antwerp, Wilrijk, Belgium

Sandra Van Aert
Electron Microscopy for Materials Science (EMAT), University of Antwerp, Antwerp, Belgium
NANOlab Center of Excellence, University of Antwerp, Antwerp, Belgium

Preface

The last thirty years have witnessed the development of what has become generally known as quantitative electron microscopy. The initial focus was to make the data more quantitative by calibrating detectors so that the signal could be measured with respect to the incident electron intensity. Images, diffraction patterns, and electron energy-loss spectra were all of concern. This approach has accompanied the development of detector technology from early digitization of photographic films, to charge-coupled devices (CCD), annular detectors, and today's direct detection devices (DDD). A parallel line of research was devoted to making the data analysis more quantitative, by using image processing and statistical means to compare experimental results with simulations, or by extracting quantitative information from the signal. This in turn led to the development of new experimental techniques, more adapted to quantitative methods. The ultimate aim has always been to determine characteristics of the instrument, sample or phenomenon under investigation.

In this respect, the Antwerp school has been exemplary. We are therefore delighted to introduce this volume of AIEP, which is devoted to the development of quantitative statistical analysis, by Sandra Van Aert and her colleagues Arnold den Dekker, Annick De Backer, and Jarmo Fatermans. They show how atoms can be counted in crystalline materials by carefully analyzing the intensity of atomic columns recorded in high-resolution images and image reconstructions. The measurement of atomic column positions is equally addressed. They attack the difficult problem of detecting and counting atoms in small clusters and even show how the three-dimensional atomic structure of nanocrystals can be determined. The statistical methods they have developed are of relevance to a much wider field of investigation than electron microscopy. We hope that readers will find this comprehensive account as informative and enjoyable as we did ourselves.

Martin Hÿtch
Peter W. Hawkes

CHAPTER ONE

Introduction

Annick De Backer[a,b,d], Jarmo Fatermans[a,b,c,d], Arnold J. den Dekker[c], and Sandra Van Aert[a,b,*]
[a]Electron Microscopy for Materials Science (EMAT), University of Antwerp, Antwerp, Belgium
[b]NANOlab Center of Excellence, University of Antwerp, Antwerp, Belgium
[c]imec-Vision Lab, University of Antwerp, Wilrijk, Belgium
*Corresponding author. e-mail address: sandra.vanaert@uantwerpen.be

Contents

Materials science: discovery and design of new materials

In the last decades, materials science has evolved from macro- to micro- to nanoscience. In this field of science, there is an ever-increasing interest in the development of nanostructured materials with novel properties requiring new synthesis techniques. With respect to this research, the paradigm of the traditional materials science tetrahedron is highly relevant (Yang & Tarascon, 2012). The materials paradigm is shown in Fig. 1.1 and illustrates the relationship between material's synthesis, structure, properties, and performance. Materials characterization resulting in the availability of precise structural information is a crucial point in understanding the interrelationships in this paradigm. In combination with recent developments in nanotechnology, where one is nowadays able to construct nanostructures with a well-chosen and controlled structure, an evolution from describing toward understanding and designing materials may be realized (Olson, 1997, 2000). In addition, designing a final application requires a synergistic combination from the research results of various individual subfields to solve technological challenges (Yang & Tarascon, 2012). Examples of such novel nanostructured materials having the attention from researchers are

[d] Both authors contributed equally to this work.

Advances in Imaging and Electron Physics, Volume 217
ISSN 1076-5670
https://doi.org/10.1016/bs.aiep.2021.01.001

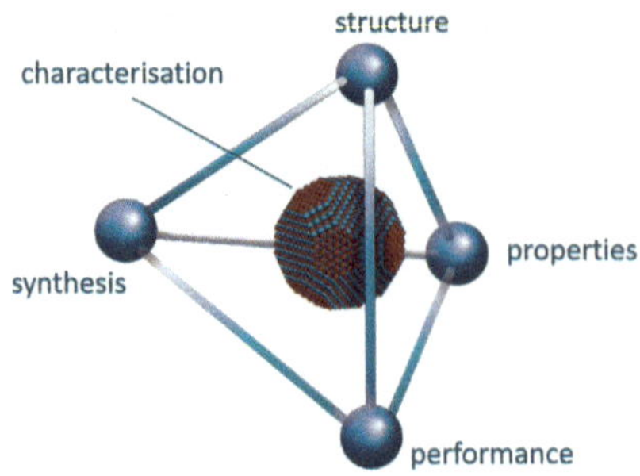

Figure 1.1 Paradigm of materials science represented in the form of a tetrahedron.

carbon nanotubes, fullerenes, metal nanoparticles, magnetic materials, ceramics, semi-conductor compounds, and catalytic nanoparticles (Klabunde, 2002; Liz-Marzàn & Kamat, 2003). The properties of these materials at the nano- or atomic scale change often significantly as compared to the properties of their bulk counterparts (Huang & Xu, 2010; Nebel, 2003; Talapatra et al., 2005). Moreover, the properties of many nanomaterials are significantly modified by extremely small changes in the local structure. Since the properties of materials are directly related to the atomic structure, the combination of theoretical and experimental studies is indispensable in order to understand the relation between structure and properties, being one of the links in the materials paradigm. On the one hand, theoretical studies allow one to calculate properties of materials with a known atomic structure (Botti et al., 2007). On the other hand, experimental characterization methods should be able to determine the structure at the atomic scale of newly synthesized materials. The combination of both approaches provides a unique link between theory and experiment if precise structural information is available for the nanomaterial of interest.

Through the years, different characterization techniques have been developed in order to provide precise structural information. X-ray and neutron diffraction can reveal the lattice structure at high resolution (Zanchet et al., 2001). However, X-rays only interact weakly with matter and they provide an average result from diffraction, instead of local structure information. This technique is therefore only suitable for periodic systems whereas nanomaterials are usually aperiodic. Scanning probe techniques, like scanning tunneling microscopy and atomic force microscopy, can be used to investigate materials at a local scale with atomic resolution (Wiesendanger, 1994). However, these techniques restrict investigations to surface information. Atomic-resolution (scanning) transmission electron microscopy ((S)TEM), on the other hand, can be used to determine both the local and the internal structure of materials at the atomic scale, because of the strong interaction of electrons with small volumes of matter (Henderson, 1995).

In a world evolving from macro- to micro- to nanoscience, these electron microscopy images need to be interpreted quantitatively instead of visually in terms of local atomic structure and chemical composition. This requires the electron microscope to evolve from a predominantly structure-oriented tool into a physical measurement instrument. Over the past two decades, the focus in (S)TEM research has therefore moved gradually from obtaining a better resolution to improving the precision with which structural and chemical information can be extracted from (S)TEM data (Krivanek et al., 2010; Urban, 2008; Van Aert, den Dekker, Van Dyck, et al., 2002a; Van Dyck et al., 2003).

Research objective

The main objective of our work is to push aberration-corrected (scanning) transmission electron microscopy toward precise measurements of unknown structure parameters using advanced statistical techniques. Although the resolution of state-of-the-art instruments has greatly improved by optimizing the lens design, equally fundamental changes in the image-processing and acquisition methods are required in order to have the instrument performing at the limits of its capabilities. Ultimately, the precision with which atomic positions have to be measured should be of the order of 1 pm (Kisielowski, Principe, et al., 2001; Muller, 1999; Springborg, 2000). In addition, the chemical composition of nanostructures has to be discerned for atom types having a difference in atomic number equal to 1. In order to extract precise structural information from atomic-resolution electron microscopy images, advanced statistical techniques will extensively be used in this work. In the first part of this work, the focus will be mainly on the methodology for a quantitative analysis of (S)TEM data. In the second part, the possibility to count the number of atoms in a projected atomic column will be investigated. The third part of this work will introduce how to detect atomic columns from noisy images. Achieving high-precision measurements based on these methodologies is of great fundamental value since ab-initio calculations require precise structural and chemical information as an input in order to compute the properties of nanostructures.

1.1. Quantitative atomic-resolution electron microscopy

Atomic-resolution (scanning) transmission electron microscopy

The principle of transmission electron microscopy (TEM) is schematically illustrated in Fig. 1.2(a). More details on TEM can be found in (De Graef, 2003; Van Dyck, 2002; Williams & Carter, 2009). In TEM, the object is illuminated by a parallel incident electron beam. The interaction between the electrons of the plane wave and the object determines a complex electron wave function at the exit plane of the object, i.e. the so-called exit wave. If the object is oriented along a main zone axis and the distance between neighboring columns is not too small, a one-to-one correspondence exists between the exit wave and the projected object structure (Geuens & Van Dyck, 2002; Van Dyck & Chen, 1999b). The exit wave is magnified by a set of apertures and magnetic lenses. The wave propagates to the image plane where it forms the so-called image wave. Since the lenses are not perfect, the exit wave is distorted by lens aberrations such as spherical aberration, defocus, and chromatic aberration while propagating to the image plane. Therefore, the image wave is more difficult to interpret in terms of structure information. Finally, the image wave is usually recorded with a charge coupled device (CCD) detecting the electron counts reaching this camera. During this process, only the intensity is recorded. As a result, the phase of the image wave is lost. Image formation in TEM is a coherent imaging process. This means that contrast reversals can occur due to lens aberrations and multiple elastic scattering.

An alternative imaging mode is referred to as scanning transmission electron microscopy (STEM) for which the details can be found in (Nellist & Pennycook, 2000; Pennycook & Nellist, 2011). In this mode, the electron beam is focused to a fine probe that is scanned across the specimen in a two-dimensional (2D) raster, as illustrated in Fig. 1.2(b). For each probe position, a so-called convergent beam electron diffraction pattern is formed in the back focal plane, where, typically, an annular detector is located. The signals of the electrons reaching the detector are integrated and displayed as a function of the probe position. Different detector geometries are nowadays available (Hovden & Muller, 2012; Shibata et al., 2010; Yang et al., 2015). Depending on the collection range of the detector, the atomic number Z-dependence of the contrast and the pixel signal-to-noise ratio (SNR) will be different. Even more flexibility in the choice of annu-

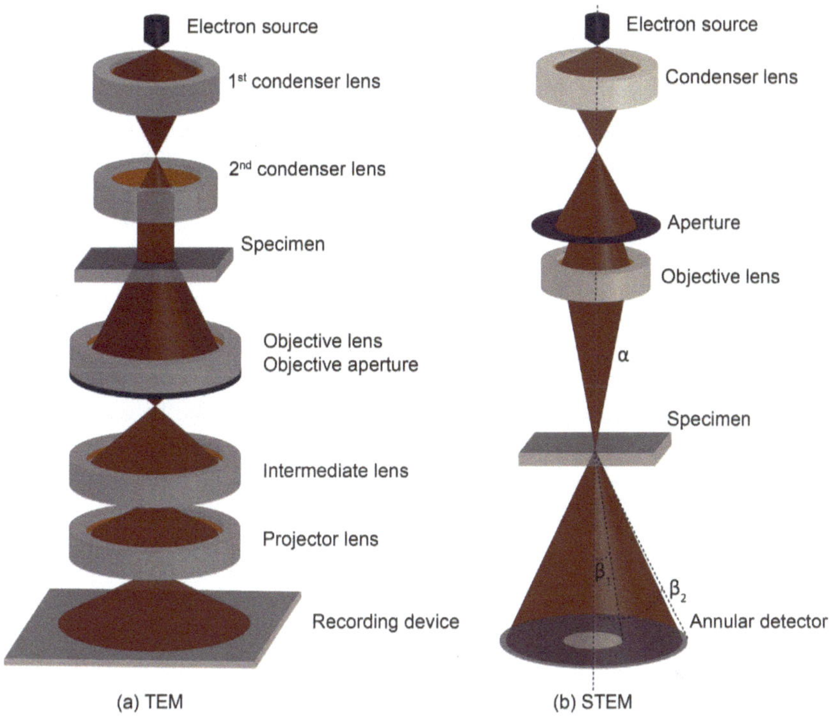

Figure 1.2 General schematic of a TEM and STEM instrument. (a) A plane wave is illuminating the object after which an image is formed using a set of electromagnetic lenses. (b) An electron beam with convergence angle α is scattered by the specimen and collected by an annular detector with inner and outer angle β_1 and β_2, respectively.

lar ranges is provided by using a direct-electron detector which allows to acquire the full diffraction pattern for every probe position (Faruqi & McMullan, 2018; Mir et al., 2017; Ryll et al., 2016; Sagawa et al., 2017; Tate et al., 2016). This results in a four-dimensional (4D) dataset, consisting of 2D diffraction patterns recorded at every 2D probe position, from which virtual STEM images can be computed with freely chosen inner and outer detector angles. For example, an annular dark-field (ADF) STEM image is obtained if the inner detector angle β_1 in Fig. 1.2(b) is larger than the convergence angle α. If the inner collection angle β_1 is much larger than the probe-convergence angle α, the imaging mode is referred to as high-angle annular dark-field (HAADF) STEM. In this case, only electrons scattered to high angles, mainly due to an inelastic scattering process, are collected. Image formation by ADF STEM is an incoherent imaging process. Images are directly interpretable in terms of structure since the image contrast depends on the atomic number of the atoms or the number of atoms in an atomic column, i.e. Z-contrast imaging. Another possible detector configuration is that of an axial detector where the outer detector angle β_2 is smaller than the convergence angle α. This results in bright-field (BF) STEM images, which are mainly formed by elastically scattered electrons.

Ever since the construction of the first electron microscope, a lot of effort has been made to improve the image resolution. After the introduction of aberration-corrected electron optics, a point resolution of about 75 to 100 pm in TEM and sub-50 pm in STEM can be obtained with the newest generation instruments at moderate acceleration voltages (Erni et al., 2009; Haider, Rose, et al., 1998; Haider, Uhlemann, et al., 1998; Jia et al., 2004; Krivanek et al., 2015; MacLaren & Ramasse, 2014; Ricolleau et al., 2012; Sawada et al., 2009; Walther & Ross, 2013). The point resolution represents the smallest detail directly interpretable from the experimental images. Therefore, these instrumental improvements significantly simplified visual interpretation.

In TEM, the point resolution can be further improved by the use of higher voltages. Recently, a point resolution of 43 pm has been reported for an aberration corrected 1.2 MV cold field-emission transmission electron microscope (Akashi et al., 2015). However, such high voltages result in radiation damage, which is problematic for the analysis of light-element structures and catalytic nanoparticles. In TEM also another resolution measure can be considered, the so-called information limit. The information limit represents the smallest detail that is transferred from the object to the image, or in other words the smallest detail that can be resolved by the instrument. The information limit is inversely proportional to the highest spatial frequency transferred from the object to the image. By using image-processing techniques the interpretable resolution can be pushed down from the point resolution to the information limit. For example, methods that reconstruct the exit wave from a focal series of images achieve this goal (Haigh et al., 2009; Hsieh et al., 2004; Kirkland et al., 1995; Lichte, 1986; Op de Beeck et al., 1996; Van Dyck et al., 1993). Ideally, the exit wave is free from imaging artefacts which means that the visual interpretation is considerably enhanced as compared to the originally recorded images, especially in the case of thin objects. In this manner, imaging with a resolution of 50 pm is possible at moderate acceleration voltages (Alem et al., 2011; Haider et al., 2010; Kisielowski et al., 2008; Kisielowski, Hetherington, et al., 2001).

Based on classical resolution criteria, it is often concluded that the directly interpretable resolution from a single image is higher in incoherent ADF STEM as compared to TEM, given that the same imaging system is used in both cases. Lord Rayleigh proved in 1896 that incoherent imaging has a factor of two higher resolution than coherent imaging (Rayleigh, 1896). In accordance with his proof, the achievement of sub-Å resolution was achieved earlier in STEM than in TEM at intermediate acceleration voltages (Batson et al., 2002; Nellist et al., 2004). In addition, directly interpretable contrast in STEM is often desirable. However, since resolution is not unambiguously defined, a more objective criterion should be used instead to evaluate different imaging systems. If images are interpreted quantitatively, a measure for the statistical precision is appropriate (Van Aert et al., 2006).

From qualitative toward quantitative imaging in electron microscopy

In recent years, electron microscopy has been evolving from qualitative imaging toward quantitative imaging. Along with this evolution, statistical analysis methods have been introduced in order to extract quantitative measurements from atomic-resolution (S)TEM images (den Dekker et al., 2005; Van Aert et al., 2005). Using statistical parameter estimation theory, unknown structure parameters can be measured with high accuracy and precision from experimental images. In this theory, the (S)TEM image is considered as a data plane from which the unknown structure parameters need to be estimated. The starting point is the availability of a parametric model describing the expectations of the experimental measurements. For atomic-resolution (S)TEM images, the projected atomic columns can be assumed to be Gaussian functions peaked at the atomic column positions. The unknown parameters are estimated by fitting this model to the experimental images using a criterion of goodness of fit, quantifying the similarity between the experimental images and the model. The model parameters are estimated using an iterative optimization scheme using the least squares estimator. In Bals et al. (2006); De Backer et al. (2011); Houben et al. (2006); Jia et al. (2008); Klingstedt et al. (2012); Urban (2008); Van Aert et al. (2012), quantitative analyzes have successfully been applied for 2D atomic column position measurements with a precision in the picometer range from TEM images or reconstructed exit waves. Additionally, ADF STEM has also become a popular technique for quantifying image intensi-

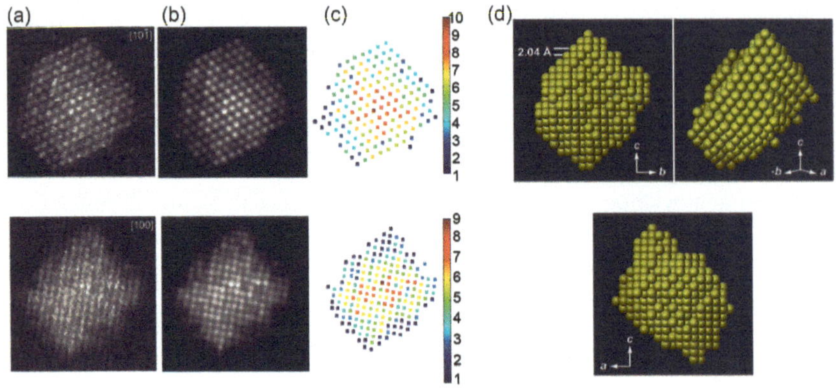

Figure 1.3 Atomic imaging of crystalline nanoparticles in three dimensions. (a) Experimental HAADF STEM images of nanosized Ag clusters embedded in an Al matrix in [10$\bar{1}$] zone-axis orientation and [100] zone-axis orientation. (b) Refined models of the images in (a). (c) Number of Ag atoms per column. (d) The computed 3D reconstruction of the Ag nanocluster viewed along three different directions. Reproduced from Van Aert et al. (2011) with permission from Nature Publishing Group.

ties (Huijben et al., 2006; Kim et al., 2012; Kimoto et al., 2010; Van Aert, Chang, et al., 2009; Yankovich et al., 2014).

The examples so far mainly achieved a precise characterization of the atomic column positions in two dimensions. Ultimately, locating the individual atoms with a precision of 1 picometer in three dimensions should be aimed for. In Van Aert et al. (2011), the level of quantification in STEM was therefore raised one step further by combining the as-described model-based fitting with another set of advanced statistical methods. In this analysis, the authors aimed for a precise quantification of the number of atoms in the third dimension from 2D projection images. This was the first achievement of a three-dimensional (3D) reconstruction of a nanoparticle at the atomic scale. The 3D reconstruction of an Ag particle embedded in an Al matrix resulting from atom counts obtained from HAADF STEM images of the same particle viewed along two different zone axes is illustrated in Fig. 1.3.

The aim in quantitative electron microscopy is to estimate the structure parameters as precisely as possible. Ultimately, this precision is only limited by the presence of noise. Due to noise, the pixel values in the experimental images will fluctuate randomly from experiment to experiment. These pixel values or observations can therefore be modeled as random variables, characterized by a joint probability density function (PDF) for continuous observations or, in the case of discrete observations, such as Poisson counting results, by a joint probability function (PF). The parametric imaging model introduced above describes the expectations of these observations. Use of the concept of Fisher information allows one to derive an expression for the highest attainable precision with which the structure parameters of the object under study can be estimated in an unbiased manner (van den Bos & den Dekker, 2001). This expression defines a lower bound on the variance of any unbiased estimator of these parameters. This bound, which is known as the Cramér-Rao lower bound (CRLB), can be derived from the P(D)F. The CRLB on the variance can be attained asymptotically when applying the maximum likelihood estimator. The CRLB is generally a function of the object parameters, the microscope parameters, and the electron dose. It provides quantitative insight into what precision might be achieved from the available image(s). It also provides insight into the sensitivity of the precision to the parameter values. An important application for which the CRLB can be used is statistical experiment design. Experiment design is the selection of free variables in an experiment to improve the precision of the estimated parameters. By calculating the CRLB as a function of the microscope settings, these experimental settings can be optimized so as to attain the highest precision (Bettens et al., 1999; den Dekker et al., 1999, 2001; Van Aert, den Dekker, & Van Dyck, 2004; Van Aert, den Dekker, van den Bos, et al., 2002, 2004; Van Aert, den Dekker, Van Dyck, et al., 2002a, 2002b; Van Aert & Van Dyck, 2001; Van den Broek et al., 2011). The approach for experiment design also provides the possibility to decide if new instrumental developments result in significantly higher attainable precisions.

So far, studies on the precision of atomic-scale measurements from (S)TEM considered the estimation of the position of atoms or atomic columns in projection (Bettens et al., 1999; den Dekker et al., 1999, 2005, 2001; Van Aert, den Dekker, van den Bos, et al., 2002; Van Aert et al., 2005; Van Dyck et al., 1997, 2003). In this work, the problem of identifying the number of atoms along the third dimension in each atomic column will also be considered. The number of atoms is a positive integer and therefore the number of atoms is a so-called restricted or discrete parameter. As a consequence, the P(D)F of the observations is not continuously differentiable with respect to the number of atoms which is a condition for the CRLB. Hence, the CRLB is not defined. Therefore, an alternative solution using the principles of detection theory will be proposed for determining the optimal experiment design (De Backer, De wael, et al.,

2015; den Dekker et al., 2013; Gonnissen et al., 2014; Kay, 2009). In this framework, the atom-counting problem can be formulated as a statistical hypothesis test, where each hypothesis corresponds to the assumption of a specific number of atoms in an atomic column. Statistical detection theory provides the tools to decide between these hypotheses and to compute the experimental settings minimizing the probability to assign an incorrect hypothesis.

1.2. Atomic-scale quantification in 3D

From 2D toward 3D imaging in electron microscopy

Besides an evolution from qualitative to quantitative analysis of (S)TEM images, there is also an evolution from 2D imaging toward 3D imaging, since 2D projection images may result in an incomplete characterization of the material's structure. Atomic-resolution (S)TEM images provide a 2D projection image of a 3D object and 2D projection images are sufficient for the investigation of periodic structures. However, a lot of interesting nanostructures are asymmetric in 3D. Different techniques to retrieve the 3D atomic structure have been suggested in the field of electron microscopy. The most known technique is electron tomography (Frank, 1992; Midgley & Dunin-Borkowski, 2009). A series of 2D projection images at different tilt angles, i.e. at incremental degrees of rotation around the specimen, is then used to obtain a 3D reconstruction. There are several limitations for a successful 3D reconstruction using electron tomography. First, the specimen needs to be stable under illumination of the electron beam during the acquisition of the tilt series. Second, the so-called projection requirement should be fulfilled (Hawkes, 1992). This means that the intensity in a projection image must be a monotonic function of a physical quantity of the object, like e.g. the thickness. Third, due to the design of the electron microscope, the range of tilt angles is often limited. This is the so-called missing wedge and causes artifacts in the 3D reconstruction (Midgley & Weyland, 2003). Finally, dedicated alignment and reconstruction algorithms are required for the computation of a 3D reconstruction of the imaged object. Originally, 3D reconstructions using electron tomography could only be obtained at the nanoscale (Bals et al., 2007; Biermans et al., 2010; Ke et al., 2010). Since then, the limits of electron tomography have been pushed beyond 3D characterization at the nanoscale in order to resolve atomic positions in a 3D reconstruction. In order to avoid beam damage a limited number of high-resolution images is favorable. As illustrated in Fig. 1.3, Van Aert et al. could locate the atomic positions for the first time in three dimensions from two atomic-resolution projection images using discrete tomography (Van Aert et al., 2011). Shortly after, a reconstruction technique based on compressive sensing has been developed for a 3D reconstruction at the atomic scale (Goris, Van den Broek, et al., 2012). This technique has successfully been applied at the atomic scale using ADF STEM for e.g. determination of the surface facets in Au nanorods (Goris, Bals, et al., 2012) and 3D elemental mapping in a bimetallic nanocrystal (Goris et al., 2013). Meanwhile, also other approaches using electron tomography resulted in atomic resolution in 3D (Belz et al., 2018; Chen et al., 2013; De Backer et al., 2017; Goris et al., 2015; Xu et al., 2015; Zhou et al., 2020).

A second possible technique for 3D structure characterization is depth sectioning by ADF STEM (Alania et al., 2017; van Benthem et al., 2005, 2006). With a focal series of images in an aberration corrected STEM, it is possible to slice through the object. In this manner, it is possible to detect the depth of heavy single impurity or dopant atoms. This technique is very promising for measuring 3D surface morphology, thickness, top/surface atomic structures, point defects in bulk materials, and even depth-sensitive spectroscopy such as energy dispersive X-ray spectroscopy at atomic dimensions (Ishikawa et al., 2015). Despite of revolutionary advances in aberration correctors, the depth resolution is nowadays still insufficient to reach atomic-depth resolution, limiting the practical applicability. A closely related technique is scanning confocal electron microscopy which requires adaptations to the electron optics (Nellist et al., 2006). This confocal mode can be obtained in a combined TEM/STEM instrument with two aberration correctors and by inserting an additional aperture in the image plane. A better depth resolution is then obtained, since out-of-focus rays are rejected. Different confocal modes are now available, but atomic-depth resolution has not been reported so far (Alania et al., 2017; Hashimoto et al., 2009, 2010; Wang et al., 2011; Wang, Behan, et al., 2010).

A third possible method for a 3D reconstruction at the atomic scale is provided by an advanced analysis of the exit wave in TEM. The exit wave is closely related to the object structure and can be reconstructed using a series of TEM images acquired at different focal values. The complex values at the atomic column positions of the reconstructed exit wave can be used to measure the thickness and to determine the surface profile using a so-called Argand plot (Wang, Chen, et al., 2010, 2012). In this plot, the real and imaginary values for each pixel value of the exit wave at the atomic column position are plotted as coordinates in a complex plane. The pixel values of columns with different numbers of atoms ideally form a circle sampled in discrete steps. Different heights of the atomic columns are observed along another circular curve. Combining the information of both curves enables one to determine the number of atoms and the surface profile. In other words, this method aims to invert the dynamical scattering between the electron beam and the atoms. The dynamical scattering can also be inverted by recasting the scattering process as an artificial neural network in order to retrieve the 3D object potential as proposed by Van den Broek and Koch (Van den Broek & Koch, 2012, 2013). An alternative method for the interpretation of the exit wave in terms of the 3D position of the individual atoms has been developed by Van Dyck et al. (Van Dyck et al., 2012). In this so-called 'Big Bang' tomography, the reconstructed exit wave is considered as a superposition of spherical waves that have been scattered by the individual atoms of the object, assuming that the atoms are point-like objects. The phase of these spherical waves is proportional to the distance of propagation. A linear fit allows then to position each individual atom along the third dimension.

A fourth possible method suggests that it is possible to determine the 3D shape of a crystal at atomic resolution from a single TEM image (Jia et al., 2014). In this approach, the peak intensities in the TEM image are interpreted in terms of atom type, composition, number of atoms, and relative position along the third dimension. This is a promising method for the determination of the 3D structure of very thin, light-element, radiation-sensitive crystals. So far, it has not yet been shown that this technique is widely applicable to heavier crystals or crystals having a thickness of more than 10 atoms.

Lastly, there is another technique which aims to retrieve information on the third dimension at the atomic scale. This method relies on quantifying the number of atoms in nanoparticles from a single ADF STEM image. Atomic-resolution ADF STEM has become a popular technique for this purpose since the observed intensities are highly sensitive to the atom type and number of atoms. The first results on this approach have been presented by Li et al. (Li et al., 2008). However, the method presented there is only applicable for regularly shaped models in combination with a priori knowledge and detailed image simulations. Other early approaches rely on calibration standards only applicable to a particular sample under study (Song et al., 2010; Young et al., 2008). Erni et al. attempted for the first time to count the number of atoms atomic column-by-atomic column by measuring the absolute differences in intensity of neighboring atomic columns (Erni et al., 2003). More generally acknowledged methods for counting the number of atoms in each projected atomic column are presented by LeBeau et al. and Van Aert et al. (LeBeau et al., 2010; Van Aert et al., 2011). The latter has already been quoted as an example of a quantitative analysis and has been illustrated in Fig. 1.3. These two atom-counting methods will be discussed in more detail in the next paragraph. The output of these atom-counting methods can be used as an input for a 3D reconstruction. By combining counting results obtained from images under different viewing directions, the 3D space can be reconstructed at an atomic level using so-called discrete tomography. In Van Aert et al. (2011), counting results of two different projections have been used for discrete tomography to obtain a 3D reconstruction of an Ag nanoparticle embedded in an Al matrix. The same approach has been used to reconstruct the core of a PbSe-CdSe nanoparticle (Bals et al., 2011). Atom-counting results have also been combined with ab-initio calculations to study the dynamical behavior of ultrasmall Ge clusters (Bals et al., 2012) and have been helpful to understand the wire structure and bonding at the atomic scale of quantum gold nanowires (Kundu et al., 2014). More recently, atom counts are used to create atomistic models along with prior knowledge on the crystal (Altantzis et al., 2019; De Backer et al., 2017; Geuchies et al., 2016; Jones et al., 2014). Energy minimization is then used to relax the nanoparticle's 3D structure.

Nanoparticle atom counting

In general, two different approaches exist for counting the number of atoms in a projected atomic column from a single ADF STEM image. These atom-counting methods are either based on image simulations or based on statistics and allow for retrieving information on the third dimension at the atomic scale. The method of LeBeau et al. relies on a simulation-based approach (LeBeau et al., 2010). The image intensities are then interpreted by a direct comparison of experimental image intensities and simulated intensities resulting from detailed numerical programs. This requires the intensity variations to be placed on an absolute scale by normalizing the measured image intensities to the incident electron beam (LeBeau et al., 2008; Rosenauer et al., 2009). This comparison provides atomic column-by-atomic column measurements of the composition or number of atoms (LeBeau et al., 2010; Martinez, Rosenauer, et al., 2014; Rosenauer et al.,

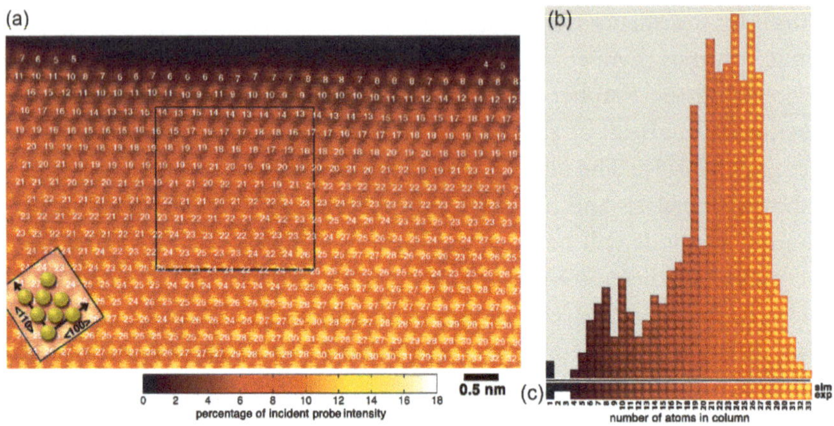

Figure 1.4 Standardless atom counting in STEM. (a) HAADF STEM image of a wedge-shaped gold file viewed along [110]. The intensity maxima correspond to gold atom columns and the white labels near the lower right of each atom column indicate the number of atoms contained in that column. The image intensities are shown on an absolute scale relative to the incident beam intensity (see scalebar). (b) Histogram of all the columns in (a) binned by the number of atoms they contain. (c) Atom column images extracted from simulation (top) and experiments (bottom) after averaging all the experimental columns in each bin shown in (b). Adapted with permission from LeBeau et al. (2010). Copyright (2010) American Chemical Society.

2009). In this manner, LeBeau et al. obtained atom counts for a thin wedge-shaped gold film viewed along the [110]-directions as illustrated in Fig. 1.4.

The approach of Van Aert et al. for atom counting, illustrated in Fig. 1.3, is a statistics-based approach (Van Aert et al., 2011). In combination with advanced statistical methods, the number of atoms in a single element atomic column can be counted with single-atom sensitivity (De Backer, Martinez, et al., 2015; De Backer et al., 2013; Van Aert et al., 2011, 2013). Nanoparticle atom counting is then achieved by quantitatively interpreting the total intensity of electrons of each atomic column scattered toward the ADF detector, i.e. the so-called scattering cross-section (E et al., 2013; Isaacson et al., 1979; Retsky, 1974; Singhal et al., 1997; Van Aert, Verbeeck, et al., 2009). In Fig. 1.5, this approach is illustrated for the Ag particle embedded in the Al matrix shown in Fig. 1.3. The scattering cross-sections estimated from the projected STEM image can be presented in the form of a histogram as in Fig. 1.5(b). Ideally, this histogram would consist of isolated components, where each component is generated by a set of columns having the same number of atoms projected on top of each other, that is in the direction perpendicular to the recorded image. In practice,

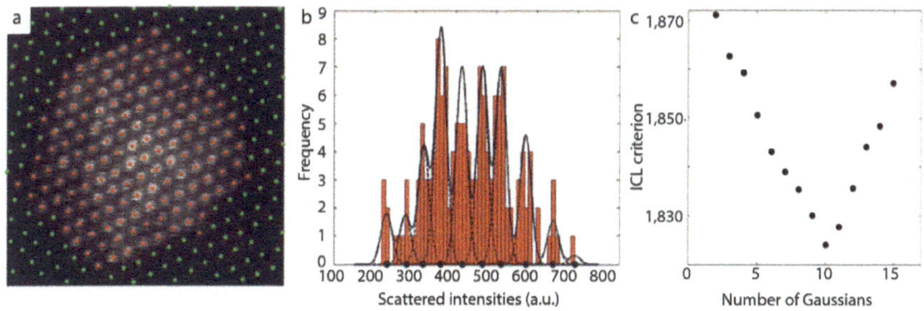

Figure 1.5 Analysis of scattering cross-sections. (a) Columns containing Ag are indicated by red markers placed at the estimated positions. (b) Histogram of the scattering cross-sections of the Ag columns. The black solid curve shows the estimated mixture model; the individual components are shown as dashed curves. (c) The Integrated Classification Likelihood criterion evaluated as a function of the number of Gaussians in a mixture model. Reproduced from Van Aert et al. (2011) with permission from Nature Publishing Group.

however, the components are smeared out owing to a combination of experimental detection noise and residual instabilities. Therefore, these results cannot directly be interpreted in terms of number of atoms. By evaluation of the so-called Integrated Classification Likelihood criterion (Fig. 1.5(c)) in combination with Gaussian mixture model estimation, the number of statistically significant components and their respective locations can be determined. From the estimated locations of the components, the number of atoms in each atomic column can be quantified. The atom-counting results are shown in Fig. 1.3(c).

1.3. Atom column detection

Model selection

In quantitative data analysis, not only precise measurements are important, but also statistical model selection plays an essential role. This is no different for electron microscopy. For example, the aforementioned ICL criterion used in nanoparticle atom counting is a model-selection technique. Also when it comes to identifying relevant atomic columns from electron microscopy images, model selection can be applied. More information about model-selection procedures and their applications can be found in Burnham and Anderson (2002); Claeskens (2016); Claeskens and Hjort (2008); Konishi and Kitagawa (2008); McQuarrie and Tsai (1998);

Zucchini (2000). Model selection consists of the task of selecting a statistical parametric model from a set of candidate models which most closely describes the underlying process that generated the data. The constructed model should be complex enough to explain relations that are hidden within the data, but, on the other hand, simple enough so that no overinterpretation of the data occurs. Statistical model selection provides a framework for handling this balance. It supports the search for good parametric models and for determining which variables are relevant and need to be included when analyzing data.

There are some general concepts related with statistical model selection (Claeskens & Hjort, 2008). Firstly, in almost all situations it is not possible to find the strictly correct or true parametric model that generated the data. This true model is often simply too complex and is almost always unknown. In practice, one works with a slightly simpler model which is almost as good. Secondly, an important concept in statistics is the balance and interplay between bias and variance (Geman et al., 1992; Sammut & Webb, 2011). When a model consists of parameters that need to be estimated from the data, the bias-variance trade-off manifests itself in the form of balancing simplicity against complexity (Myung, 2000). A simple parametric model, with for example a low number of estimated parameters, is often associated with low variability and large bias. Increasing the number of parameters, though, increases the complexity of the model which leads to a higher degree of variability but smaller bias. The key of statistical model selection lies in this balance where both overfitting and underfitting preferably need to be avoided (Burnham & Anderson, 2002; Everitt & Skrondal, 2010; Lever et al., 2016). Overfitting refers to the fact that a model consists of too many parameters, hereby overinterpreting the data. Underfitting, on the other hand, is quite the opposite where a model has too few parameters, which disables it from capturing relevant information. In addition, statistics often relies on the principle of parsimony, or Ockham's razor, which may be of relevance as well in other domains. It states that when several theories exist to describe some phenomenon, the simplest one consistent with the empirical evidence is preferred (Gauch, 2003; Hoffman et al., 1997; Myung & Pitt, 1997; Schaffer, 2015; Sivia & Carlile, 1992). In statistical modeling, this statement could be translated by claiming that only parameters that really matter should be included in the selected parametric model. Furthermore, data analysis is dependent on the scientific context and for what purpose it is used. In statistics, it can also be that certain quantities or functions of parameters are more important than others. Interestingly, model-selection methods have been developed that are aimed on focusing specifically on the most important quantities (Claeskens & Hjort, 2003; Hjort & Claeskens, 2006). An important note on model selection is that different selection methods might end up offering different results, even when the data and list of candidate models is the same. These conflicting recommendations do not necessarily form a contradiction, but indicate the importance of understanding the working principle of model-selection methods and in what context and for what aim they are relevant. Lastly, it is noted that most model-selection strategies work by assigning a score to each of the candidate models. In certain cases, it might be advantageous to perform a multi-modal approach where all fitted models are averaged in a weighted manner according to their scores (Stoica et al., 2004).

An important application of model-selection procedures is to detect and extract relevant features or peaks from noisy data (Knuth et al., 2015). Noise is a nuisance and the goal is to find the underlying signal in possibly heavily distorted data. Model selection may, for example, be used to estimate the degree of a polynomial from a noisy data sample (Mana et al., 2014). In addition, model selection has been used to determine the model order of a Gaussian mixture model of the spatial sensitivity function of a light sensor used in a robotics application (Malakar et al., 2013). Further examples concern exoplanet detection by determining the importance of various photometric effects in an exoplanetary system (Knuth et al., 2012; Placek et al., 2014) and force-field selection in biomolecular structure determination from nuclear magnetic resonance data (Habeck, 2011). In X-ray photoelectron spectroscopy, model selection has been applied to detect spectral peaks from noisy data (Armstrong & Hibbert, 2009). In order to highlight the importance and difficulty of correctly detecting peaks from noisy data, an example from experimental molecular spectroscopy has been shown in Fig. 1.6. The spectrum of Fig. 1.6(a) has been obtained by examining the sample being placed in a beam of neutrons, allowing the recording of a diffraction pattern that provides information of the structure of the material under investigation. The spectrum, though, consists of very noisy data which hampers a visual determination of the locations of spectral peaks that are present. The technique of model selection can be successfully applied to this spectrum, answering the question of how many peaks the spectrum consists (Monserrat et al., 2015; Sivia & Carlile, 1992; Sivia et al., 1992, 1993). Fig. 1.6(b) shows the best estimate of the amplitudes and positions of the excitations peaks. The analysis by model selection has revealed that there are 9 peaks present in the data shown in Fig. 1.6(a). This clearly shows

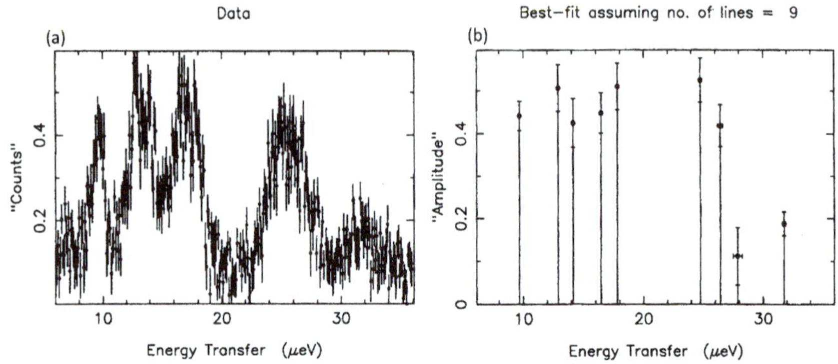

Figure 1.6 (a) Experimental neutron diffraction spectrum of lutidine. (b) Best estimate of peaks, including amplitudes and positions, extracted from (a). (Reprinted from Sivia and Carlile (1992) with the permission of AIP Publishing.)

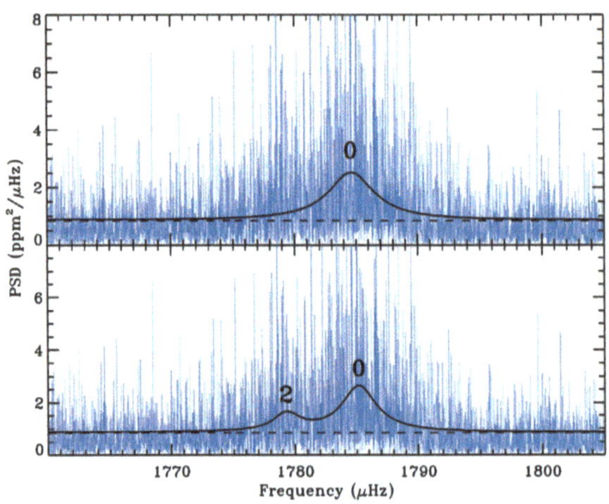

Figure 1.7 Detection of oscillation peaks from the PSD of star KIC 9139163. Upper panel: model fit consisting of one peak with angular degree equal to 0. Lower panel: favored model fit assessing the presence of two peaks with angular degrees 2 and 0. (Credit: Corsaro and De Ridder (2014). Reproduced with permission ©ESO.)

the importance of model selection since it is not straightforward to detect these peaks from only visually interpreting the data of Fig. 1.6(a).

Another example of the importance of model selection is illustrated in Fig. 1.7 which shows the power spectral density (PSD) of a star. The example stems from the domain of asteroseismology where stellar physics and stellar evolution are investigated through oscillations that occur in stars (García, 2013). Since oscillations are related to the internal structure of a star, information can be gained about a star's interior. From the data in Fig. 1.7, oscillation peaks need to be reliably detected in order to be able to extract stellar structural information. Two competing models are shown in Fig. 1.7 where the upper one contains only one oscillation peak, whereas the lower one contains two peaks. By applying model selection, it has been shown that the model containing two peaks is strongly favored (Corsaro & De Ridder, 2014). As for the example of neutron diffraction in Fig. 1.6, a visual interpretation of the noisy data in Fig. 1.7 is insufficient to reliably detect the possible presence of peaks. Therefore, a quantitative model-selection method is required which is able to disentangle signal from noise.

Model selection in materials science and technology

Model selection is not restricted to the examples described above, but can play an important role as well in materials science. For example, model-selection techniques are used for nanoparticle atom counting, which has been mentioned in the previous section. Moreover, also for reliably and directly detecting the presence of atoms of nanomaterials, model selection can be applied. This can play an important role in, for example, determining the locations of atoms at the edge of a nanoparticle whose physical properties may depend strongly on its exact surface morphology. Also for ultrasmall nanoclusters, the detection of single atoms is crucial. Such atomic clusters can be thought of as fundamental building blocks leading to metamaterials with physical and chemical properties that are not available in nature (Binns, 2001; Claridge et al., 2009). The growth mechanisms for small nanoclusters are much more exotic than for bulk materials and, therefore, they possess more complex structures, providing the need of applying model-selection techniques to study their exact atomic arrangements. Model selection in materials science may also be applied to specifically detect light-element atomic columns that are present in materials. Light atoms, such as hydrogen, lithium, and oxygen, play a key-role in modern technology. In hydrogen-storage applications, hydrogen is used as a medium to store energy (Schlapbach & Zuttel, 2001; von Colbe et al., 2019). For example, energy created by intermittent energy sources, such as wind and solar energy, can be stored during periods of high availability and low demand. As such, the fluctuating nature of intermittent energy sources can be addressed. An important application of lithium is that of lithium batteries, which are commonly used for portable electronics and electric vehicles. As compared to other types of batteries, lithium batteries are light weight and possess higher energy densities (Du Pasquier et al., 2003; Lee et al., 2011; Tarascon & Armand, 2001). In order to develop lithium-based batteries with improved charging rate, capacity, and life time, it is essential to study the movement of lithium ions during charging and discharging processes (Levi et al., 1999; Morcrette et al., 2003; Nishimura et al., 2008). Furthermore, oxygen appears in a wide range of technologically interesting materials. Oxide materials have been reported to possess a whole range of fascinating phenomena (MacLaren & Ramasse, 2014; Martin et al., 2010). For example, at oxide interfaces novel electrical transport properties may arise (Ohtomo & Hwang, 2004; Ohtomo et al., 2002; Seidel et al., 2009; Shah et al., 2010; Yu et al., 2010). In all of the aforementioned examples, it is crucial that light-element atomic columns can be detected in order to understand the role of the light atoms on the physical and chemical properties of the material.

Electron beam damage

Nowadays, due to instrument developments, (S)TEM enables to routinely image specimens at atomic resolution with high signal-to-noise ratio (SNR). This is only viable for specimens that can withstand high-energy electron irradiation meaning that they do not show significant signs of damage due to the electron beam. For biological samples and for samples which are composed predominantly of low atomic number elements, though, large radiation doses have a destructive effect (Egerton et al., 2004; Tach & Tach, 1971). The primary damage mechanisms from exposure to high-energy electrons are knock-on damage and radiolysis (Ugurlu et al., 2011; Williams & Carter, 2009). For both mechanisms, inelastic scattering occurs in which part of the incident electron energy is transferred during the collision event. Knock-on damage manifests itself as the displacement of atoms in the sample. Hereby, the incoming electrons transfer enough energy to even eject atoms from the specimen. In general, knock-on damage is a significant damaging mechanism for inorganic samples, but is less significant for biological samples. By using lower incident electron energies, knock-on damage can be mitigated when the energy transferred during inelastic scattering events is below the threshold energy for displacement of a certain kind of atom. Decreasing the incident energy of the imaging electrons decreases the likelihood of a knock-on damage event, but can increase the frequency of radiolysis (Csencsits & Gronsky, 1987), which is an important damaging mechanism for biological samples. Here, damage arises due to ionizations by the incident electrons interacting with valence electrons and breaking chemical bonds.

In order to limit the occurrence of beam damage due to the electron beam, beam-sensitive materials, such as light-element nanostructures, need to be investigated with a limited incoming electron dose, roughly of the order of around 10^3–10^5 $e^-/Å^2$ for individual atomic-resolution STEM images. Several techniques are available to mitigate beam damage on the sample in different imaging modes. A first example is the possibility of acquiring TEM images at cryogenic temperatures since low temperatures increase the tolerance to radiation damage (Chiu et al., 1987; Knapek & Dubochet, 1980; Newmark, 1982). Imaging specimens at liquid nitrogen or liquid helium temperatures falls within the domain of transmission electron cryomicroscopy (CryoTEM) (Adrian et al., 1984). Here, a specimen holder is used which maintains the specimen at cryogenic temperatures. This allows one to study specimens prepared in vitreous, or amorphous, ice, which is advantageous to image individual molecules or molecular assemblies in the field of structural biology (Callaway, 2015; Li, Baker, et al., 2009; Wolf et al., 2014), since it preserves the internal structure of small tissue or cells, or to image materials that are prone to sublimation at room temperature (Levin et al., 2017). Due to advances in detector technology and software algorithms, CryoTEM has enabled structure determination with angstrom resolution of molecules including enzymes, proteins, ribosomes and viruses (Bartesaghi et al., 2015; Campbell et al., 2015; Cheng, 2018; Dellisanti, 2015; Khoshouei et al., 2017; Merk et al., 2016; Sirohi et al., 2016; Vonck & Mills, 2017). It is noted that such resolutions are achieved by combining many individual particle images (Frank, 1992). As a result, the resolution of the reconstruction of a particle depends on the homogeneity of the sample and on the accuracy of the image alignment. Alternatively, in order to reduce beam-damage effects, electron diffraction can be used, where bright spots corresponding to repetitive crystal spacings in Fourier or diffraction space are recorded (Downing & Li, 2001; Hadermann et al., 2012, 2011; Karakulina et al., 2018). Since unit cell information within the crystal is amplified, enough signal can be generated in low-dose conditions (Unwin & Henderson, 1975). In real-space imaging, using a lower acceleration voltage limits the effect of knock-on damage (Egerton et al., 2004; Girit et al., 2009). In addition, in STEM, the effects of beam damage can be further limited by reducing the pixel dwell time and beam current while recording the image (Buban et al., 2010). Other interesting options to mitigate beam damage are to physically limit the amount of electrons reaching the specimen by compressed sensing (Béché et al., 2016; Binev et al., 2012; Saghi et al., 2015), where the electron dose is reduced by measuring image intensities for a few pixels only, or to use alternative scanning strategies in STEM (Velazco et al., 2020). In electron tomography, the total dose can be reduced by acquiring the series of images in a fast manner (Koneti et al., 2019; Migunov et al., 2015; Vanrompay et al., 2019), or by limiting the number of images in the series to only a few number of projections, for example by performing discrete tomography (Herman & Kuba, 2007).

Since there is a direct relationship between the incident dose and detected signal, imaging specimens in STEM with a limited incoming electron dose of around 10^3–10^5 $e^-/Å^2$ results in images exhibiting low intensities of the order of 10^2 detected electrons/$Å^2$. Recognizing structural features or even atomic columns from such images by just a visual inspection is unreliable and might lead to biased structural information. Usually, for beam-stable materials, the number of atomic columns can be determined visually due to the high incoming electron dose, typically of the order of more than 10^6–10^8 $e^-/Å^2$ (Williams & Carter, 2009), that can be used to image these materials. This results into images containing a high number of detected electrons, roughly of the order of 10^4 electrons/$Å^2$, leading to images exhibiting high quality. This is illustrated in Fig. 1.8(a) for an atomic-resolution image of $SrTiO_3$ where the atomic columns can be easily recognized. For beam-sensitive and light-element materials, though, the incoming electron dose should be limited in order to avoid beam damage, resulting into low-quality images. Visual inspection of such images may lead to biased results. Figs. 1.8(b), (c) and (d) show images recorded with a limited incoming electron dose of the order of 10^4–10^5 $e^-/Å^2$ of $SrTiO_3$, graphene, and a small Ge cluster, respectively (Fatermans et al., 2018). Although some features can be recognized from these images, it is not straightforward to visually indicate the presence or absence of atomic columns, especially when no prior knowledge about the material's structure is available. Therefore, often imaging processing techniques are necessary in order to interpret low-dose images and to obtain reliable structural information. Several noise-reduction techniques exist which are aimed to remove noise from images and to highlight relevant features (Vieira et al., 2013; Westin et al., 2000). Nowadays, deep-learning techniques making use of neural networks are also being developed to remove noise or other image artefacts from images and to perform image restoration (Madsen et al., 2018). In case of a series of low-dose images, a higher signal can be

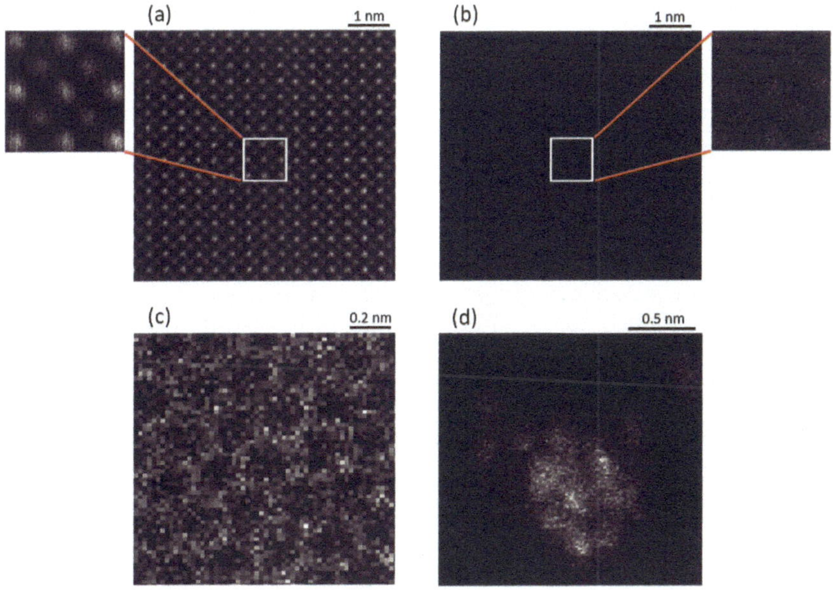

Figure 1.8 (a) Atomic-resolution electron microscopy image of $SrTiO_3$ using a high incoming electron dose of almost 10^6 $e^-/Å^2$. (b), (c), and (d) Images of $SrTiO_3$, graphene, and a small Ge cluster, respectively, using a limited incoming electron dose of the order of 10^4–10^5 $e^-/Å^2$, resulting into a detected electron dose of the order of 10^2 $e^-/Å^2$.

obtained by combining the individual images in the series by using image-registration or image-alignment algorithms (Goshtasby, 2005; Jones et al., 2015). An alternative way is to use template matching where parts of an image are matched with a template which can be a smaller area of the image or a separate image (Altantzis et al., 2016; Brunelli, 2009). A drawback of the techniques of image registration and template matching is that only average structure information can be obtained, losing information on the local level which might have an important effect on the physical properties of nanomaterials. Alternatively, also statistical techniques are available for detecting atoms or the number of atoms in a column and to recover information about the structure of the material under investigation (De Backer, De wael, et al., 2015; Fatermans et al., 2018; Gonnissen et al., 2014, 2016; Kramberger & Meyer, 2016; Kramberger et al., 2017; Meyer et al., 2014; Van Aert et al., 2019).

1.4. This work

The outline of this work is as follows. In the first part of this work, quantitative analysis methods for different types of atomic-resolution electron microscopy images are introduced and discussed. In Chapter 2, the basic principles of statistical parameter estimation theory will first be reviewed. These principles will then also be applied. To begin with, it has been investigated how to measure atomic column positions as accurately and precisely as possible using a focal series of images (De Backer et al., 2011). In theory, it is expected that the precision would considerably improve using a maximum likelihood estimator based on the full series of focal images. As such, the theoretical lower bound on the variances of the unknown atomic column positions can be attained. However, this approach is numerically demanding. Therefore, maximum likelihood estimation has been compared with the results obtained by fitting a model to the reconstructed complex exit wave or to the phase of the reconstructed complex exit wave rather than to the full series of images.

Since ADF STEM has become a popular technique for quantitative structure characterization, the accuracy and the precision with which measurements for the atomic column positions and scattering cross-sections from ADF STEM images can be estimated, have been investigated as well (De Backer et al., 2016). The observations for ADF STEM are counting events, which are modeled as Poisson distributed variables. The expectation model of the observations is a superposition of Gaussian peaks. Under these assumptions, the CRLB on the variance for the positions and scattering cross-sections can be computed. In practice, the expectation model is fitted using the well-known least squares estimator, which is identical to the maximum likelihood estimator for independent and identically normally distributed pixel values. However, this assumption about the statistical distribution of the pixel values is not always valid, especially not at low electron doses, needed to avoid radiation damage in beam-sensitive matter. By means of image simulations, it has been investigated whether the variance on the least squares estimates of scattering cross-sections and atomic column positions attains the CRLB. This study reveals if quantitative ADF STEM analysis would benefit when applying the maximum likelihood estimator which has optimal statistical properties.

Quantitative structure characterization is not limited to exit wave reconstruction or ADF STEM, but can also be applied to simultaneously acquired ADF and annular bright-field (ABF) STEM images (Fatermans et al., 2020). In order to do so, a novel parametric model is proposed to extract accurate and precise measurements from combined ABF-ADF STEM image data. Such a simultaneous acquisition from these detector geometries is of relevance since this technique combines the improved light-element visualization of ABF STEM with the easily interpretable image contrast of ADF STEM.

In Chapter 3, an efficient algorithm for the estimation of the atomic column positions and intensities from (S)TEM images will be introduced. The earlier used implementation of the least squares estimator limits the area of a (S)TEM image that quantitatively can be analyzed. Therefore, an efficient algorithm that solves some practical problems has been introduced in the StatSTEM software package (De Backer et al., 2016). This new algorithm has been applied to several experimental studies to show that the proposed approach works in practice.

The second part of this work discusses the possibility to count the number of atoms of mono-type crystalline nanostructures from atomic-resolution ADF STEM images. The methodology for atom counting discussed in detail in Chapter 4, has been proposed for the first time in Van Aert et al. (2011). In that paper, the statistics-based atom-counting method has been proven to result in reliable counts for the number of atoms in a metal nanocluster embedded into a stabilizing matrix with the same crystal nanostructure. Although, the results presented there are very promising, a thorough study on the possibilities and inherent limitations of the method is indispensable (De Backer et al., 2013). This study has been carried out in detail in this work. Only because of such a study, it can be discovered whether the statistics-based atom-counting method gives reliable counting results for nanoparticles of arbitrary shape and size. Once the possibilities and limitations are known, the atom-counting method has been applied to several challenging nanostructures (De Backer, Martinez, et al., 2015; Van Aert et al., 2013). Experimental atom-counting results have been validated by using detailed image simulations and electron tomography.

In Chapter 5, the optimal experiment design will be derived for atom counting. Since the aim in a quantitative atom-counting analysis is to extract as precisely as possible the number of atoms in a projected atomic column from the experimental images, it will be shown how even higher precisions for the estimates of the number of atoms can be obtained by statistical experiment design (De Backer, De wael, et al., 2015). Designing the optimal experiment for nanoparticle atom counting can greatly enhance the reliability of the atom-counting results, which is of great importance when studying beam-sensitive materials. Often visual interpretability is used as a criterion to optimize the experiment. However, the dose efficiency is not taken into account in such an analysis. The dose efficiency is defined as the ratio of the number of detected electrons to the number of incident electrons. This ratio is affected by e.g. the detector geometry. For the often used HAADF STEM detector in STEM imaging, the inner detector angle β_1 is much larger than the probe convergence angle α. This results in high resolution and contrast at the expense of dose efficiency. The aim is to optimize the design in terms of precision with which the number of atoms can be counted. Therefore, the principles of detection theory have been used to optimize the experiment design, which corresponds in this work to deriving the optimal STEM detector for atom counting. This technique of hypothesis testing can also be used to determine the minimally required electron dose in order to attain a pre-specified precision.

The third part of this work treats the concept of atom column detection from STEM images (Fatermans et al., 2018). Often, in order to obtain structural information with high accuracy and precision by image-quantification procedures in atomic-resolution (S)TEM, it is important that the atomic columns in the image can be recognized. As discussed before, visually determining the number of columns is not straightforward for materials which are sensitive to the electron beam. This is due to the use of a limiting incoming electron dose which causes images to exhibit low intensities, hampering visual interpretation. Therefore, a new method is developed which brings both image quantification and atom detection together in one framework. For this, in Chapter 6, parameter estimation theory is combined with model-order selection (Fatermans et al., 2019). Hereby, besides estimating the unknown structure parameters of the expectation model, also the number of parameters need to be determined, most closely representing the underlying process that generated the experimental observations. Typical model-selection methods perform a trade-off between high goodness of fit and low model complexity, since model fit can be easily improved by increased model complexity without necessarily bearing any interpretable relationship with the underlying data-generating process. In this work, it will be shown that for both ADF STEM and simultaneously acquired ABF-ADF STEM, model selection can be used to automatically and objectively determine the number of atomic columns which are present in the image data (Fatermans et al., 2020, 2018). This method is referred to as the maximum a posteriori (MAP) probability rule (Fatermans et al., 2019). As a result, atomic columns and even single atoms can be reliably detected and the most probable atomic structure of unknown nanomaterials can be determined. This is especially relevant for analyzing the structure of beam-sensitive nanomaterials, as such materials need to be imaged in STEM with limited incoming electron doses, typically varying from around 10^3 to around 10^5 $e^-/Å^2$ leading to images consisting of low intensities, which makes visual inspection unreliable. The validity and usefulness of this method will be demonstrated to experimental and simulated images of samples of different shape, size, and atom type.

In principle, atom detection from STEM images can be performed with any model-selection criterion. Therefore, the performance of the MAP probability rule in correctly detecting atoms from ADF STEM images is compared to that of other model-selection criteria in Chapter 7 (Fatermans et al., 2019). It will be shown that the MAP probability rule takes into account more dimensions of model complexity as compared to other commonly used model-selection criteria. This allows for a superior performance to detect the correct number of atomic columns from ADF STEM images. In addition, the MAP probability rule offers a more flexible way to detect atoms from images than other criteria. This is due to the fact that the prior information can be tuned, resulting into a different value for the complexity of the model under consideration. Moreover, by using the MAP probability rule, it is clear what prior knowledge has been taken into account during the analysis, which is not always straightforward for other model-selection criteria. Besides detecting atomic columns from STEM images, the MAP probability rule also offers a way to evaluate the relation between STEM image-quality measures, such as SNR and contrast-to-noise ratio (CNR), and atom detectability, which is defined as the probability to detect an atom from an image (Fatermans et al., 2019). In general, atom detectability is indeed related to image quality, since, for example, one expects the detectability of atomic columns in an image to increase with increasing image quality. In this manner, the probability to detect an atomic column from STEM image data, which can be provided by the MAP probability rule, is correlated with the quality of the image. As such, for STEM images, a new image-quality measure, the so-called integrated CNR (ICNR), will be proposed that better correlates with atom detectability than conventional image-quality measures (Fatermans et al., 2020, 2019).

Finally, in Chapter 8, conclusions will be drawn and future perspectives are considered.

Acknowledgments

This project has received funding from the European Research Council (ERC) under the European Union's Horizon 2020 research and innovation programme (Grant Agreement No. 770887 and No. 823717 ESTEEM3). The authors acknowledge financial support from the Research Foundation Flanders (FWO, Belgium) through Projects No. G.0502.18N, G.0267.18N, G.0368.15N, and a postdoctoral grant to A. D. B. The authors acknowledge funding from the University of Antwerp through a TOP project.

CHAPTER TWO

Statistical parameter estimation theory: principles and simulation studies

Annick De Backer[a,b,d], Jarmo Fatermans[a,b,c,d], Arnold J. den Dekker[c], and Sandra Van Aert[a,b,*]

[a]Electron Microscopy for Materials Science (EMAT), University of Antwerp, Antwerp, Belgium
[b]NANOlab Center of Excellence, University of Antwerp, Antwerp, Belgium
[c]imec-Vision Lab, University of Antwerp, Wilrijk, Belgium
*Corresponding author. e-mail address: sandra.vanaert@uantwerpen.be

Contents

2.1. Introduction

In this chapter, the basic principles of statistical parameter estimation theory will be introduced (den Dekker et al., 2005; Van Aert et al., 2005;

[d] Both authors contributed equally to this work.

Advances in Imaging and Electron Physics, Volume 217
ISSN 1076-5670
https://doi.org/10.1016/bs.aiep.2021.01.002

van den Bos, 2007; van den Bos & den Dekker, 2001). These principles are widely applicable in different fields of science where unknown parameters need to be measured from an experiment. Often indirect techniques are used to measure physical parameters since the unknown parameters are hidden in the available set of observations. The measurement of an unknown parameter always occurs in the presence of fluctuations in the observations, limiting the precision with which the unknown parameters can be determined. Statistical parameter estimation theory and statistical optimal experiment design aim for the highest possible precision in the determination of the unknown parameters.

For high-resolution (scanning) transmission electron microscopy (HR(S)TEM) images, the observations are characterized by a set of pixel values from which unknown structure parameters such as, the atomic positions, atomic displacements, atom types, and/or number of atoms, have to be estimated. The experimental HR(S)TEM images fluctuate randomly about their expectation values due to the presence of noise. Therefore, the observations are modeled as stochastic variables. By definition, a stochastic variable is characterized by a probability (density) function, whereas a set of stochastic variables is characterized by the joint probability (density) function. The joint probability (density) function defines the expectation values of the observations and the fluctuations of the observations about the expectation values. The expectation values are described by the expectation model, i.e. a parametric model containing the unknown structure parameters.

Using the joint probability (density) function, which depends on the unknown structure parameters, the Fisher information matrix can be defined for continuous parameters. This concept can be used to quantitatively express the amount of information about the unknown structure parameters available in the observations. The Fisher information allows us to derive the attainable precision, i.e. the lowest possible variance, with which a continuous parameter can be estimated unbiasedly from a set of observations assumed to obey a certain distribution. This is the so-called Cramér-Rao lower bound (CRLB) giving the lower bound on the variance of the unbiased estimates of the parameters. It is important to notice that this lower bound is independent of the estimation method used. This lower bound therefore represents the intrinsic limit on the precision. The joint probability (density) function can also be used to define the maximum likelihood estimator that achieves the CRLB asymptotically, i.e. for an increasing number of observations. The CRLB can therefore be used to quantify the precision of structure parameters.

It will be shown that the concept of attainable precision can be a powerful tool in order to decide how the structure parameters of interest should be estimated from the available set of observations. Here, it has been investigated how to measure atomic column positions as accurately and precisely as possible using a focal series of HRTEM images, which is often recorded for the reconstruction of the wave function at the exit plane of the object (De Backer et al., 2011). This exit wave is closely related to the object structure, has a better signal-to-noise ratio (SNR) than the individual HRTEM images, and is free from residual lens aberrations. Using simulations, it can be investigated whether the reconstructed complex exit wave contains the same amount of information concerning the atomic column positions as compared to the full series of focal images, and which amount of information can be retrieved from the phase of the reconstructed complex exit wave.

In practice, the least squares estimator is often used in order to derive unknown structure parameters from the available set of observations. It can be shown that the maximum likelihood estimator is identical to the well-known uniformly weighted least squares estimator if the observations are independent and identically normally distributed about their expectations (van den Bos & den Dekker, 2001). However, it is assumed that HR(S)TEM images are statistically independent counting results which can be modeled as Poisson distributed random variables. The Poisson distribution tends to be a normal distribution for increasing expectations of the image pixel values. However, this assumption about the statistical distribution of the pixel values is not always valid, especially not at low electron doses which are needed to avoid radiation damage in beam-sensitive matter. Therefore, it is important to verify whether the lower bound on the variance of the structure parameters is attained when applying the least squares estimator to the Poisson distributed observations for different electron doses (De Backer et al., 2016). The accuracy and precision of the thus estimated structure parameters of interest of annular dark-field (ADF) STEM images, i.e. the atomic column positions and scattering cross-sections, will be evaluated. This evaluation will tell us if it can be concluded that the structure parameters of the ADF STEM image are estimated unbiasedly with the highest possible precision.

Imaging by STEM, though, is not limited to the well-established imaging mode of ADF imaging, which allows to obtain images with high resolution and easily interpretable image contrast and which is strongly dependent on the atomic number (Jesson & Pennycook, 1995; Nellist & Pennycook, 2000). For certain materials, it is important to be able to directly visualize the light-element atomic columns, as their exact locations might influence the physical properties of the material. In this case, visualization of light-element atomic columns from ADF STEM images is challenging since light elements only scatter electrons weakly to high detector angles leading to low intensities in ADF images. Direct visualization of light elements, though, has been enabled by the annular bright-field (ABF) mode in STEM where an annular detector spanning a range within the illumination cone of the electron beam is used (Findlay et al., 2009; Okunishi et al., 2009). A simultaneous acquisition of both ABF and ADF STEM images, by using ABF and ADF detectors operating at the same time, is an interesting option to visualize atoms of a large range of atomic numbers for studying and interpreting materials at the atomic scale consisting of both light and heavy atoms. Interestingly, the principles of statistical parameter estimation theory are applicable to both ADF and ABF image data. In order to quantify simultaneously acquired ABF and ADF STEM images by using statistical parameter estimation theory, an extension of the commonly used parametric models in STEM is required (Fatermans et al., 2020). Hereby, it is important to take the effect of specimen tilt into account, since small tilts from the crystal zone axis affect especially ABF STEM intensities (Brown et al., 2017; Gao et al., 2018; Zhou et al., 2016).

This chapter is organized as follows. In Section 2.2, the framework of statistical parameter estimation theory will be discussed in detail. This includes the parametric statistical model of the observations, the derivation of the Fisher information and CRLB, i.e. the attainable precision, and the derivation of the maximum likelihood estimator. Section 2.3 shows how the attainable precision is used as a measure to theoretically examine the performance of different approaches for the estimation of the atomic column positions from a focal series of HRTEM images. In Section 2.4, the concepts of statistical parameter estimation theory are used to show how unbiased estimates with the highest possible precision can be obtained for the parameters of interest from an ADF STEM image. In Section 2.5, a new parametric model is developed in order to quantitatively analyze ABF and ADF STEM images simultaneously. Section 2.6 consists of conclusions.

2.2. Theory

In this section, the principles of statistical parameter estimation theory will be introduced. More details and references on this theory can be found in van den Bos (2007). A key component of statistical parameter estimation theory is a parametric statistical model describing the observations. In Subsection 2.2.1, a parametric statistical model will be proposed to model HR(S)TEM images. In Subsection 2.2.2, it will be shown how an expression for the attainable precision of the parameter estimates, i.e. the CRLB, can be derived. In Subsection 2.2.3, the maximum likelihood estimator will be derived from the parametric statistical model describing the observations.

2.2.1 Parametric statistical model

Due to the inevitable presence of noise, sets of observations made under the same conditions will differ from experiment to experiment. Therefore, the observations are modeled as stochastic variables. A set of pixel observations can be represented by a row vector of length $W = K \times L$ with $k = 1, \ldots, K$ and $l = 1, \ldots, L$:

$$\mathbf{w} = (w_{11}, w_{12}, w_{21}, \ldots, w_{KL})^T \tag{2.1}$$

where w_{kl} corresponds to the stochastic variable for the pixel in an HR(S)TEM image at position (k, l). Each set of observations is characterized by a joint probability (density) function:

$$p_{\mathbf{w}}(\boldsymbol{\omega}) = p(\omega_{11}, \omega_{12}, \omega_{21}, \ldots, \omega_{KL}) = p(\boldsymbol{\omega}). \tag{2.2}$$

This joint probability (density) function describes the probability that the observation $\mathbf{w}$ is equal to $\boldsymbol{\omega}$. The expectation value μ and variance σ^2 of a continuous stochastic variable are defined by the joint probability density function:

$$\mu = \mathbb{E}[w_{kl}] = \int_{-\infty}^{\infty} \omega_{kl} p(\boldsymbol{\omega}) d\boldsymbol{\omega} \tag{2.3}$$

$$\sigma^2 = \text{var}(w_{kl}) = \mathbb{E}\left[(w_{kl} - \mu)^2\right] = \int_{-\infty}^{\infty} (w_{kl} - \mu)^2 p(\boldsymbol{\omega}) d\boldsymbol{\omega}. \tag{2.4}$$

For discrete variables, the integration needs to be replaced by a summation. Statistical parameter estimation theory uses a model that describes the expectation values of the observations as a function of the sought physical parameters. The expectation value of the pixel (k, l) is then given by:

$$\mathbb{E}\left[w_{kl}\right] = f_{kl}(\boldsymbol{\theta}) \tag{2.5}$$

where $\boldsymbol{\theta} = (\theta_1, \theta_2, \ldots, \theta_R)^T$ equals an $R \times 1$ vector with the unknown parameters.

An explicit expression for the joint probability function in Eq. (2.2) can be derived by taking into account knowledge about the statistical properties of the errors in the experimental measurements. The statistical description of the observations here is valid for TEM and STEM images. Since these images are formed by counting electrons scattered to the detector, the pixel values are inevitably subject to Poisson noise causing each observed image pixel value to be Poisson distributed (Haight, 1967; Mood et al., 1974; van den Bos & den Dekker, 2001). Besides Poisson noise, also other noise sources arise related to the detector system and electronics (Ishikawa, Lupini, et al., 2014). Nevertheless, the statistical Poisson noise associated with the particle nature of electrons impinging onto the detector remains the dominant noise source in (S)TEM imaging (Seki et al., 2018). Hereby, the time between subsequent single electron impacts fits a Poisson random process (Sang & LeBeau, 2016), confirming the validity of modeling the observations in (S)TEM as Poisson distributed random variables. This results into the following probability distribution (Herrmann, 1997; Papoulis & Pillai, 2002):

$$\frac{(\lambda_{kl})^{\omega_{kl}}}{\omega_{kl}!} \exp\left(-\lambda_{kl}\right), \tag{2.6}$$

where λ_{kl} denotes the expected value of the stochastic variable w_{kl} at the pixel (k, l) which on his turn equals the expectation model:

$$\mathbb{E}\left[w_{kl}\right] = \mathrm{var}\left(\omega_{kl}\right) = \lambda_{kl} = f_{kl}(\boldsymbol{\theta}). \tag{2.7}$$

The joint probability function for a set of pixel observations which are modeled as Poisson distributed random variables equals the product of the probabilities given by Eq. (2.6), assuming that the observations of the pixels are statistically independent:

$$p(\boldsymbol{\omega}) = \prod_{k=1}^{K} \prod_{l=1}^{L} \frac{(\lambda_{kl})^{\omega_{kl}}}{\omega_{kl}!} \exp\left(-\lambda_{kl}\right). \tag{2.8}$$

The unknown parameters are included through the expectation values λ_{kl} described by a suitable parametric model as noted in Eq. (2.7). For

HR(S)TEM images, the intensity is often assumed to be sharply peaked at the atomic column positions. Therefore, HR(S)TEM images are often modeled as a superposition of Gaussian peaks. The expectation of the image at pixel (k, l) at position (x_k, y_l) can then be described as:

$$f_{kl}(\boldsymbol{\theta}) = \zeta + \sum_{i=1}^{I} \sum_{m_i}^{M_i} \eta_{m_i} \exp\left(-\frac{\left(x_k - \beta_{x_{m_i}}\right)^2 + \left(y_l - \beta_{y_{m_i}}\right)^2}{2\rho_i^2} \right) \tag{2.9}$$

with ζ a constant background, ρ_i the column type dependent width of the Gaussian peak, η_{m_i} the column intensity of the m_ith Gaussian peak, and $\beta_{x_{m_i}}$ and $\beta_{y_{m_i}}$ the x- and y-coordinate of the m_ith atomic column, respectively. The index i refers to a particular column type and the index m_i refers to the mth column of column type i. The indices in the summation of Eq. (2.9) can be simplified in the case of a mono-type crystalline nanostructure since in that case only one column type is present. The unknown parameters of the imaging model of Eq. (2.9) are given by the parameter vector:

$$\boldsymbol{\theta} = \left(\beta_{x_{1_1}}, \ldots, \beta_{x_{M_I}}, \beta_{y_{1_1}}, \ldots, \beta_{y_{M_I}}, \rho_1, \ldots, \rho_I, \eta_{1_1}, \ldots, \eta_{M_I}, \zeta\right)^T. \tag{2.10}$$

The proposed theoretical parametric model should model the image intensities accurately, in order to extract reliable quantitative measurements from HR(S)TEM images. This model, which consists of a superposition of Gaussian peaks, as described by Eq. (2.9), is preferred above other more complicated models. Complicated models describe the image intensities in detail using a more physics-based approach, i.e. modeling the electron-object interaction and the transfer of the electrons through the microscope. However, for such models, the number of unknown parameters will be higher, since e.g. unknown microscope parameters should then be included in the parametric model. Although, in general, a higher number of unknown parameters results in a better fit, the estimates of the model parameters are less precise as compared to using a simpler model with less parameters. Besides, the structure parameters of interest, such as atomic column positions and scattering cross-sections, are sufficiently well described by the simpler model given by Eq. (2.9). Moreover, the model given by Eq. (2.9) has been validated using different statistical model assessment methods (den Dekker et al., 2005; Van Aert et al., 2005). The assumption of Gaussian peaks describing the atomic columns in HR(S)TEM image data is mainly based on the source size distribution being well approximated

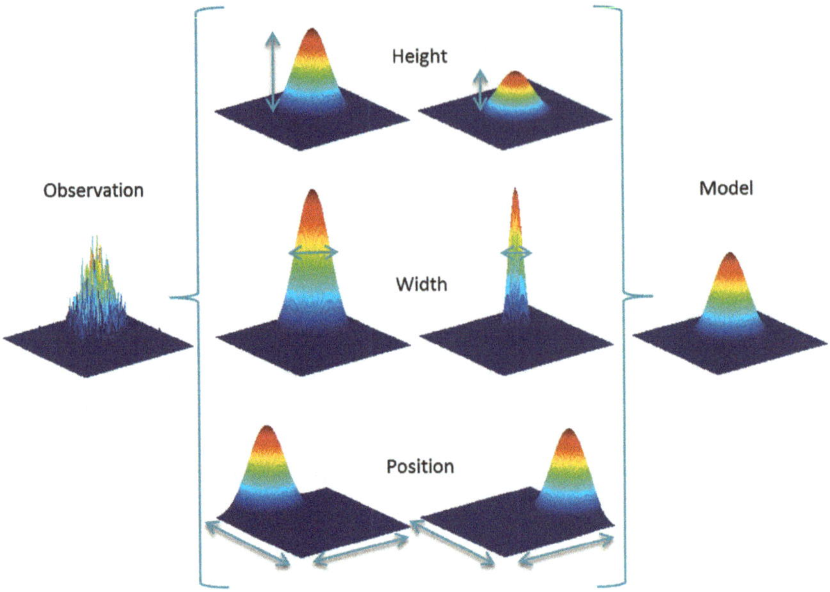

Figure 2.1 Visualization of modeling a Gaussian peak to an observed atomic column in HR(S)TEM by adjusting the height, width and position.

as a Gaussian distribution (Klenov et al., 2007; Mory et al., 1985; Nellist & Rodenburg, 1994) and on the fact that the projected potential of an atomic column is well described by considering the 1s-state (Broeckx et al., 1995; Van Aert, den Dekker, Van Dyck, et al., 2002b; Van Dyck & Chen, 1999a), which is also well approximated by a Gaussian function (Geuens & Van Dyck, 2002). As such, the atomic columns in atomic-resolution (S)TEM images can be well described by a parametric model of Gaussian peaks. Fig. 2.1 visualizes the working principle of model-based parameter estimation, where the height, width, and position of a Gaussian peak in the model are adjusted in such a way in order to find the best match with an observed Poisson-disturbed atomic column.

2.2.2 Fisher information and Cramér-Rao lower bound

In this subsection, the parametrized probability function of the observations, which is derived in Section 2.2.1, will be used to define the Fisher information matrix and to derive the CRLB on the variance, i.e. the attainable precision with which the unknown parameters can be estimated unbiasedly from the observations. Also an expression for the CRLB of a function of the parameters will be given. For example, this can be the distance between two atomic column positions which depends on the position coordinates of the atomic columns. More details about Fisher information and the CRLB can be found in Frieden (1998); van den Bos (1982); van den Bos and den Dekker (2001).

Fisher information

The joint probability (density) function depends on the unknown parameters. This dependence can be expressed as $p(\boldsymbol{\omega}; \boldsymbol{\theta})$ and can be used to define an $R \times R$ covariance matrix. This covariance matrix is the so-called Fisher information matrix given by:

$$F_{\boldsymbol{\theta}} = -\mathbb{E}\left[\frac{\partial^2 \log p(\mathbf{w}; \boldsymbol{\theta})}{\partial \boldsymbol{\theta}^2}\right] \tag{2.11}$$

where $p(\boldsymbol{\omega}; \boldsymbol{\theta})$ equals the joint probability (density) function of the stochastic variables $\mathbf{w}$. The expression between square brackets represents the Hessian matrix of $\log p(\boldsymbol{\omega}; \boldsymbol{\theta})$. The Fisher information is a measure for the physical fluctuation of the observations.

For the observed pixel values in an electron microscopy image, where $p(\boldsymbol{\omega}; \boldsymbol{\theta})$ is given by Eq. (2.8), it follows from Eqs. (2.7), (2.8), and (2.11) that the (r, s)th element of $F_{\boldsymbol{\theta}}$ equals:

$$F_{rs} = \sum_{k=1}^{K} \sum_{l=1}^{L} \frac{1}{\lambda_{kl}} \frac{\partial \lambda_{kl}}{\partial \theta_r} \frac{\partial \lambda_{kl}}{\partial \theta_s}. \tag{2.12}$$

Cramér-Rao lower bound

Using the concept of Fisher information, a lower bound on the variance, i.e. the highest possible precision, can be defined for unbiased estimators. This lower bound is the so-called Cramér-Rao lower bound (CRLB). The Cramér-Rao inequality, with $\hat{\boldsymbol{\theta}}$ any unbiased estimator of $\boldsymbol{\theta}$, equals:

$$\operatorname{cov}\left(\hat{\boldsymbol{\theta}}, \hat{\boldsymbol{\theta}}\right) \geq F_{\boldsymbol{\theta}}^{-1}, \tag{2.13}$$

with

$$\operatorname{cov}\left(\hat{\boldsymbol{\theta}}, \hat{\boldsymbol{\theta}}\right) = \mathbb{E}\left[\left(\hat{\boldsymbol{\theta}} - \boldsymbol{\theta}\right)\left(\hat{\boldsymbol{\theta}} - \boldsymbol{\theta}\right)^T\right]. \tag{2.14}$$

This inequality expresses that the difference of the matrix $\operatorname{cov}\left(\hat{\boldsymbol{\theta}}, \hat{\boldsymbol{\theta}}\right)$ and $F_{\boldsymbol{\theta}}^{-1}$ is positive semi-definite. Since the diagonal elements of $\operatorname{cov}\left(\hat{\boldsymbol{\theta}}, \hat{\boldsymbol{\theta}}\right)$ represent the variances of $\hat{\theta}_1, \hat{\theta}_2, \ldots, \hat{\theta}_R$ and since the diagonal elements of a positive semi-definite matrix are nonnegative, these variances are larger than or equal to the corresponding diagonal elements of $F_{\boldsymbol{\theta}}^{-1}$:

$$\mathrm{var}\left(\hat{\theta}_r\right) \geq \left[F_{\boldsymbol{\theta}}^{-1}\right]_{rr}. \tag{2.15}$$

The matrix $F_{\boldsymbol{\theta}}^{-1}$ is called the Cramér-Rao lower bound on the variance of $\hat{\boldsymbol{\theta}}$. The CRLB is independent of the estimation method, but depends on the statistical properties of the observations and the hypothetical true values of the parameters.

The general expression for the CRLB for vector functions of the vector parameters is of importance when one is interested in a new parameter which is a function of the proper parameters. Let $\boldsymbol{\gamma}(\boldsymbol{\theta}) = (\gamma_1(\boldsymbol{\theta}), \gamma_2(\boldsymbol{\theta}), \ldots, \gamma_G(\boldsymbol{\theta}))^T$ be a $G \times 1$ vector which depends on the $R \times 1$ parameter vector $\boldsymbol{\theta} = (\theta_1, \theta_2, \ldots, \theta_R)^T$. If $\hat{\boldsymbol{\gamma}}$ is an unbiased estimator, then it can be shown that:

$$\mathrm{cov}\left(\hat{\boldsymbol{\gamma}}\right) \geq \frac{\partial \boldsymbol{\gamma}(\boldsymbol{\theta})}{\partial \boldsymbol{\theta}^T} F_{\boldsymbol{\theta}}^{-1} \frac{\partial \boldsymbol{\gamma}(\boldsymbol{\theta})^T}{\partial \boldsymbol{\theta}} \tag{2.16}$$

where $\frac{\partial \boldsymbol{\gamma}(\boldsymbol{\theta})}{\partial \boldsymbol{\theta}}$ is the $G \times R$ Jacobian matrix of which the elements are given by:

$$\begin{pmatrix} \frac{\partial \gamma_1(\boldsymbol{\theta})}{\partial \theta_1} & \cdots & \frac{\partial \gamma_1(\boldsymbol{\theta})}{\partial \theta_R} \\ \vdots & \cdots & \vdots \\ \frac{\partial \gamma_G(\boldsymbol{\theta})}{\partial \theta_1} & \cdots & \frac{\partial \gamma_G(\boldsymbol{\theta})}{\partial \theta_R} \end{pmatrix}.$$

The matrix $\mathrm{cov}\left(\hat{\boldsymbol{\gamma}}\right)$ is the $G \times G$ covariance matrix of the estimator $\hat{\boldsymbol{\gamma}}$. It should be noticed that the CRLB can only be computed when the joint probability (density) function of the observations is known. If the CRLB depends on the true values of the parameters, which are often not known, the highest possible precision can still be quantified when substituting the real values by the estimated values (Van Aert et al., 2005).

2.2.3 Maximum likelihood estimator

In the previous subsection, the concepts of Fisher information and CRLB have been introduced. The Fisher information matrix contains information about the fluctuations of the observations. Furthermore, different estimators will show different precisions. In this subsection, it is discussed how the maximum likelihood estimator (MLE) of the parameters of the statistical expectation model may be derived from the parametrized joint probability (density) function. The MLE is an estimator that achieves the CRLB asymptotically, i.e. for an increasing number of observations.

Definition

The MLE is derived following the steps below:

1. The joint probability (density) function depends on the unknown parameters $\boldsymbol{\theta} = (\theta_1, \theta_2, \ldots, \theta_R)^T$ through the statistical expectation model:

$$p(\boldsymbol{\omega}; \boldsymbol{\theta}).$$

2. The independent variables $\boldsymbol{\omega} = (\omega_{11}, \omega_{12}, \omega_{21}, \ldots, \omega_{KL})^T$ are replaced by the observations $\mathbf{w} = (w_{11}, w_{12}, w_{21}, \ldots, w_{KL})^T$:

$$p(\mathbf{w}; \boldsymbol{\theta}).$$

3. The true parameters $\boldsymbol{\theta} = (\theta_1, \theta_2, \ldots, \theta_R)^T$ are replaced by the independent variables $\mathbf{t} = (t_1, t_2, \ldots, t_R)^T$:

$$L(\mathbf{t}) := p(\mathbf{w}; \mathbf{t})$$

where $L(\mathbf{t})$ is the likelihood function.

4. The MLE $\hat{\boldsymbol{\theta}}_{ML}$ of the parameters $\boldsymbol{\theta}$ are defined by the values of $\mathbf{t}$ that maximize the likelihood function:

$$\hat{\boldsymbol{\theta}}_{ML} = \arg\max_{\mathbf{t}} L(\mathbf{t}) = \arg\max_{\mathbf{t}} \log L(\mathbf{t})$$

where $\log L(\mathbf{t})$ is the log-likelihood function.

The joint probability (density) function evaluated at the maximum likelihood estimates $p(\boldsymbol{\omega}; \hat{\boldsymbol{\theta}}_{ML})$ generates observations around or equal to $\mathbf{w}$ with a larger probability than the joint probability (density) function evaluated with any other set of parameters $p(\boldsymbol{\omega}; \mathbf{t})$ (Myung, 2003).

Properties

The MLE has some important properties (van den Bos, 2007):

- *Invariance property*: if $\hat{\boldsymbol{\theta}}_{ML} = \left(\hat{\theta}_1, \hat{\theta}_2, \ldots, \hat{\theta}_R\right)^T$ is the MLE for the parameters $\boldsymbol{\theta} = (\theta_1, \theta_2, \ldots, \theta_R)^T$ and $\boldsymbol{\gamma}(\boldsymbol{\theta})$ is a function of the parameters $\boldsymbol{\theta}$, then $\hat{\boldsymbol{\gamma}}\left(\hat{\boldsymbol{\theta}}_{ML}\right)$ is the MLE of $\boldsymbol{\gamma}(\boldsymbol{\theta})$.

For an increasing number of observations, the most important properties are the following ones:

- *Consistency*: The MLE is consistent meaning that the MLE converges toward the true value of the parameters.
- *Asymptotic efficiency*: The MLE is asymptotically most precise, i.e., the MLE attains asymptotically the CRLB:

$$\operatorname{cov}\left(\hat{\boldsymbol{\theta}}_{ML}\right) \rightarrow CRLB.$$

- *Asymptotic normality*: The probability density function of the MLE asymptotically tends to be a normal distribution having the true parameters $\boldsymbol{\theta}$ as expectation values and the CRLB matrix as covariance matrix:

$$\hat{\boldsymbol{\theta}}_{ML} \overset{a}{\sim} \mathcal{N}(\boldsymbol{\theta}; CRLB).$$

Maximum likelihood estimator for independent normally distributed observations

For an HR(S)TEM image, the observed image pixel values can be assumed to be statistically independent electron counting results, and therefore, they are Poisson distributed. This has led to the expression given by Eq. (2.8) for the joint probability function. For an increasing number of electron counts per pixel, i.e. an increasing expectation value λ_{kl}, the Poisson distribution tends to be a normal distribution with both mean and variance equal to λ_{kl} (Papoulis & Pillai, 2002). This approximation is justified if the magnitude of the observations is large compared to the square root of this magnitude (Koster et al., 1987). In Fig. 2.2, the correspondence between the Poisson distribution and normal distribution with equal mean and variance has been illustrated for increasing values of the mean.

The MLE estimator can be calculated for statistically independent identical normally distributed observations having expectation value $\mathbb{E}[w_{kl}] = f_{kl}(\boldsymbol{\theta})$ and variance $\operatorname{var}(w_{kl}) = \sigma^2$, following the steps described in the beginning of this section. If the contrast in the images is low, the deviations of the observations from their expectations may be supposed to be identically distributed, i.e. $\sigma_{kl} \approx \text{cst} = \sigma$ (Miedema et al., 1994). Then, the joint probability density function as a function of the parameters $\boldsymbol{\theta}$ is given by:

$$p(\boldsymbol{\omega}) = \prod_{k=1}^{K} \prod_{l=1}^{L} \frac{1}{\sqrt{2\pi}\sigma} \exp\left(-\frac{\left(\omega_{kl} - f_{kl}(\boldsymbol{\theta})\right)^2}{2\sigma^2}\right). \tag{2.17}$$

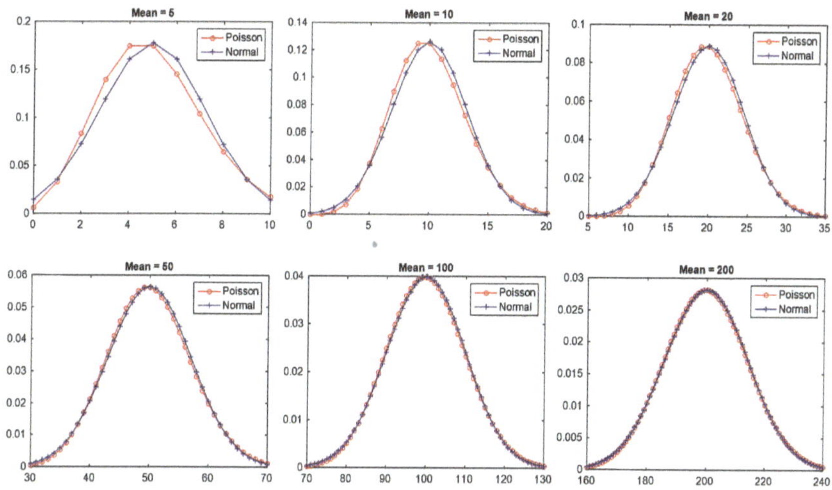

Figure 2.2 Comparison of Poisson and normal distribution with equal mean and variance for increasing mean.

The log-likelihood function is given by:

$$\log L(\mathbf{t}) = -\frac{W}{2}\log(2\pi) - W\log(\sigma) - \frac{1}{2}\sum_{k=1}^{K}\sum_{l=1}^{L}\left(\frac{w_{kl} - f_{kl}(\mathbf{t})}{\sigma}\right)^2. \quad (2.18)$$

The MLE is then given by:

$$\hat{\boldsymbol{\theta}}_{ML} = \arg\max_{\mathbf{t}} \log L(\mathbf{t}) = \arg\min_{\mathbf{t}} \sum_{k=1}^{K}\sum_{l=1}^{L}\left(w_{kl} - f_{kl}(\mathbf{t})\right)^2. \quad (2.19)$$

This means that for independent identically normally distributed observations the MLE estimator equals the uniformly weighted least squares estimator (LSE). The uniformly weighted least squares estimates $\hat{\boldsymbol{\theta}}_{LS}$ of the unknown parameters $\boldsymbol{\theta}$ are given by the values $\mathbf{t}$ that minimize the uniformly weighted least squares criterion. In Chapter 3, an efficient algorithm for the least squares estimation of the parameters of the statistical expectation model for HR(S)TEM images will be discussed in detail.

More generally, if the condition of normally distributed observations with equal variances is dropped and one considers the deviations of the observations from their expectations to be independent, normally distributed with a different standard deviation σ_{kl} for each pixel, the joint probability function becomes:

$$p(\boldsymbol{\omega}) \approx \prod_{k=1}^{K}\prod_{l=1}^{L} \frac{1}{\sqrt{2\pi\sigma_{kl}^2}} \exp\left(-\frac{(\omega_{kl}-f_{kl}(\boldsymbol{\theta}))^2}{2\sigma_{kl}^2}\right). \tag{2.20}$$

Then, the log-likelihood function and the MLE equal

$$\log L(\boldsymbol{t}) = -\frac{W}{2}\log(2\pi) - \sum_{k=1}^{K}\sum_{l=1}^{L}\log(\sigma_{kl}) - \frac{1}{2}\sum_{k=1}^{K}\sum_{l=1}^{L}\left(\frac{w_{kl}-f_{kl}(\boldsymbol{t})}{\sigma_{kl}}\right)^2, \tag{2.21}$$

and

$$\hat{\boldsymbol{\theta}}_{ML} = \arg\max_{t} \log L(\boldsymbol{t}) = \arg\min_{t} \sum_{k=1}^{K}\sum_{l=1}^{L}\left(\frac{w_{kl}-f_{kl}(\boldsymbol{t})}{\sigma_{kl}}\right)^2, \tag{2.22}$$

respectively. The term at the right-hand side of Eq. (2.22) is a weighted least squares estimator, being a generalization of the aforementioned uniformly weighted least squares estimator.

In this section, it has been discussed how the parameters of a statistical expectation model can be determined using the maximum likelihood estimator. Furthermore, the lower bound on the variance of the parameters to be estimated was theoretically described. In the next sections, the precision and accuracy with which the unknown parameters are estimated from HR(S)TEM images will be evaluated and the variance of the estimated parameters will be compared with the theoretical lower bound. First, it will be shown in Section 2.3 how the tools described in this section can help to decide how to estimate the atomic column positions as precisely as possible from a focal series of images. Second, the accuracy and precision of the parameters of interest estimated from ADF STEM images will be investigated in Section 2.4. In Section 2.5, a new parametric model is developed in order to quantitatively analyze ABF and ADF STEM images simultaneously.

2.3. Exit wave reconstruction: precise measurements of atomic column positions

In this section, the tools of statistical parameter estimation theory will be applied for precise measurements of the atomic column positions from a reconstructed exit wave. More details on this can be found in De Backer et al. (2011). The reconstructed exit wave is a complex wave. Often, only the reconstructed phase is employed in a quantitative analysis, although it can be expected that a better precision can be obtained when using the full complex exit wave or the originally required focal series images. The precision of the atomic column position estimates obtained from the reconstructed phase and from the complex exit wave will therefore be compared with the highest attainable precision, i.e. the CRLB, calculated based on the full data set of the focal series. First, a brief introduction is given on exit wave reconstruction. Next, the estimation of the atomic column positions for exit wave reconstruction and the attainable precision for the atomic column position estimates are described, before the results of a simulation study are discussed.

2.3.1 Brief introduction on exit wave reconstruction

The phase information in TEM is lost due to the image recording process, which is a limiting factor since the phase contains significant information on the positions of the atomic columns. In order to retrieve this phase information, the imaging process can be inverted numerically. In this manner, the complete wave function at the exit plane of the object can be computed. A commonly applied technique is to use a series of images recorded at different focal values (Allen et al., 2004a; Allen & Oxley, 2001; Coene & Janssen, 1992; Coene et al., 1996; Hsieh et al., 2004; Kirkland & Meyer, 2004; Meyer et al., 2002; Op de Beeck et al., 1996; Thust et al., 1996; Van Dyck & Op de Beeck, 1990). In this manner, both the amplitude and phase information of the exit wave are recovered. In addition, a better SNR is achieved and residual lens aberrations are eliminated (Erni et al., 2010; Houben et al., 2006; Kirkland & Meyer, 2004; Meyer et al., 2002; Thust et al., 1996). The reconstructed exit wave is closely related to the object structure. In combination with quantitative methods, atomic column positions can be measured with a precision of a few picometers (Bals et al., 2006; Jia, Mi, Faley, et al., 2009; Jia et al., 2008; Urban, 2008; Van Aert, Chang, et al., 2009; Xu et al., 2005).

The ultimate goal is to determine structure parameters with the highest possible precision and accuracy. In the case of a focal series, this means that the parameters need to be determined using the maximum likelihood estimator applied to the originally recorded focal images, which is numerically demanding. Typically, only the phase information of the reconstructed exit wave is used to determine the atomic column positions using a parametric model consisting of a superposition of Gaussian peaks (Bals et al., 2006;

Houben et al., 2006; Jia, Mi, Urban, et al., 2009; Klingstedt et al., 2012; Van Aert, Chang, et al., 2009). In principle, more precise results might be expected if the reconstructed amplitude is also taken into account using a physics-based parametric model using the channeling theory (Geuens & Van Dyck, 2002, 2005; Van Aert et al., 2007; Van Dyck & Chen, 1999b; Van Dyck & Op de Beeck, 1996). In the current work, a real space model-based approach for exit wave reconstruction based on channeling theory is introduced in order to investigate the precision that can ultimately be obtained for the atomic column positions. By means of image simulations of an ideal Au [100] sample, which are subject to counting noise, different parameter estimation methods have been compared. It has been investigated what precision can be obtained by using the phase information of the reconstructed exit wave only or by using the full complex exit wave for different specimen thicknesses. This precision has been compared with the attainable precision, i.e. the CRLB, that could in principle be obtained asymptotically by applying the maximum likelihood estimator to the full series of focal images. Although our simulation experiments represent an idealistic case, they serve as a reference for model-based exit wave reconstruction and statistical parameter estimation theory of atomic column positions. The simulations should answer the question whether the use of only the phase information in a quantitative analysis is justified and if the analysis would benefit if the amplitude would be used in addition.

2.3.2 Model-based exit wave reconstruction

In order to investigate the precision with which the atomic column positions can be measured, a model-based exit wave reconstruction technique is considered here. The image intensities of the focal series are backpropagated to the zero-focus plane to reconstruct the wave at the exit face of the specimen. Next, the set of backpropagated images is averaged. This real space approach for exit wave reconstruction is very intuitive and understandable. In most reconstruction methods, the only aim is to reconstruct the complex exit wave as such. However, since it is ultimately the purpose to derive also the crystal structure from the exit wave, we want to go one step further and include also the channeling theory for the electron interaction already in the exit wave. Therefore, this model-based approach for exit wave reconstruction is conceptually different from the existing focal series reconstruction methods (Allen et al., 2004a; Allen & Oxley, 2001; Coene & Janssen, 1992; Coene et al., 1996; Hsieh et al., 2004; Kirkland & Meyer,

2004; Meyer et al., 2002; Op de Beeck et al., 1996; Thust et al., 1996; Van Dyck & Op de Beeck, 1990).

The channeling theory describes the quantum mechanical interaction of the electrons with the object and provides a closed-form analytical expression for the exit wave which is parametric in the structure parameters (Geuens & Van Dyck, 2002, 2005; Van Dyck & Chen, 1999b; Van Dyck & Op de Beeck, 1996). The complex exit wave of an object consisting of atomic columns described in the channeling theory consists of a vacuum wave and an interaction wave:

$$\begin{aligned}\psi_{ex}(\mathbf{r}, z) &= 1 + \psi_{int}(\mathbf{r}, z) \\ &= 1 + \sum_{i=1}^{I} \sum_{m_i=1}^{M_i} c_{m_i} \phi_{1s,m_i}(\mathbf{r} - \boldsymbol{\beta}_{m_i}) \left[\exp\left(-i\pi \frac{E_{1s,m_i}}{E_0} \frac{1}{\lambda} z \right) - 1 \right],\end{aligned} \tag{2.23}$$

where $\mathbf{r} = (x, y)^T$ is a two-dimensional vector in the exit plane of the object, perpendicular to the incident beam direction, E_0 denotes the incident electron energy, λ the electron wavelength, and z the object thickness. Lastly, $\phi_{1s,m_i}(\mathbf{r} - \boldsymbol{\beta}_{m_i})$ is the eigenfunction of the lowest energy bound state with energy E_{1s,m_i}. The corresponding excitation coefficients c_{m_i} can be found in Van Dyck and Op de Beeck (1996). The 1s-state of an atom column m_i at position $\boldsymbol{\beta}_{m_i} = \left(\beta_{x_{m_i}}, \beta_{y_{m_i}} \right)^T$ can be approximated by (Geuens & Van Dyck, 2002):

$$\phi_{1s,m_i}(\mathbf{r} - \boldsymbol{\beta}_{m_i}) = \frac{1}{a_{m_i}\sqrt{2\pi}} \exp\left(-\frac{\left(x - \beta_{x_{m_i}}\right)^2 + \left(y - \beta_{y_{m_i}}\right)^2}{4a_{m_i}^2} \right), \tag{2.24}$$

where a_{m_i} is the column dependent width. The interpretation of Eq. (2.23) is as follows. Each atomic column acts as a channel in which the wave function oscillates periodically with depth (Van Aert et al., 2007; Van Dyck & Op de Beeck, 1996). This periodicity is related to the average mass density of an atomic column and is given by

$$D_{1s} = \left| \frac{2E_0\lambda}{E_{1s,m_i}} \right|. \tag{2.25}$$

This distance D_{1s} is called the extinction distance. The wave function in the image plane is given by the exit wave convoluted with a point spread function accounting for the effect of defocus ϵ and spherical aberration C_s (Van Dyck, 2002):

$$\psi_{\text{im}}(\mathbf{r}, z; C_s, \epsilon) = 1 + \psi_{\text{int}}(\mathbf{r}, z) * p_{C_s}(\mathbf{r}; C_s) * p_\epsilon(\mathbf{r}; \epsilon). \tag{2.26}$$

The intensity in the image plane is given by the modulus square of the image wave (Van Dyck, 2002):

$$I(\mathbf{r}, z; C_s, \epsilon) = |\psi_{\text{im}}(\mathbf{r}, z; C_s, \epsilon)|^2. \tag{2.27}$$

Other aberration effects are neglected in the simulations as they can be numerically corrected in the reconstruction (Allen et al., 2004a, 2004b; Coene et al., 1996; Kirkland & Meyer, 2004; Meyer et al., 2002).

In order to reconstruct the exit wave, the image intensities of the focal series are backpropagated. This backpropagation is performed by a convolution with the inverse point spread function of the defocus $p_\epsilon^{-1}(\mathbf{r}; \epsilon) = p_\epsilon^*(\mathbf{r}; \epsilon)$, which equals the point spread function of the defocus with the opposite focal value $p_\epsilon(\mathbf{r}; -\epsilon)$. In this manner, a compensation is obtained for the effect of the defocus. The same procedure can be applied for spherical aberration resulting in the following expression:

$$\Psi_{\text{rec}}(\mathbf{r}, z) = \langle I(\mathbf{r}, z)\rangle = \frac{1}{M}\sum_{m=1}^{M} I(\mathbf{r}, z; C_s, \epsilon) * p_\epsilon(\mathbf{r}, -\epsilon_m) * p_{C_s}(\mathbf{r}; -C_s). \tag{2.28}$$

The focal value ϵ_m corresponds to the focal value of an image of the focal series. Other aberrations, like incoherence effects present in the experimental images can be corrected by including them in the backpropagation by the appropriate inverse point spread function (Allen et al., 2004a; de Jong & Van Dyck, 1993). It can be shown that the reconstructed exit wave using Eq. (2.28) will show good agreement with the true input exit wave (De Backer et al., 2011). This agreement enables us to use the parametric model from the channeling theory given by Eq. (2.23) in a statistical parameter estimation procedure applied to the reconstructed exit wave.

In the next subsection, the least squares estimation procedure when using the phase of the exit wave only or when using the full complex exit wave will be discussed.

2.3.3 Estimation of atomic column positions

Least squares estimator using the phase of the reconstructed exit wave

It is generally known that the phase of the exit wave is peaked at the atomic column positions. Accordingly, an empirical model is often used in which the phase is modeled as a superposition of Gaussian functions which are peaked at the atomic column positions (Bals et al., 2006; Houben et al., 2006; Jia, Mi, Urban, et al., 2009; Van Aert, Chang, et al., 2009). Therefore, the expectation at the pixel (k, l) at position (x_k, y_l) of the phase of the exit wave can be described by the model of Eq. (2.9):

$$f_{kl}^{\mathrm{p}}(\boldsymbol{\theta}) = \zeta + \sum_{i=1}^{I} \sum_{m_i}^{M_i} \eta_{m_i} \exp\left(- \frac{\left(x_k - \beta_{x_{m_i}}\right)^2 + \left(y_l - \beta_{y_{m_i}}\right)^2}{2\rho_i^2} \right) \tag{2.29}$$

where the superscript p refers to the phase. The parameters of this model

$$\boldsymbol{\theta} = \left(\beta_{x_{1_1}}, \ldots, \beta_{x_{M_I}}, \beta_{y_{1_1}}, \ldots, \beta_{y_{M_I}}, \rho_1, \ldots, \rho_I, \eta_{1_1}, \ldots, \eta_{M_I}, \zeta\right)^T \tag{2.30}$$

are estimated in the least squares sense. The uniformly weighted least squares estimated $\hat{\boldsymbol{\theta}}$ are given by the values of $\mathbf{t}$ that minimize the uniformly weighted least squares criterion:

$$\hat{\boldsymbol{\theta}} = \arg \min_{\mathbf{t}} \sum_{k=1}^{K} \sum_{l=1}^{L} \left(w_{kl} - f_{kl}^{\mathrm{p}}(\mathbf{t})\right)^2 \tag{2.31}$$

with w_{kl} the value of the reconstructed phase at the pixel (k, l).

Least squares estimator using the phase and amplitude of the reconstructed exit wave

In order to attain a higher precision, especially for thicker specimens, all the information available in the reconstructed exit wave, i.e. phase and amplitude information, can be used to estimate the unknown parameters. The implementation for the complex exit wave is fully equivalent to the procedure for the phase, although more extended, and is based on the channeling theory. The complex exit wave is given by Eq. (2.23). This physics-based parametric model for the complex exit wave can be rewritten and separated in a real $\mathcal{R}$ and imaginary $\mathcal{I}$ part:

$$f_{kl}^{\mathrm{c}}(\boldsymbol{\theta}) = \mathcal{R}(\boldsymbol{\theta}) + i\mathcal{I}(\boldsymbol{\theta})$$

$$= \left[\zeta_{\mathcal{R}} + \sum_{i=1}^{I} \sum_{m_i=1}^{M_i} \eta_{m_i}^{\mathcal{R}} \exp\left(-\frac{\left(x_k - \beta_{x_{m_i}}\right)^2 + \left(y_l - \beta_{y_{m_i}}\right)^2}{2\rho_i^2} \right) \right]$$

$$+ i \left[\zeta_{\mathcal{I}} + \sum_{i=1}^{I} \sum_{m_i}^{M_i} \eta_{m_i}^{\mathcal{I}} \exp\left(-\frac{\left(x_k - \beta_{x_{m_i}}\right)^2 + \left(y_l - \beta_{y_{m_i}}\right)^2}{2\rho_i^2} \right) \right] \tag{2.32}$$

where the superscript c refers to the complex wave. The effect of the thickness of the object is incorporated in the parameters $\eta_{m_i}^{\mathcal{R}}$ and $\eta_{m_i}^{\mathcal{I}}$. The vector containing the unknown parameters of the complex model is:

$$\boldsymbol{\theta} = \left(\beta_{x_{1_1}}, \ldots, \beta_{x_{M_I}}, \beta_{y_{1_1}}, \ldots, \beta_{y_{M_I}}, \right.$$
$$\left. \rho_1, \ldots, \rho_I, \eta_{1_1}^{\mathcal{R}}, \ldots, \eta_{M_I}^{\mathcal{R}}, \eta_{1_1}^{\mathcal{I}}, \ldots, \eta_{M_I}^{\mathcal{I}}, \zeta_{\mathcal{R}}, \zeta_{\mathcal{I}}\right)^T. \tag{2.33}$$

Minimization of the uniformly weighted least squares criterion yields the estimated parameters:

$$\hat{\boldsymbol{\theta}} = \arg\min_{\mathbf{t}} \sum_{k=1}^{K} \sum_{l=1}^{L} \left[\left(w_{kl}^{\mathcal{R}} - \mathcal{R}(\mathbf{t})\right)^2 + \left(w_{kl}^{\mathcal{I}} - \mathcal{I}(\mathbf{t})\right)^2 \right] \tag{2.34}$$

where $w_{kl}^{\mathcal{R}}$ and w_{kl}^{I} represent the real and imaginary parts of the reconstructed complex exit wave, respectively.

2.3.4 Attainable precision

It is expected that the lower bound on the variance of the unknown parameters can be attained by applying the maximum likelihood estimator to the originally recorded focal images, which is, however, unpractical and computationally demanding. In order to derive this Cramér-Rao lower bound on the variance of the unknown parameters, the model describing the expectations of the original observations of the focal series is needed. The expectation model for the image intensity of an image of the recorded focal series used for this purpose equals:

$$f_{kl}^{\mathrm{i},m}(\boldsymbol{\theta}) = N \frac{I(\mathbf{r}, z; C_s, \epsilon_m)}{I_{\mathrm{norm}}} \Delta x \Delta y \tag{2.35}$$

where the superscript i refers to the image intensity and m to the mth image of the focal series. This result defines the number of electrons expected to be found at a pixel (k, l) taking into account the electron dose.

In this equation, N denotes the total number of incident electrons and Δx and Δy represent the sampling distances in the x- and y-directions, and $I(\mathbf{r}, z; C_s, \epsilon_m)$ denotes the intensity in the image plane given by Eq. (2.27). The normalization factor I_{norm} in Eq. (2.35) is given by:

$$I_{\text{norm}} = \int I(\mathbf{r}, z; C_s, \epsilon_m) d\mathbf{r} \tag{2.36}$$

where the integral extends over the whole field of view. For HRTEM images in a focal series, the observations w_{kl}^m are assumed to be statistically independent having a Poisson distribution with the following joint probability density function equivalent to Eq. (2.8) (den Dekker et al., 2005; Van Aert et al., 2005):

$$p(\boldsymbol{\omega}) = \prod_{m=1}^{M} \prod_{k=1}^{K} \prod_{l=1}^{L} \frac{\left(\lambda_{kl}^m\right)^{\omega_{kl}^m}}{\omega_{kl}^m!} \exp\left(-\lambda_{kl}^m\right) \tag{2.37}$$

where λ_{kl}^m is the expectation of a pixel value, i.e. $\mathbb{E}[w_{kl}] = \lambda_{kl}^m = f_{kl}^{\text{i},m}(\boldsymbol{\theta})$. The Fisher information matrix for this set of Poisson distributed observations then becomes (equivalent to Eq. (2.12))

$$F_{rs} = \sum_{m=1}^{M} \sum_{k=1}^{K} \sum_{l=1}^{L} \frac{1}{\lambda_{kl}^m} \frac{\partial \lambda_{kl}^m}{\partial \theta_r} \frac{\partial \lambda_{kl}^m}{\partial \theta_s}. \tag{2.38}$$

The CRLB on the variance is given by the inverse of the Fisher information matrix. The derivatives in Eq. (2.38) are calculated with respect to the parameters $\boldsymbol{\theta}$ in the expectation model given by Eq. (2.35), i.e. the positions, the width, and the height of the atomic columns and a constant background. The diagonal elements of the inverse Fisher information matrix define lower bounds on the parameters $\boldsymbol{\theta}$. The thus obtained numbers will be compared with the variance of the structure parameters estimated from the phase of the reconstructed exit wave and estimated from the reconstructed complex exit wave. The exit wave is reconstructed using the method described in Subsection 2.3.2. This comparison is first carried out for an isolated atomic column and next for an assembly of atom columns. Furthermore, the comparison is done for different specimen thicknesses. For thin specimens for which the weak phase object approximation (WPOA) is valid, most object information will be contained in the phase of the reconstructed exit wave. The WPOA is valid if (Geuens & Van

Table 2.1 Object and microscope parameters for the simulation of the focal series.

Parameter	**Symbol**	**Value**
Atomic number	Z	79
Column dependent width	a_{m_i} (Å^2)	0.13
1s state energy	E_{1s,m_i} (eV)	−210.8
Interatomic distance	d (Å)	4.08
Extinction distance	D_{1s} (nm)	5.7
Extinction distance	D_{1s} (atoms)	14
Accelerating voltage	V_0 (kV)	300
Mean focal value	ϵ (nm)	−75
Focal step	γ (nm)	2.5
Spherical aberration constant	C_s (mm)	0.5
Number of focal images	M	21
Pixel size	$\Delta x = \Delta y$ (Å)	0.1
Number of pixels x- and y-direction	$K = L$	1024
Electron dose	N_d (e$^-$Å^{-2})	7854

Dyck, 2005; Van Dyck & Op de Beeck, 1996):

$$\left|E_{1s,n}\right| < \frac{E_0\lambda}{\pi z}. \tag{2.39}$$

This can be rewritten in terms of the extinction distance D_{1s} given by Eq. (2.25), i.e. the WPOA is valid if the specimen thickness is smaller than $0.16D_{1s}$.

2.3.5 Simulations of Au [100]

Focal series for different thicknesses are simulated of an isolated Au [100] atomic column and of a 7 × 7 lattice of projected Au [100] atomic columns, modeled using the channeling theory. The microscope parameters for the simulation of the focal series are shown in Table 2.1; the value of C_s is arbitrarily chosen nonzero. For Au [100], the extinction distance D_{1s} equals 5.7 nm, i.e. about 14 atoms. This means that the WPOA is valid for a specimen thickness less than two atoms. However, it should be noted that some deviations can be expected from Eq. (2.25) for very heavy atoms.

The complex exit wave is reconstructed using Eq. (2.28) for 100 different Poisson distributed noise realizations of a focal series calculated from the expectation model given by Eq. (2.35). The unknown parameters of the parametric models for the phase of the reconstructed exit wave and

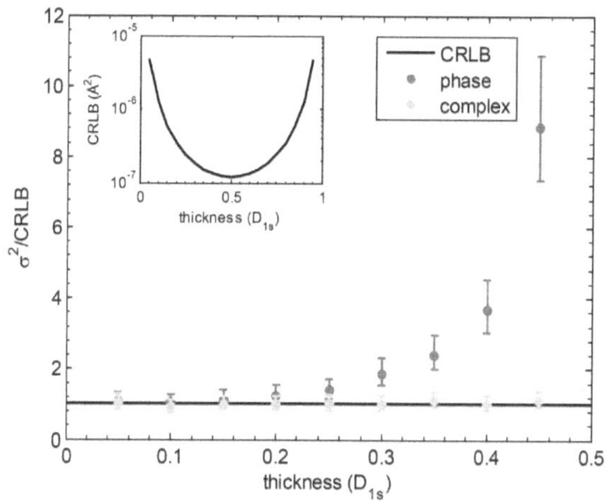

Figure 2.3 Attainability of the CRLB for the estimated column position coordinates as a function of the object thickness for an isolated Au [100] atomic column. The error bars show the 95% confidence intervals for the estimated variances. The inset shows the CRLB as a function of the object thickness. Reprinted from De Backer et al. (2011), Copyright (2011), with permission from Elsevier.

the reconstructed complex exit wave, described by Eqs. (2.29) and (2.32), respectively, are estimated in the least squares sense. The sample mean and sample variance are computed from the 100 estimates and compared with the true values and the CRLB on the variance of the position parameters, respectively.

Isolated atomic column

The results of the analysis of the attainability of the CRLB for an isolated column as a function of the object thickness are presented in Fig. 2.3. The figure shows the ratio of the estimated variance to the calculated CRLB for the x- and y-position coordinates. The CRLB on the variance of the position parameters also depends on the object thickness as is shown in the inset of Fig. 2.3. The attainable precision for the position parameters is highest when the thickness equals half the extinction distance, $0.5D_{1s}$. Here, the electrons are most strongly localized at the atomic column positions implying that the exit wave is peaked maximally at the atomic positions. From Fig. 2.3, it can be concluded that the lower bound on the variance is attained for all thicknesses when the column position coordinates are estimated from the complex reconstructed exit wave using the parametric model given by the channeling theory (Eq. (2.32)). This result is also valid for thicknesses larger than $0.5D_{1s}$ and will show a periodic behavior since the exit wave varies periodically with depth. By estimating the position coordinates from the phase of the reconstructed exit wave using the model of Gaussian peaks of Eq. (2.29), the CRLB is attained only when the WPOA is valid, i.e. if the thickness is smaller than $0.16D_{1s}$. It is clear that for larger object thicknesses, for which the WPOA is no longer valid, estimating the position coordinates from the complex reconstructed exit wave is necessary in order to attain the highest possible precision.

Assembly of atomic columns

To evaluate the effect of neighboring atomic columns, a 7×7 lattice of projected Au [100] atomic columns is modeled. For an assembly of atomic columns two different thicknesses are considered: $z_1 = 0.1D_{1s}$ corresponding to a specimen thickness of only one atom and $z_2 = 0.4D_{1s}$ corresponding to a specimen thickness of five atoms. The WPOA is valid for z_1 but not for z_2. The positions of the central 5×5 atomic columns are considered in the analysis in order to avoid boundary effects at the outermost atomic columns of the simulations. The results for the attainability of the CRLB for both thicknesses z_1 and z_2 are presented in Fig. 2.4. The ratio of the estimated variance of the x- and y-coordinates to the CRLB are plotted for the 25 atomic columns. As expected for the thin object, the variances on the position parameters correspond to the theoretical lower bound on the variance for both estimating the position coordinates from the phase information only and from the full complex exit wave. For the thicker object, where the WPOA is no longer valid, the CRLB can only be attained by estimating the position coordinates from the complex model for the exit wave to the reconstructed exit wave. If the position parameters are determined based on the full complex exit wave, the precision is approximately three to four times better in terms of the variance than when determining the atomic column positions from the phase of the reconstruction. This result is in agreement with the results of the isolated atomic column.

Discussion

In our evaluation of the precision with which the positions of the atomic columns can be estimated from (i) a focal series of images, (ii) using the complete reconstructed complex exit wave (i.e. amplitude and phase), or (iii) using only the phase of the reconstructed complex exit wave, the simplified channeling theory has been used to describe the dynamical scattering of the electrons through the object. Obviously, this approximation

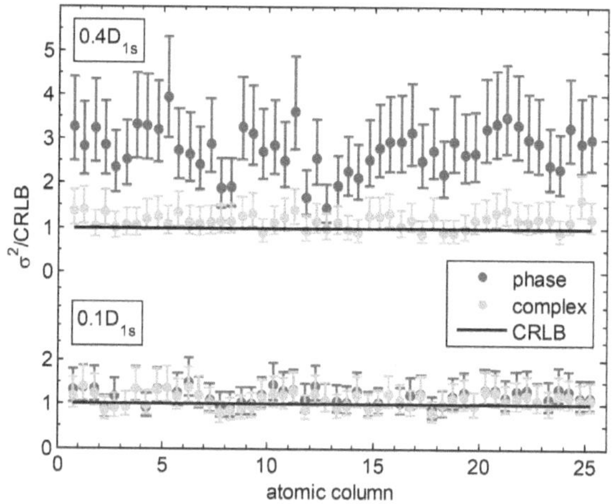

Figure 2.4 Attainability of the CRLB for 25 position estimates of Au [100] atomic columns. The bottom of the graph shows the result for thickness $z_1 = 0.1D_{1s}$ and the top of the graph shows the result for thickness $z_2 = 0.4D_{1s}$. Each atomic column position corresponds to two estimated ratios in the graph, the x-coordinate and y-coordinate, resulting in 50 data points on the horizontal axis. The error bars show the 95% confidence intervals for the estimated variances. Reprinted from De Backer et al. (2011), Copyright (2011), with permission from Elsevier.

is of limited validity. However, the analysis presented here focuses on the question whether the use of only the phase information in a quantitative analysis for atomic column measurements is justified or if the analysis would benefit if the amplitude would be used in addition. The simplified model-based channeling theory is very useful for this purpose and the results are independent of the approximations made. Furthermore, it should be noted that, even when using a more accurate simulation approach, in practice the lower bound and estimated variances will be typically larger, since their actual values critically depend on the image contrast. A reduction of contrast will indeed result in a loss of precision (Van Aert, Chang, et al., 2009). Phonon scattering and different experimental conditions, like the stability of the specimen and the microscope, atomic thermal vibrations, lattice configuration, residual microscope aberrations, time dependent incoherence may contribute to a loss of image contrast.

Summary

In this section, we investigated the precision with which the atomic columns can be measured from an exit wave reconstructed from a focal series of images. Although one might have expected that the CRLB could only be attained asymptotically when deriving the maximum likelihood estimates from the full series of focal images, it has been demonstrated that the reconstructed complex exit wave contains the same amount of information on the positions parameters. Estimating the position parameters from the complex exit wave using the parametric model from the channeling theory thus results in the highest possible precision. This precision cannot be obtained when using the phase information only except for thin specimens where the weak phase approximation is valid, that is, thinner than $0.16D_{1s}$ with D_{1s} the extinction distance. For thicker specimens, an improvement of a factor of about four in terms of the variance with which the atomic column positions can be estimated is observed, when using the full complex exit wave in the estimation procedure. From this study, we can conclude that it is therefore recommendable to use the full complex exit wave to determine quantitative information about the atomic column positions.

In the next section, the concepts of statistical parameter estimation theory will be used to evaluate the accuracy and precision of the measurements of the atomic column positions and the scattering cross-sections from ADF STEM images.

2.4. ADF STEM: accurate and precise measurements of the scattering cross-sections and atomic column positions

Since ADF STEM images show Z-contrast and are sensitive to the number of atoms in the atomic columns, ADF STEM has become a popular technique for chemical composition determination and atom counting because it provides directly interpretable images. Statistical model-based estimation procedures are required in this field, because it is impossible to e.g. directly count the number of atoms by means of visual interpretation only. Statistical parameter estimation theory enables one to extract quantitative information, such as chemical composition (Martinez, Rosenauer, et al., 2014; Van Aert, Verbeeck, et al., 2009) or the number of atoms (Van Aert et al., 2011, 2013). In Chapter 4 of this work, a statistics-based method for atom counting will be discussed in detail and will be applied to different nanostructures. Here, the aim is to count the number of atoms with single atom sensitivity. The parameters of the imaging model are therefore interpreted in terms of total intensities of scattered electrons by the atomic columns. These so-called scattering cross-sections are further explored in order to count the number of atoms that are present in an atomic column. The goal is then to estimate these scattering cross-sections and atomic column positions unbiasedly with the highest possible precision. Therefore, it is important to assess the accuracy and precision of the estimated parameters. This has been discussed in detail in De Backer et al. (2016). Furthermore, it should be noticed that it is assumed that the pixel values of the ADF STEM images are statistically independent counting results which can be modeled as Poisson distributed random variables. For the estimation of the unknown parameters of the expectation model, the least squares estimator is employed, which equals the maximum likelihood estimator in the case of independent normally distributed random variables as derived in Subsection 2.2.3. The Poisson distribution tends to be a normal distribution for increasing expectation values for the pixel values. However, this assumption about the statistical distribution of the pixel values is not always valid, especially not at low electron doses, needed to avoid radiation damage in beam-sensitive matter. In order to evaluate the accuracy and precision of the estimated parameters, use will be made of detailed ADF STEM multislice simulations. This will be done for the scattering cross-section and atomic column positions obtained using different noise realizations of the ADF STEM multislice simulation. The sample means of the repetitive noise realizations will be compared with the true values of the parameters in order to verify the accuracy and the sample variances will be compared with the CRLB.

First, model-based parameter estimation will be reformulated for ADF STEM images in Subsection 2.4.1. Here, also the parameter of interest for the interpretation of the ADF STEM image intensities, i.e. the scattering cross-section, will be introduced. Next, the attainable precision for the position parameters and scattering cross-sections will be defined in Subsection 2.4.2. Then, the ADF STEM simulations will be discussed briefly in Subsection 2.4.3 before the results of the simulation study for the assessment of the accuracy and precision of the parameters of interest will be shown in Subsection 2.4.4.

2.4.1 Model-based parameter estimation in ADF STEM

As a first step in the quantitative analysis of ADF STEM images, such as the statistics-based atom-counting analysis for mono-type crystalline nanostructures which will be presented in Chapter 4, the ADF STEM images will be considered as a data plane from which the structure parameters have to be determined with the highest possible precision and accuracy. Therefore, the expectation model of Eq. (2.9) assuming mono-type atomic columns will be used:

$$f_{kl}(\boldsymbol{\theta}) = \zeta + \sum_{n=1}^{N} \eta_n \exp\left(-\frac{\left(x_k - \beta_{x_n}\right)^2 + \left(y_l - \beta_{y_n}\right)^2}{2\rho^2}\right). \tag{2.40}$$

The unknown parameters are thus given by the parameter vector:

$$\boldsymbol{\theta} = \left(\beta_{x_1}, \ldots, \beta_{x_N}, \beta_{y_1}, \ldots, \beta_{y_N}, \rho, \eta_1, \ldots, \eta_N, \zeta\right)^T. \tag{2.41}$$

The parametric model of Eq. (2.40) is fitted to the experimental data using the least squares criterion quantifying the similarity between the experimental images and the model:

$$\hat{\boldsymbol{\theta}} = \arg\min_{\mathbf{t}} \sum_{k=1}^{K} \sum_{l=1}^{L} \left(w_{kl} - f_{kl}(\mathbf{t})\right)^2. \tag{2.42}$$

The volume under a peak above a background reflects the total intensity of electrons scattered toward the ADF detector for every atomic column. These are the so-called scattering cross-sections which can be used to quantify chemical composition or number of atoms in a projected atomic column from an ADF STEM image. These scattering cross-sections can be calculated as follows (Van Aert, Verbeeck, et al., 2009):

$$\hat{V}_n = 2\pi \hat{\eta}_n \hat{\rho}^2. \tag{2.43}$$

The precision and accuracy of the estimated atomic column positions $\hat{\beta}_{x_n}$ and scattering cross-sections $\hat{V}_n$ will be evaluated in the next paragraphs.

2.4.2 Attainable precision

It is expected that the lower bound on the variance of the unknown parameters, i.e. the atomic column positions and the scattering cross-sections, can asymptotically be attained when applying the maximum likelihood estimator. Here, the least squares estimator of Eq. (2.42) is used, although the observations are modeled as independent Poisson distributed variables. The Poisson distribution tends to be a normal distribution, for an increasing expectation value λ_{kl} as discussed in Section 2.2.3. This means, that when the detected electron dose is high, the uniformly weighted least squares estimator equals maximum likelihood estimator. However, for low doses this approximation is not valid. Therefore, the attainability of the lower bound on the variance of atomic column position and scattering cross-section will be investigated for different incident electron doses.

The Fisher information matrix for ADF STEM images, which are a set of Poisson distributed observations equals (Eq. (2.12)):

$$F_{rs} = \sum_{k=1}^{L} \sum_{l=1}^{L} \frac{1}{\lambda_{kl}} \frac{\partial \lambda_{kl}}{\partial \theta_r} \frac{\partial \lambda_{kl}}{\partial \theta_s}. \tag{2.44}$$

The diagonal elements of the inverse Fisher information matrix define lower bounds on the parameters $\boldsymbol{\theta}$. Following Eq. (2.16), the expression for the CRLB on the variance of the scattering cross-sections, which is a function of the parameters $\boldsymbol{\theta}$ is given by:

$$\sigma_{V_n}^2 = \frac{\partial V_n}{\partial \boldsymbol{\theta}} F_{\boldsymbol{\theta}}^{-1} \left(\frac{\partial V_n}{\partial \boldsymbol{\theta}} \right)^T \tag{2.45}$$

where V_n is the scattering cross-section for the nth atomic column in the ADF STEM image given by Eq. (2.43). The elements of the Fisher information matrix are given by Eq. (2.44). The vector $\frac{\partial V_n}{\partial \boldsymbol{\theta}}$ is a $(3N+2)$-vector consisting of zeros except for the $(2N+1)$th and $(2N+1+n)$th elements. These elements are equal to:

$$\begin{aligned} \left[\frac{\partial V_n}{\partial \boldsymbol{\theta}} \right]_{1,2N+1} &= \frac{\partial V_n}{\partial \rho} = 4\pi \eta_n \rho \\ \left[\frac{\partial V_n}{\partial \boldsymbol{\theta}} \right]_{1,2N+1+n} &= \frac{\partial V_n}{\partial \eta_n} = 2\pi \rho^2. \end{aligned}$$

Explicit numbers for the CRLB on the variance of the atomic column positions and the scattering cross-sections can be computed by substituting the values of the true parameters into Eqs. (2.44) and (2.45). However, since the observations are Poisson distributed noise realizations of ADF STEM images that were simulated using the STEMsim software (Rosenauer & Schowalter, 2008), only the true parameters of the atomic column positions are available. Therefore, the elements F_{rs} and $\sigma_{V_n}^2$ may only be estimated. This will be done by substituting estimated values for the parameters ρ and η_n for the true parameters. These estimates for the true parameters of ρ and η_n are obtained by fitting the model given by Eq. (2.40) to the detailed STEMsim model for the ADF STEM image, i.e. in absence of noise, which will be described in the next subsection. The estimated Fisher information

Table 2.2 Settings used for absorptive potential multislice simulations of Pt in [100] and [110] zone axes.

Parameter	Symbol	Value
Acceleration voltage	V_0 (kV)	300
Defocus	ϵ (nm)	−8.8741
Spherical aberration	C_s (mm)	0.04
Spherical aberration of 5th order	C_5 (mm)	10
Objective aperture angle	α (mrad)	21.1
Spatial incoherence of source	FWHM (Å)	0.7
Number of unit cells per supercell	$N_a \times N_b$	9×9
HAADF inner collection angle	β_1 (mrad)	60
HAADF outer collection angle	β_2 (mrad)	190
Maximum specimen thickness [100]	$z_{[100]}$ (nm)	29.4
Maximum specimen thickness [110]	$z_{[110]}$ (nm)	20.8
Debye-Waller factor	B (Å^2)	0.384
Pixel size [100]	$\Delta x = \Delta y$ (Å)	0.1569
Pixel size [110]	$\Delta x = \Delta y$ (Å)	0.1121

matrix and estimated lower bounds on the variances will be denoted as $\hat{F}_\theta$ and $\hat{\sigma}^2$, respectively.

2.4.3 HAADF STEM simulation study

The analysis presented here will make use of multislice simulations under the absorptive potential approximation (Ishizuka, 2002) because of their suitability to describe electron-sample interaction for thin samples. The ADF STEM images are calculated using the STEMsim software (Rosenauer & Schowalter, 2008). The absorptive potential approximation is computationally less demanding while it still describes the image intensities properly up to 50 nm thickness of the sample (Rosenauer et al., 2009). We simulated a Pt structure in [100] and [110] zone axis up to 75 atoms thickness, i.e. ≈ 30 nm and ≈ 21 nm, respectively. The simulated images were convolved with a Gaussian function to account for spatial incoherence. The simulation settings are summarized in Table 2.2. The Debye-Waller factor in the table describes the atomic vibrations. This Debye-Waller factor is temperature dependent and has been parametrized for several elemental crystals and compounds with zinc-blende structure by Gao and Peng (1999).

Using the model described by Eq. (2.40), the scattering cross-sections and atomic column positions are estimated from the simulated images. As presented in Martinez, Rosenauer, et al. (2014), fitting a constant background to a simulation is not needed. In experimental images, the background considers signals that do not originate from the sample under study, such as the sample support or grid. The simulated images do not include these contributions. Consequently, a background with value 0 was considered for this analysis.

The analysis of simulated images for two different zone axes is intended to consider different atomic column spacings, having a different effect on the cross-talk of the electron beam through the sample.

2.4.4 Results

Detailed ADF STEM simulations are available for an increasing thickness from 1 to 75 atoms for Pt in the [100] and [110] orientation as described in the previous subsection. In order to evaluate the attainability of the CRLB, different values for the incident electron dose N_d ($e^-/\text{Å}^2$) are considered:

$$\lambda_{kl} = N_p f_{kl} = N_d \Delta x \Delta y f_{kl} \tag{2.46}$$

where λ_{kl} equals the expectation model including the electron dose, f_{kl} equals the detailed ADF STEM simulation, N_p equals the incident electron dose per pixel, $\Delta x = \Delta y$ equals the pixel size, and N_d the incident electron dose per area. Atomic-resolution imaging is still possible with doses of the order of $N_d = 10^2$ $e^-\text{Å}^{-2}$ for low-angle annular dark-field (LAADF) STEM according to Buban et al. (2010). Therefore, electron doses of $N_d = 5 \cdot 10^2$ $e^-\text{Å}^{-2}$ and $N_d = 5 \cdot 10^5$ $e^-\text{Å}^{-2}$ are used in the simulation study based on HAADF STEM images. For each thickness, 100 noise realizations are simulated for the two different doses. Next, the noise realizations are normalized with respect to the incident electron dose, i.e. divided by the incident electron dose placing the noise realizations on the same intensity scale as the detailed image simulations f_{kl} from STEMsim. The detailed STEM simulations and corresponding noise realizations are illustrated in Fig. 2.5.

From the noise realizations, the parameters of the parametric model given by Eq. (2.40) are estimated. The parameters that are evaluated here are the normalized scattering cross-section and the x-coordinate of the central atomic column in the simulated ADF STEM image. Only the x-coordinate is evaluated since the x- and y-coordinate show exactly the same behavior for this symmetric simulation. In order to evaluate the accuracy of the estimates, the sample means have been compared with the reference values. For the atomic column positions, the reference value equals the true value

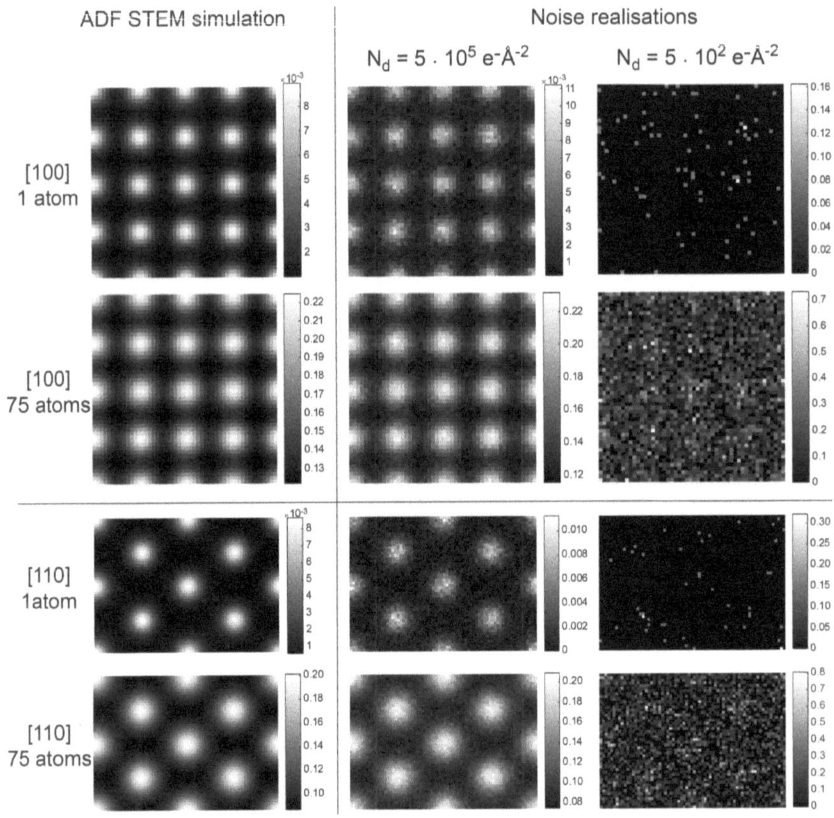

Figure 2.5 Detailed ADF STEM simulations for [100] and [110] orientation for 1 atoms and 75 atoms thickness. The corresponding normalized noise realizations for an incoming electron dose of $N_d = 5 \cdot 10^2$ e$^-$Å^{-2} and $N_d = 5 \cdot 10^5$ e$^-$ Å^{-2} are also shown.

of the atomic column position from the input structure of the STEM simulation described in the previous paragraph. For the scattering cross-sections, the values estimated from the detailed ADF STEM simulation in absence of Poisson distributed noise are taken as a reference. The $(1 - \alpha) \times 100\%$ confidence intervals of the sample mean of the x-coordinate $\bar{\beta}_x$ and the scattering cross-section $\bar{V}$ of the central atom are given by:

$$\left[\bar{\beta}_x - t_{1-\frac{\alpha}{2},N-1}\frac{s_{\beta_x}}{\sqrt{N}}, \bar{\beta}_x + t_{1-\frac{\alpha}{2},N-1}\frac{s_{\beta_x}}{\sqrt{N}}\right] \tag{2.47}$$

$$\left[\bar{V} - t_{1-\frac{\alpha}{2},N-1}\frac{s_V}{\sqrt{N}}, \bar{V} + t_{1-\frac{\alpha}{2},N-1}\frac{s_V}{\sqrt{N}}\right] \tag{2.48}$$

with s_{β_x} and s_V the square root of the sample variance and $t_{1-\frac{\alpha}{2},N-1}$ equals the $1-\frac{\alpha}{2}$-quantile of the Student's t distribution with $N-1$ degrees of freedom, where N equals the sample size.

In order to evaluate the precision of the estimates, the sample variances are compared with the estimated lower bounds on the variances. The $(1-\alpha)\times 100\%$ confidence intervals on the sample variance of the x-coordinate $s^2_{\beta_x}$ and on the sample variance of the scattering cross-section s^2_V are given by:

$$\left[\frac{(N-1)s^2_{\beta_x}}{\chi^2_{\frac{\alpha}{2},N-1}}, \frac{(N-1)s^2_{\beta_x}}{\chi^2_{1-\frac{\alpha}{2},N-1}}\right] \tag{2.49}$$

$$\left[\frac{(N-1)s^2_{V}}{\chi^2_{\frac{\alpha}{2},N-1}}, \frac{(N-1)s^2_{V}}{\chi^2_{1-\frac{\alpha}{2},N-1}}\right] \tag{2.50}$$

with $s^2_{\beta_x}$ and s^2_V the sample variances for β_x and V, respectively, and $\chi^2_{1-\frac{\alpha}{2},N-1}$ and $\chi^2_{\frac{\alpha}{2},N-1}$ are the $1-\frac{\alpha}{2}$- and $\frac{\alpha}{2}$-quantiles of the χ^2 distribution with $N-1$ degrees of freedom, respectively.

The results of the study of the accuracy of the scattering cross-sections and position coordinates are shown in Figs. 2.6 and 2.7 for electron doses of $N_d = 5\cdot 10^5$ e$^-$Å^{-2} and $N_d = 5\cdot 10^2$ e$^-$Å^{-2}, respectively. The sample means from the estimates resulting from the 100 noise realizations are plotted together with the true values of the scattering cross-sections and x-coordinate of the central atomic column. It is clear that unbiased estimates are obtained for both the scattering cross-sections as well as the x-coordinate of the central atomic column. For an incident electron dose of $N_d = 5\cdot 10^5$ e$^-$Å^{-2}, the 95% confidence intervals of the scattering cross-sections are too small to be visible in Fig. 2.6. It is clear that the confidence intervals on the sample means become larger when reducing the incident electron dose to $N_d = 5\cdot 10^2$ e$^-$Å^{-2}. The estimates for the scattering cross-sections and the x-coordinates of the central atomic column are still accurate for this lower incident electron dose.

The results for the precision of the estimated scattering cross-sections and position coordinates are shown in Figs. 2.8 and 2.9 for electron doses of $N_d = 5\cdot 10^5$ e$^-$Å^{-2} and $N_d = 5\cdot 10^2$ e$^-$Å^{-2}, respectively. In these figures, the sample variances from the estimates of the 100 noise realizations are plotted together with the theoretical lower bound on the variance, the CRLB. Two main conclusions can be drawn from these Figs. 2.8 and 2.9. First, the CRLB allows us to evaluate the precision as a function of the number of atoms for the scattering cross-sections and position coordinates.

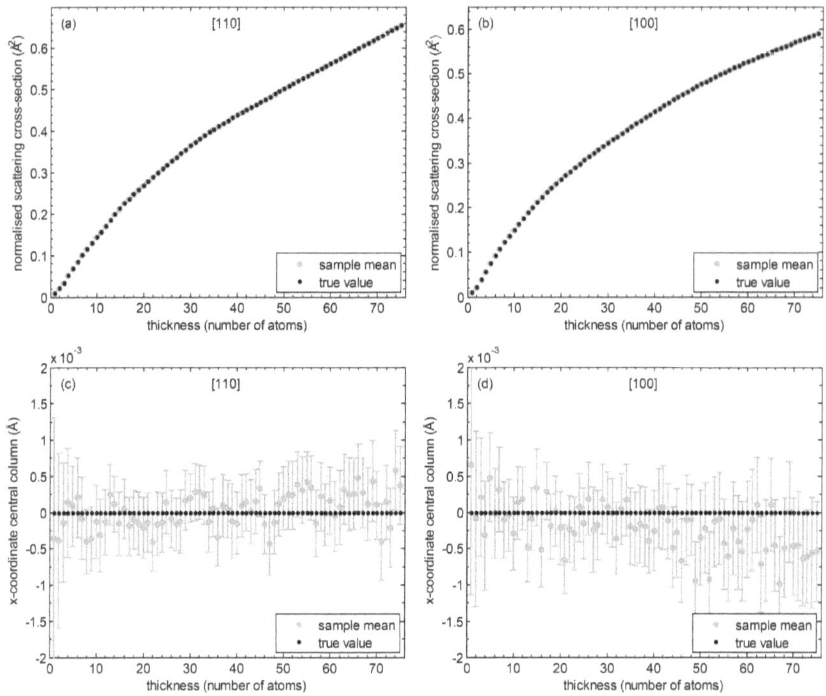

Figure 2.6 Accuracy of the estimated scattering cross-sections and position coordinates for an incoming electron dose of N_d=$5 \cdot 10^5$ e$^-$Å^{-2}. The true values and the sample mean together with the 95% confidence interval are shown for the (a) scattering cross-sections of Pt [110], (b) scattering cross-sections of Pt [100], (c) *x*-coordinate of the central atomic column of Pt [110], and (d) *x*-coordinate of the central atomic column of Pt [100].

The variance on the scattering cross-section increases with increasing number of atoms in a projected atomic column. The variance of the position coordinate of the atomic columns decreases up to 20 atoms. Beyond this thickness, the probe will be delocalized and the variance on the position coordinates increases. Second, the attainability of the CRLB can be evaluated from these figures. For an incident electron dose of $N_d = 5 \cdot 10^5$ e$^-$Å^{-2} in Fig. 2.8, the CRLB is attained for the whole range of number of atoms. For the reduced electron dose of $N_d = 5 \cdot 10^2$ e$^-$Å^{-2} in Fig. 2.9, the CRLB on the variance of the scattering cross-sections and *x*-coordinate of the central atomic column is not attained for very small number of atoms. This result is not unexpected since the detected number of electrons is very low in this case. Indeed, with a pixel size of 0.1121 Å for Pt [110], the incident

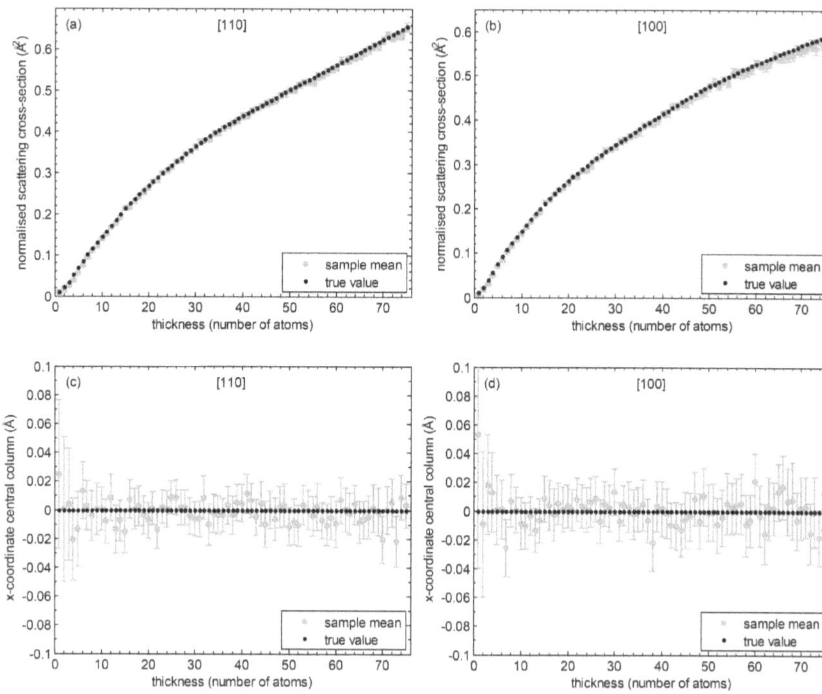

Figure 2.7 Accuracy of the estimated scattering cross-sections and position coordinates for an incoming electron dose of N_d=5 · 10^2 e$^-$Å^{-2}. The true values and the sample mean together with the 95% confidence interval are shown for the (a) scattering cross-sections of Pt [110], (b) scattering cross-sections of Pt [100], (c) *x*-coordinate of the central atomic column of Pt [110], and (d) *x*-coordinate of the central atomic column of Pt [100].

electron dose per pixel equals approximately 6 e$^-$/pixel; for Pt [100], with a pixel size of 0.1569 Å, the incident electron dose per pixel equals approximately 12 e$^-$/pixel. Since the maximal pixel intensity for single atom thickness expressed as a fraction of the incident electron beam equals approximately 0.01, the probability that an electron is detected will be very low. This is also clear from the noise realizations with $N_d = 5 \cdot 10^2$ e$^-$Å^{-2} in Fig. 2.5. In this case, the noise realization does not show the shape of the atomic columns anymore. Because of the prior knowledge of the presence of atomic columns, an attempt to estimate the model parameters is made, although the model estimation starts to fail here. For practical applications requiring the same low incident electron dose, the detected dose should be increased, which can be done by increasing the pixel size

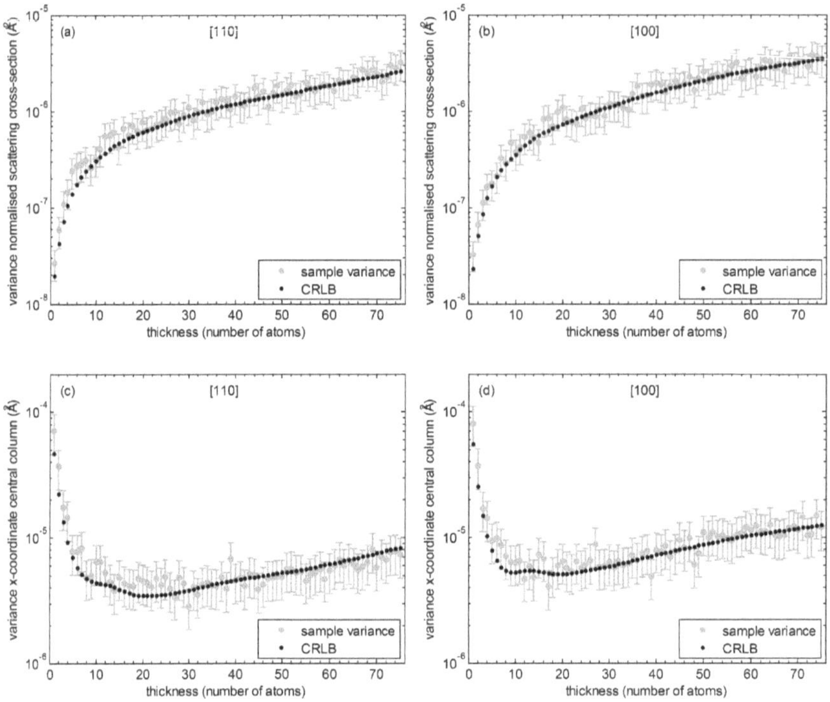

Figure 2.8 Precision of the estimated scattering cross-sections and position coordinates for an incoming electron dose of N_d=$5 \cdot 10^5$ $e^{-}Å^{-2}$. The Cramér-Rao lower bound and the sample variances together with the 95% confidence interval are shown for the (a) scattering cross-sections of Pt [110], (b) scattering cross-sections of Pt [100], (c) *x*-coordinate of the central atomic column of Pt [110], and (d) *x*-coordinate of the central atomic column of Pt [100].

and/or decreasing the inner detector angle. Nevertheless, this simulation study shows that the lower bound on the variance is attained for the estimation of the scattering cross-sections and position coordinates from ADF STEM images which are assumed to have independent Poisson distributed pixel values, using the least squares estimator. This result is of great importance, since this means that the estimated scattering cross-sections, which will be extensively used as an input for the atom-counting method, are obtained with the highest possible precision. Furthermore, from Figs. 2.8 and 2.9, it can also be derived that for an increasing electron dose, the variance decreases which leads to a more precise estimation. The dependence of precision on the incident electron dose is an unavoidable effect in parameter estimation.

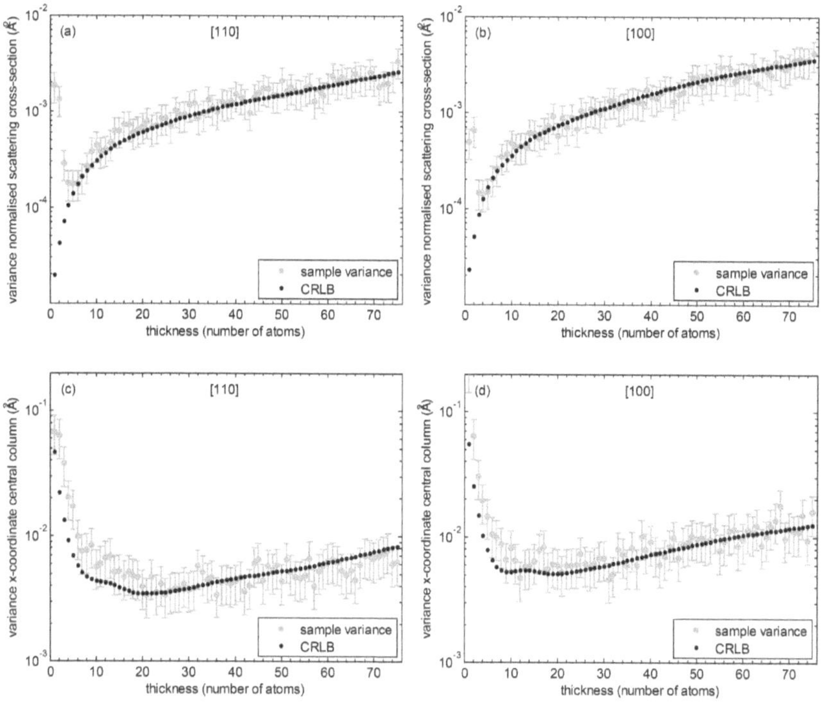

Figure 2.9 Precision of the estimated scattering cross-sections and position coordinates for an incoming electron dose of N_d=$5 \cdot 10^2$ $\mathbf{e^{-}Å^{-2}}$. The Cramér-Rao lower bound and the sample variances together with the 95% confidence interval are shown for the (a) scattering cross-sections of Pt [110], (b) scattering cross-sections of Pt [100], (c) *x*-coordinate of the central atomic column of Pt [110], and (d) *x*-coordinate of the central atomic column of Pt [100].

It is noted that in this analysis only Poisson noise has been taken into account, whereas in real experiments environmental or instrumental disturbances of acoustic, mechanical, or electromagnetic nature can cause additional distortions in images recorded by STEM (Jones & Nellist, 2013; von Harrach, 1995). Scan noise in STEM originates from unwanted lateral movements of the electron probe wile scanning the specimen (Muller et al., 2006; Recnik et al., 2005). It manifests itself as the slicing of atoms or atomic columns in STEM images (Buban et al., 2010). Scan noise may have a considerable effect on the precision that can be attained of quantities of interest (Van Aert et al., 2019). In addition, also the effect of sample or stage drift may contribute to distortions as images can appear warped. Image-reconstruction methods have been developed to correct microscopy images, mitigating the occurrence of scan and drift distortions (Anguiano & Aguilar, 1999; Berkels et al., 2014; Jones & Nellist, 2013; Jones et al., 2015; Ophus et al., 2016; Salmons et al., 2010; Sang & LeBeau, 2013).

2.5. ABF-ADF STEM: simultaneous model fitting

In Subsection 2.2.1, the concepts of a parametric statistical model have been discussed in detail. There, it has been mentioned through Eq. (2.9) that for atomic-resolution STEM images the projected atomic columns can be described as a superposition of Gaussian peaks (De Backer et al., 2016). In the most general case of assuming a different width for each estimated Gaussian peak, the expectation $f_{kl}(\boldsymbol{\theta})$ at the pixel (k, l) at the position (x_k, y_l) can be written as:

$$f_{kl}(\boldsymbol{\theta}) = \zeta + \sum_{n=1}^{N} \eta_n \exp\left(- \frac{(x_k - \beta_{x_n})^2 + (y_l - \beta_{y_n})^2}{2\rho_n^2} \right) \tag{2.51}$$

where ζ is a constant background, and ρ_n, η_n, β_{x_n}, and β_{y_n} are the width, the height and x- and y-coordinates of the nth atomic column described by a Gaussian peak, respectively, and N is the total number of atomic columns. The unknown parameters of this expectation model are represented by the parameter vector:

$$\boldsymbol{\theta} = (\beta_{x_1}, \ldots, \beta_{x_N}, \beta_{y_1}, \ldots, \beta_{y_N}, \rho_1, \ldots, \rho_N, \eta_1, \ldots, \eta_N, \zeta)^T. \tag{2.52}$$

Parameter estimation theory in atomic-resolution STEM is not restricted to the model given by Eq. (2.51). In Subsection 2.4.1, Eq. (2.40) shows the expectation model for mono-type atomic columns assuming Gaussian peaks with equal widths. Which parametric model to use depends on the material under investigation, but it is noted that even if the material consists of columns of different atom types, the model assuming Gaussian peaks with equal widths, given by Eq. (2.40), is acceptable since the width of an atomic column is mainly determined by the finite source size and to a lesser extent by the atom type (LeBeau & Stemmer, 2008).

For analyzing materials containing both light and heavy atoms, an interesting option is to acquire ABF and ADF STEM images simultaneously. Hereby, two annular detectors are used. One collects the electrons scattered toward high scattering angles, whereas the other is placed within the illumination cone of the electron probe. This technique combines the improved light-element visualization of ABF STEM with the easily interpretable image contrast of ADF STEM. In addition, due to the simultaneous acquisition, a pixel at location (x_k, y_l) possesses both an ABF and ADF intensity. In order to obtain quantitative measurements from the ABF and ADF image data, statistical parameter estimation theory may be applied using a parametric model such as, for example, the one given in Eq. (2.51). Although this methodology has been predominantly applied to quantify ADF STEM images, it can be used to describe both ABF and ADF image data. For ADF STEM images, the atomic columns are fitted by Gaussian peaks with a positive height η as columns in ADF STEM are depicted as bright spots on a dark background. Atomic columns in ABF STEM, though, are depicted as dark spots on a bright background. As a result, they are fitted by Gaussian peaks with a negative height η. Nevertheless, models such as the one given by Eq. (2.51) are not optimal for quantifying simultaneously acquired ABF and ADF STEM images since the correlation between the images is neglected. To overcome this problem, the aforementioned models for quantifying STEM images need to be extended to fit both the ABF and ADF images simultaneously. More details on simultaneous model fitting can be found in Fatermans et al. (2020). Simultaneous fitting can be achieved by assuming that the Gaussian peaks describing a particular atomic column, one with negative height for the ABF image and one with positive height for the ADF image, are located at the same position, corresponding to the actual position of the projected atomic column. When the observed STEM pixel values of the ABF and ADF image are depicted as $\boldsymbol{w}^{ABF}$ and $\boldsymbol{w}^{ADF}$, respectively, and under the assumption that each estimated Gaussian peak has a different width, the expectation model $\boldsymbol{f}_{kl}(\boldsymbol{\theta})$, describing simultaneously acquired ABF and ADF STEM images, gives the expectation of the observed pixel values $\boldsymbol{w}_{kl} = (w_{kl}^{ABF}, w_{kl}^{ADF})$ at position (x_k, y_l):

$$\begin{aligned}\boldsymbol{f}_{kl}(\boldsymbol{\theta}) &= (f_{kl}^{ABF}, f_{kl}^{ADF}) \\ &= \left(\zeta^{ABF} + \sum_{n=1}^{N} \eta_n^{ABF} exp\left(-\frac{(x_k-\beta_{x_n})^2 + (y_l-\beta_{y_n})^2}{2(\rho_n^{ABF})^2}\right),\right. \\ &\left.\zeta^{ADF} + \sum_{n=1}^{N} \eta_n^{ADF} exp\left(-\frac{(x_k-\beta_{x_n})^2 + (y_l-\beta_{y_n})^2}{2(\rho_n^{ADF})^2}\right)\right)\end{aligned} \quad (2.53)$$

where f_{kl}^{ABF} and f_{kl}^{ADF} denote the expectation models describing the ABF and ADF image, respectively. The parameters ζ^{ABF} and ζ^{ADF} describe the constant background and ρ_n^{ABF}, ρ_n^{ADF}, η_n^{ABF}, η_n^{ADF} depict the widths and heights of the nth atomic column described by two Gaussian peaks, respectively, in both the ABF and ADF images. For the model given by Eq. (2.53), the unknown parameters are represented by the parameter vector:

$$\begin{aligned}\boldsymbol{\theta} = (&\beta_{x_1}, \ldots, \beta_{x_N}, \beta_{y_1}, \ldots, \beta_{y_N}, \rho_1^{ABF}, \ldots, \rho_N^{ABF}, \rho_1^{ADF}, \ldots, \rho_N^{ADF}, \\ &\eta_1^{ABF}, \ldots, \eta_N^{ABF}, \eta_1^{ADF}, \ldots, \eta_N^{ADF}, \zeta^{ABF}, \zeta^{ADF})^T.\end{aligned} \tag{2.54}$$

However, it is well known that the observed atomic column positions in ABF STEM images are sensitive to specimen tilt (Brown et al., 2017; Gao et al., 2018; Zhou et al., 2016). This causes a deviation of the observed positions from the actual positions of the projected atomic columns. The observed locations of the atomic columns in ADF STEM images, though, are less sensitive to specimen tilt and can be considered to be reliable indicators of the true column positions even in the presence of some tilt (Findlay et al., 2010; Maccagnano-Zacher et al., 2008; So & Kimoto, 2012). Thus, due to the effect of specimen tilt, the observed column positions in ABF STEM can be shifted as compared to the observed positions in ADF STEM. Since small tilts of the electron beam with respect to the crystal zone axis can easily be present in STEM experiments, it is important to take into account the effect of the shifted observed column positions in ABF images in statistical parameter estimation theory in order to obtain accurate quantitative measurements. For this purpose, the expectation model proposed in Eq. (2.53) can be altered by allowing the ABF peak locations to deviate from the ADF peak locations along the tilting direction:

$$\begin{aligned} \boldsymbol{f}_{kl}(\boldsymbol{\theta}) &= (f_{kl}^{ABF}, f_{kl}^{ADF}) \\ &= \left(\zeta^{ABF} + \sum_{n=1}^{N} \eta_n^{ABF} exp\left(-\frac{(x_k - \gamma_{x_n})^2 + (y_l - \gamma_{y_n})^2}{2(\rho_n^{ABF})^2} \right), \right. \\ &\quad \left. \zeta^{ADF} + \sum_{n=1}^{N} \eta_n^{ADF} exp\left(-\frac{(x_k - \beta_{x_n})^2 + (y_l - \beta_{y_n})^2}{2(\rho_n^{ADF})^2} \right) \right) \end{aligned} \tag{2.55}$$

with

$$(\gamma_{x_n}, \gamma_{y_n}) = (\beta_{x_n} + \alpha_n cos\phi, \beta_{y_n} + \alpha_n sin\phi) \tag{2.56}$$

where ϕ and α_n indicate the direction and amplitude of the shift of the nth Gaussian peak in the ABF image with respect to the corresponding peak in the ADF image, respectively. The amplitude of the shift is not necessarily equal for all atomic columns in the image, but depends on atom type and orientation (Brown et al., 2017; Gao et al., 2018; Zhou et al., 2016). This is

Table 2.3 Microscope parameter values for simulation of an ABF and ADF STEM image of a 10 nm thick $SrTiO_3$ unit cell using MULTEM including 5 mrad specimen tilt.

Parameter	Symbol	Value
Acceleration voltage	V_0 (kV)	300
Defocus	ϵ (Å)	0
Spherical aberration	C_s (mm)	0.001
Spherical aberration of 5th order	C_5 (mm)	0
Semiconvergence angle	α (mrad)	22.9
ABF detector inner radius	β_1^{ABF} (mrad)	12
ABF detector outer radius	β_2^{ABF} (mrad)	22
ADF detector inner radius	β_1^{ADF} (mrad)	88
ADF detector outer radius	β_2^{ADF} (mrad)	98
Pixel size	$\Delta x = \Delta y$ (Å)	0.23
Number of scanned pixels	$K \times L$	27 × 27
FWHM of the source image	FWHM (Å)	0.7

due to the coherent image formation process in ABF STEM. In Eqs. (2.55) and (2.56), a general approach has been followed where no prior knowledge about the atom types and orientations is considered. Note that the shift of the ABF peaks only occurs along the tilting direction, as no atom position shift is observed in a direction perpendicular to this direction (Zhou et al., 2016). For the model given by Eq. (2.55) the parameter vector containing the unknown parameters is given by

$$\begin{aligned}\boldsymbol{\theta} = (&\beta_{x_1}, \ldots, \beta_{x_N}, \beta_{y_1}, \ldots, \beta_{y_N}, \rho_1^{ABF}, \ldots, \rho_N^{ABF}, \rho_1^{ADF}, \ldots, \rho_N^{ADF}, \\ &\eta_1^{ABF}, \ldots, \eta_N^{ABF}, \eta_1^{ADF}, \ldots, \eta_N^{ADF}, \alpha_1, \ldots, \alpha_N, \phi, \zeta^{ABF}, \zeta^{ADF})^T.\end{aligned} \tag{2.57}$$

In order to verify whether the proposed parametric model in Eq. (2.55) is a valid model to describe images in the presence of tilt, a simulation of a $SrTiO_3$ unit cell with a thickness of 10 nm has been performed using MULTEM (Lobato et al., 2016; Lobato & Van Dyck, 2015) with a specimen tilt of 5 mrad. The parameters of this simulation are included in Table 2.3. The simulated ADF and ABF STEM images are shown in Figs. 2.10(a) and (b), respectively. First, these images have been analyzed using the parametric model including the effect of specimen tilt given by Eq. (2.55), where equal widths for the Gaussian peaks in the ABF image, ρ^{ABF}, and ADF image, ρ^{ADF}, have been used. The estimated column positions that follow from this approach are shown in Figs. 2.10(a) and (b)

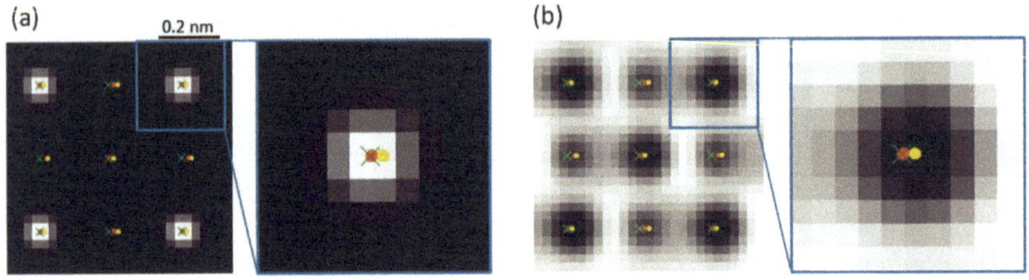

Figure 2.10 (a) Simulated ADF and (b) ABF STEM image of a 10 nm thick $SrTiO_3$ unit cell with a tilt angle of 5 mrad where the highlighted areas have been enlarged in the insets. The green crosses depict the true atomic column positions. The estimated atomic column positions obtained by performing model fitting including specimen tilt are shown by red dots, whereas for neglecting specimen tilt the estimated column positions are depicted by yellow dots. (Reprinted from Fatermans et al. (2020) with permission from Elsevier.)

by red dots. From Figs. 2.10(a) and (b), it can be seen that there is a close correspondence between the estimated column positions and the true column positions, shown by green crosses, which have been determined by the atom positions of the top plane of the tilted specimen. In case the analysis is, instead, performed using a parametric model neglecting the effect of specimen tilt, such as the one given by Eq. (2.53), the estimated column positions deviate further from the true positions. This is shown in Figs. 2.10(a) and (b) where the estimated column positions using a model without taking specimen tilt into account are shown by yellow dots. Quantitatively, the average deviation of the 9 estimated atomic column positions in the simulated image from their true column positions including the effect of specimen tilt is 8 ± 3 pm, whereas for neglecting tilt this is 18 ± 3 pm. This indicates that the estimated atomic column locations, obtained from analyzing simultaneously acquired ABF and ADF STEM image data by the parametric model proposed in Eq. (2.55) including the effect of specimen tilt, correspond better with the true atomic column locations than for analyzing the image data by a model neglecting the effect of specimen tilt.

2.6. Conclusions

In this chapter, the principles of statistical parameter estimation theory have been introduced. Within this framework, electron microscopy observations can be described by a parametric statistical model defining the expectations of the observations and the fluctuations of the observations about these expectations. Furthermore, statistical parameter estimation theory provides a theoretical lower bound, i.e. the Cramér-Rao lower bound, on the variance of any unbiased estimator which can be reached asymptotically using the maximum likelihood estimator. In practice, the least squares estimator which equals the maximum likelihood estimator for independent normally distributed random values, is used to estimate the unknown parameters of the model.

The Cramér-Rao lower bound has been used to study the amount of information on the position parameters contained in a reconstructed exit wave (De Backer et al., 2011). It has been shown that this CRLB is attained using the reconstructed complex exit wave. This precision cannot be obtained when using the phase information only, except for thin specimens where the weak phase object approximation is valid.

In order to explore the ultimate limits in ADF STEM, the precision and accuracy of the estimates for the atomic column position coordinates and scattering cross-sections have been evaluated using multislice simulations (De Backer et al., 2016). Although, the pixel values of ADF STEM images are modeled as independent Poisson distributed variables, it turns out that even for low incident electron doses, the Cramér-Rao lower bound on the variance of the estimates has been obtained. This is a remarkable result since the CRLB is expected to be attained asymptotically when using the maximum likelihood estimator and here we used the least squares estimator which only equals the maximum likelihood estimator in the case of independent normal distributed random variables. This conclusion is extremely important, since these estimates will be extensively used as an input for a statistics-based atom-counting method discussed in Chapter 4.

Furthermore, a new parametric model has been introduced for quantitatively analyzing simultaneously acquired ABF and ADF STEM images (Fatermans et al., 2020). Hereby, the effect of specimen tilt, which shifts the ABF peak locations from the true atomic column locations, has been taken into account since small tilts of the electron beam with respect to the crystal zone axis can easily be present in STEM experiments.

Concluding, based on statistical parameter estimation theory, electron microscopy images may be analyzed quantitatively and estimates for the structure parameters, such as the atomic column positions and scattering cross-sections can be obtained with the highest possible precision.

In Chapter 3, the implementation of the least squares estimator will be reviewed. A direct implementation of the least squares estimator limits the area that can be analyzed, since the estimation of all the parameters of the parametric model at the same time is computationally very intensive. Therefore, an efficient implementation of the least squares estimator is required, which enables the quantification of larger areas (De Backer et al., 2016).

Acknowledgments

This project has received funding from the European Research Council (ERC) under the European Union's Horizon 2020 research and innovation programme (Grant Agreement No. 770887 and No. 823717 ESTEEM3). The authors acknowledge financial support from the Research Foundation Flanders (FWO, Belgium) through Projects No. G.0502.18N, G.0267.18N, G.0368.15N, and a postdoctoral grant to A. D. B. The authors acknowledge funding from the University of Antwerp through a TOP project.

CHAPTER THREE

Efficient fitting algorithm

Annick De Backer[a,b], Jarmo Fatermans[a,b,c], Arnold J. den Dekker[c], and Sandra Van Aert[a,b,*]

[a]Electron Microscopy for Materials Science (EMAT), University of Antwerp, Antwerp, Belgium
[b]NANOlab Center of Excellence, University of Antwerp, Antwerp, Belgium
[c]imec-Vision Lab, University of Antwerp, Wilrijk, Belgium
*Corresponding author. e-mail address: sandra.vanaert@uantwerpen.be

Contents

3.1. Introduction

Nowadays, more and more quantitative measurements for the analysis of nanomaterials are required. This includes locating atomic column positions with a precision in the picometer range, a precise determination of the chemical composition of materials, and counting the number of atoms with single atom sensitivity. Moreover, one is interested in a quantitative analysis of larger areas preserving the highest possible precision that can be attained using statistical parameter estimation theory. Therefore, in this chapter, the implementation of the least squares estimator (LSE), which is used for the estimation of the parameters of the model describing the observations in a high-resolution (scanning) transmission electron microscopy (HR(S)TEM) image, will be reviewed.

As discussed in Chapter 2, in Subsection 2.2.3, the maximum likelihood estimator (MLE) equals the LSE for independent normally distributed vari-

Advances in Imaging and Electron Physics, Volume 217
ISSN 1076-5670
https://doi.org/10.1016/bs.aiep.2021.01.003

ables. However, direct implementation of the LSE in which all parameters are estimated at the same time is computationally very intensive and feasible only for images containing a limited number of projected atomic columns, i.e. a limited field of view. Therefore, a more efficient algorithm is proposed for obtaining the unknown structure parameters. The use of this algorithm enables us to analyze large fields of view. To provide end-users with this well-established quantification method, a user friendly program, StatSTEM, is developed which is freely available under a GNU public license (De Backer et al., 2016).

The basic idea of this algorithm is the segmentation of the image into smaller sections containing individual columns. In this manner, only the parameters corresponding to a single atomic column are estimated at the same time, instead of all the parameters of the parametric model. Two different schemes will be proposed for two different variants of the parametric models discussed in Chapter 2. In the first model, each atomic column is modeled with a Gaussian peak having a different width. This method will be referred to as 'model estimation A'. In the second model, the same width is imposed for atomic columns of the same atom type. This method will be referred to as 'model estimation B'. Different widths for each Gaussian in the parametric model corresponding to the intensity of an atomic column in the HR(S)TEM image, as in 'model estimation A', are required if no prior knowledge is available on e.g. the atomic column types. However, if prior knowledge is available, the number of parameters of the expectation model that needs to be estimated in the model can be reduced, as in 'model estimation B'. This approach enhances the precision of the parameter estimates. In addition, the estimation of the parameters of such a model will be more robust in the case of the inevitable presence of different types of noise (e.g. scanning noise in STEM images), since less parameters need to be estimated in this model. Indeed, the estimates for the parameters for 'model estimation A' do often not converge because of noise.

In Section 3.2, the efficient algorithm for the estimation of atomic column positions and intensities from HR(S)TEM images will be introduced. After this, the efficient estimation algorithm will be applied to three different nanostructures for which the analysis of a large field of view is required. In the first example in Section 3.3, estimated models will be obtained for ADF STEM images of a core-shell nanostructure recorded along different zone axes (Goris et al., 2013). The estimated models as such will be used as an input for a three-dimensional (3D) reconstruction in order to determine the position and atom type of each atom. In the next two examples in

Section 3.4, the estimated atomic column position coordinates will be used to make displacement maps of a large field of view. This will be done for five-fold-twinned decahedral Au nanoparticles in Subsection 3.4.1 (Goris et al., 2015) and for the positions of Al columns within $LaAlO_3$ in Subsection 3.4.2.

3.2. Efficient model-estimation algorithm

In this section, an efficient model estimation method for the analysis of atomic resolution (S)TEM images including images with a large field of view will be described. The general idea behind this method has been proposed in den Dekker et al. (2013). The model estimation algorithm has later further been improved and extended in De Backer et al. (2016). As compared to the scheme presented in den Dekker et al. (2013), the starting values for the width of the Gaussian peaks and the estimation of the height of the Gaussian peaks and the background within the iterative part have now been optimized. More important, a new estimation scheme has been added allowing to estimate the same width for estimated Gaussian peaks of atomic columns of the same atom type. In addition, a software package, StatSTEM, including this algorithm has been developed.

The parametric model that is implemented is the superposition of Gaussian peaks describing the peaked intensity at the atomic column positions as described in Subsection 2.2 of Chapter 2. The idea of the new algorithm is a segmentation of the image into smaller sections containing individual atomic columns without ignoring overlap between neighboring columns. Therefore, it is required that the image can be described as a superposition of the individual atomic columns. For example, incoherent imaging in ADF STEM meets this necessary condition.

In Subsection 3.2.1, 'model estimation A' with a different width assumed for each Gaussian is introduced. Next, Subsection 3.2.2 introduces 'model estimation B' where the same width is assumed for Gaussian peaks of the atomic columns of the same atom type.

3.2.1 Model estimation A: *different width for each estimated Gaussian peak of an atomic column*

When a different width is assumed for each estimated Gaussian corresponding to the intensity of an atomic column, the parametric model of Eq. (2.9) can be rewritten as:

$$f_{kl}(\boldsymbol{\theta}) = \zeta + \sum_{n=1}^{N} \eta_n \exp\left(-\frac{\left(x_k - \beta_{x_n}\right)^2 + \left(y_l - \beta_{y_n}\right)^2}{2\rho_n^2} \right), \tag{3.1}$$

with ζ a constant background, ρ_n and η_n the width and height of the nth Gaussian peak, β_{x_n} and β_{y_n} the x- and y-coordinate of the nth atomic column respectively, and N the total number of atomic columns. The unknown parameters of the expectation model are then given by:

$$\boldsymbol{\theta} = \left(\beta_{x_1}, \ldots, \beta_{x_N}, \beta_{y_1}, \ldots, \beta_{y_N}, \rho_1, \ldots, \rho_N, \eta_1, \ldots, \eta_N, \zeta\right)^T. \tag{3.2}$$

It is assumed that proper starting values are available for the parameters of the empirical model. Good starting values are necessary in order to avoid ending up in a local minimum during the model-estimation procedure. Initial values for the widths ρ_n of the Gaussian peaks are easy to provide as will be explained later and starting coordinates for the positions β_{x_n} and β_{y_n} of projected atomic columns can be given using a peak finding routine. For this purpose, the image resolution should be high enough. This means that a high-resolution image is required, where individual columns are atomically resolved. If the resolution is only just sufficient, prior knowledge on the structure, e.g. the presence of dumbbells, can still provide the starting values for the atomic column positions.

The parametric model given by Eq. (3.1) is linear in the parameters for the height of the Gaussian peaks η_n and the background ζ and is nonlinear in the parameters for the positions β_{x_n}, β_{y_n}, and the width of the Gaussian peaks ρ_n. The parameters β_{x_n}, β_{y_n}, and ρ_n are therefore the nonlinear parameters of the model and will be denoted as $\boldsymbol{\alpha}$. An initial guess for the linear parameters of the model, i.e. the heights η_n and background ζ denoted as the vector $\boldsymbol{\xi}$, is not needed, since the linear parameters in the model can be replaced by their linear least squares estimates given the values of the nonlinear parameters of the model (Lawton & Sylvestre, 1971). These linear least squares estimates are the best companion values for the linear parameters and the computation involves only a simple linear regression. In addition, this approach has a second benefit apart from the fact that no starting guesses are needed for the linear parameters. The number of parameters that needs to be estimated in an iterative procedure is reduced to the nonlinear parameters only, increasing the convergence speed of the model estimation.

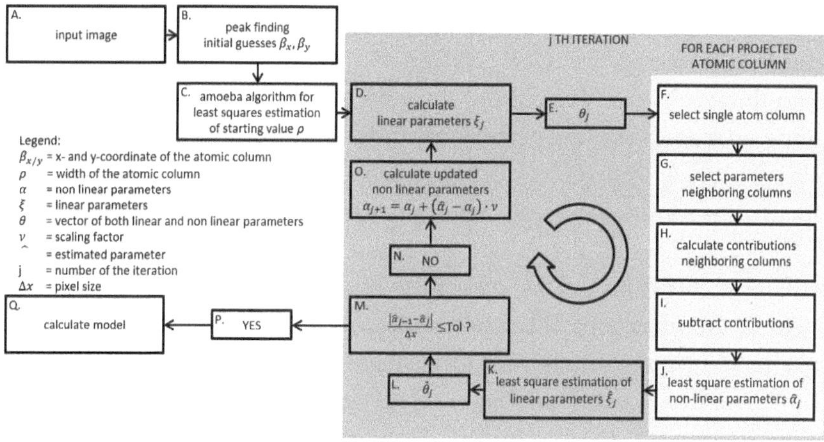

Figure 3.1 Schematic for model estimation A: algorithm for the expectation model having a different width for each estimated Gaussian peak. Reprinted from De Backer et al. (2016), Copyright (2016), with permission from Elsevier.

Schematically, the proposed algorithm for model estimation A is outlined in Fig. 3.1. In the first steps (B.) and (C.), initial guesses are obtained for the nonlinear parameters $\boldsymbol{\alpha}_1$ for the input image (A.). In step (B.), initial guesses for the x- and y-coordinates of the atomic columns in the image are obtained by a peak finding routine. Next, using these starting coordinates and a rough starting estimate of the width ρ based on the distance between the atomic columns, the starting value for the width is refined in step (C.) while fixing the values of the peak position parameters to their initial values obtained in step (B.). For this purpose, a derivative-free unconstrained nonlinear optimization method, the Nelder-Mead simplex method, is used in order to reduce the memory consumption (Lagarias et al., 1998). If a limited number of parameters needs to be estimated, then this method converges toward good estimates of the parameters. Here, a single starting value for the width of all atomic columns is estimated. For the estimation of this refined starting value for the width, the tolerance level of the Nelder-Mead simplex algorithm is set less tight. Next, the iterative part of the fitting procedure can be executed (shown in gray in Fig. 3.1). Based on the values of the nonlinear parameters $\boldsymbol{\alpha}_j$, starting values for the linear parameters $\boldsymbol{\xi}_j$ are calculated by means of linear regression (D.), where j corresponds to the iteration index. Having starting values for the linear and nonlinear parameter of the jth iteration available (E.), the actual estimation of the parameters can be done. Therefore, regions of interest containing single atomic columns are selected from the HR(S)TEM image (F.). These regions are defined based on the average distance between the atomic columns in the image. These distances can be calculated from the coordinates given by the peak finding routine. Next, the parameters of the neighboring projected atomic columns are selected (G.) in order to take overlap between neighboring Gaussian shapes into account. The values of the nonlinear and linear parameters are given by the values that result from step (B.), (C.), and (D.) in the first iteration and by the values that resulted from step (O.) and (D.) in further iterations. The contributions of these neighboring Gaussian peaks in the region of interest of the single atomic column are calculated (H.). Then, these contributions are subtracted from the image of the single atomic column (I.). Ideally, the thus obtained image only contains scattered intensities resulting from this single projected atomic column and can be used for a nonlinear least squares estimation of the nonlinear parameters of this single Gaussian (J.). Steps (F.) till (J.) are repeated for each projected atomic column in the HR(S)TEM image in order to obtain a complete set of estimated nonlinear parameters $\hat{\boldsymbol{\alpha}}_j$ for the jth iteration. Next, the best companion values for the linear parameters for the jth iteration, $\hat{\boldsymbol{\xi}}_j$ are calculated (K.). Once the parameters are estimated, convergence is evaluated (M.). If convergence is not yet obtained, the nonlinear starting values for the $(j+1)$th iteration are calculated based on the starting and estimated values for the nonlinear parameters of the jth iteration (O.). Here, the updates are modified by a scaling factor ν in order to avoid divergence on the parameters to be estimated. The scaling factor is chosen equal to 0.25. When for a large number of the atomic columns the parameters are already converged, this scaling factor is increased to 0.5. Having a new set of nonlinear parameters available, the iterative part of the fitting algorithm can be executed again from step (D.). The iterative approach allows a better modeling of the overlap between neighboring Gaussian peaks. If convergence is attained (P.), the model can be calculated based on the estimated parameters (Q.).

3.2.2 Model estimation B: *same width for estimated Gaussian peaks of atomic columns of the same atom type*

When the same width is estimated for Gaussian peaks of atomic columns of the same atom type, the expectation model and corresponding unknown structure parameters $\boldsymbol{\theta}$ are given by Eqs. (2.9) and (2.10) which are repeated

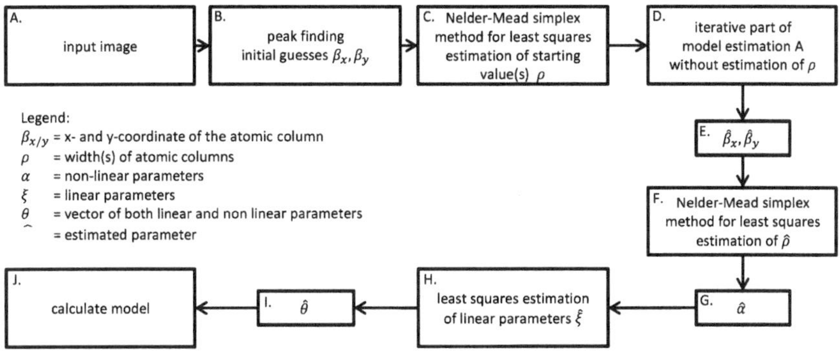

Figure 3.2 Schematic for model estimation B: algorithm for the expectation model having the same width for Gaussian peaks of atomic columns of the same atom type. Reprinted from De Backer et al. (2016), Copyright (2016), with permission from Elsevier.

here:

$$f_{kl}(\boldsymbol{\theta}) = \zeta + \sum_{i=1}^{I} \sum_{m_i}^{M_i} \eta_{m_i} \exp\left(- \frac{\left(x_k - \beta_{x_{m_i}}\right)^2 + \left(y_l - \beta_{y_{m_i}}\right)^2}{2\rho_i^2} \right), \tag{3.3}$$

with unknown parameter:

$$\boldsymbol{\theta} = \left(\beta_{x_{1_1}}, \ldots, \beta_{x_{M_I}}, \beta_{y_{1_1}}, \ldots, \beta_{y_{M_I}}, \rho_1, \ldots, \rho_I, \eta_{1_1}, \ldots, \eta_{M_I}, \zeta\right)^T. \tag{3.4}$$

The basic idea of this routine is equivalent to the method for the model having different widths for each estimated Gaussian peak. The estimation method is subdivided into the estimation of the position parameters $\beta_{x_{m_i}}$ and $\beta_{y_{m_i}}$, and the estimation of the width ρ_i for each atom type. This approach is valid since the correlation between the position parameters and the width of the Gaussian peaks equals almost zero. Also throughout this estimation routine, the best companion values for the linear parameters $\boldsymbol{\xi}$ are used in each iteration increasing the convergence of the model estimation.

A schematic is shown in Fig. 3.2. The first steps of model estimation B are similar to steps (A.), (B.), and (C.) of model estimation A. In step (B.), initial guesses are obtained for the coordinates of the projected atomic columns in input image (A.) using a peak finding routine. If different column types are present, the atomic columns are grouped by type. Next, in step (C.), refined initial guesses for the widths of the atomic columns of the same atom type are obtained using the Nelder-Mead simplex method with a less tight tolerance level applied to the full image containing all atomic columns. Since the number of different atom types in an HR(S)TEM image is limited, this method gives also for model estimation B accurate estimates in an efficient manner. Next, the iterative part of model estimation A is performed till convergence is reached (D.) without the estimation of the parameters for the width of the Gaussian peaks, i.e. with a fixed value for the width ρ_i of each atomic column type set in step (C.). In this way, estimates are obtained for the x- and y-coordinates $\beta_{x_{m_i}}$ and $\beta_{y_{m_i}}$ (E.). Next, estimates for the width ρ_i and the companion values for the linear parameters $\boldsymbol{\xi}$ need to be estimated for the fixed positions obtained from the first part of this method. A full optimization of the parameter ρ_i is done using the Nelder-Mead simplex method with a tight tolerance level applied to the full image containing all atomic columns (F.). After the optimization of the parameter ρ_i, estimates are available for all the nonlinear parameters $\hat{\boldsymbol{\alpha}}$ in the parametric model (G.). Finally, the best companion values for the linear parameters $\hat{\boldsymbol{\xi}}$ (H.) and the model based on the estimated parameters (I. and J.) can be calculated.

In this section, the implementation of the least squares estimator has been adapted enabling us to analyze large fields of view of HR(S)TEM images in a quantitative manner. As such, model-estimation procedures A or B allow one to evaluate larger nanostructures in a quantitative manner by means of a least squares estimation of the parameters of the statistical expectation models. In the next sections, this algorithm will be applied to different nanostructures in which a quantitative analysis of a large field of view is required. It will be shown that the analysis of larger fields of views opens up a variety of new possibilities for the characterization of nanomaterials. In the first example, the estimated models as such will be used to reduce scanning errors in HAADF STEM images of a core-shell (Au@Ag) nanorod. The refined models serve as an input for atomic-resolution electron tomography aiming to determine the position and atom type of each atom in a 3D reconstruction. In the second set of examples, the positions of the atomic columns are interpreted for large fields of view in order to make displacement maps of the atomic column positions.

3.3. Three-dimensional elemental mapping in bimetallic nanocrystals

The first application of the efficient algorithm described in the previous section is the estimation of models for two-dimensional (2D) projection images of a bimetallic nanocrystal as an input for electron tomography (Goris et al., 2013). A thorough understanding of the 3D structure and composition of core-shell nanostructures is indispensable to obtain a deeper insight on their physical behavior. Such 3D information can be reconstructed from 2D projection images using electron tomography. Different electron tomography techniques have already enabled the 3D characterization of a variety of nanostructures at the atomic level (Bals et al., 2011; Belz et al., 2018; Chen et al., 2013; De Backer et al., 2017; Goris et al., 2015; Scott et al., 2012; Van Aert et al., 2011; Xu et al., 2015; Zhou et al., 2020). However, these methods have all focused on the investigation of nanomaterials containing only one type of chemical element. Now, for the first time, the position and atom type of each atom in a hetero-nanostructure will be determined using a combination of statistical parameter estimation techniques and compressive sensing based tomography (Donoho, 2006; Goris, Bals, et al., 2012). As an example, the interface in core-shell Au@Ag nanorods will be investigated.

These bimetallic nanostructures are of great interest as they provide additional flexibility, stability, or even novel properties as compared to their monometallic counterparts (Cortie & McDonagh, 2011; Henglein, 2000; Hodak et al., 2000; Tedsree et al., 2011). The deposition of Ag-coated Au (Au@Ag) nanorods, in particular, has been extensively studied because of the possibility to tune their overall geometry (and optical response) by changing the size and aspect ratio of the original Au nanorods (Cardinal et al., 2010; Liu & Guyot-Sionnest, 2004; Ma et al., 2010; Park et al., 2011; Sánchez-Iglesias et al., 2010; Xiang et al., 2008; Yang & Chang, 2006). However, to understand and tailor the properties of these Au@Ag nanorods, a thorough knowledge of the structure and the chemical composition at the atomic scale, in particular at the interfaces between the metallic components, is required.

Five high-resolution high-angle annular dark-field (HAADF) STEM projection images of the same nanorod were acquired along different major zone axes as shown in Figs. 3.3(A-E). The aim is to identify the position and atom type of individual atoms from a reliable, atomic-resolution 3D reconstruction. Since the compressive sensing based electron tomography method can strongly be influenced by (scanning) noise, a reliable manner to reduce these effects is by modeling the projection images. No prior knowledge was available on the atom type of each atomic column, although it was known that the nanoparticles under study are core-shell Au@Ag particles. However, it is impossible to visually discern for each atomic column

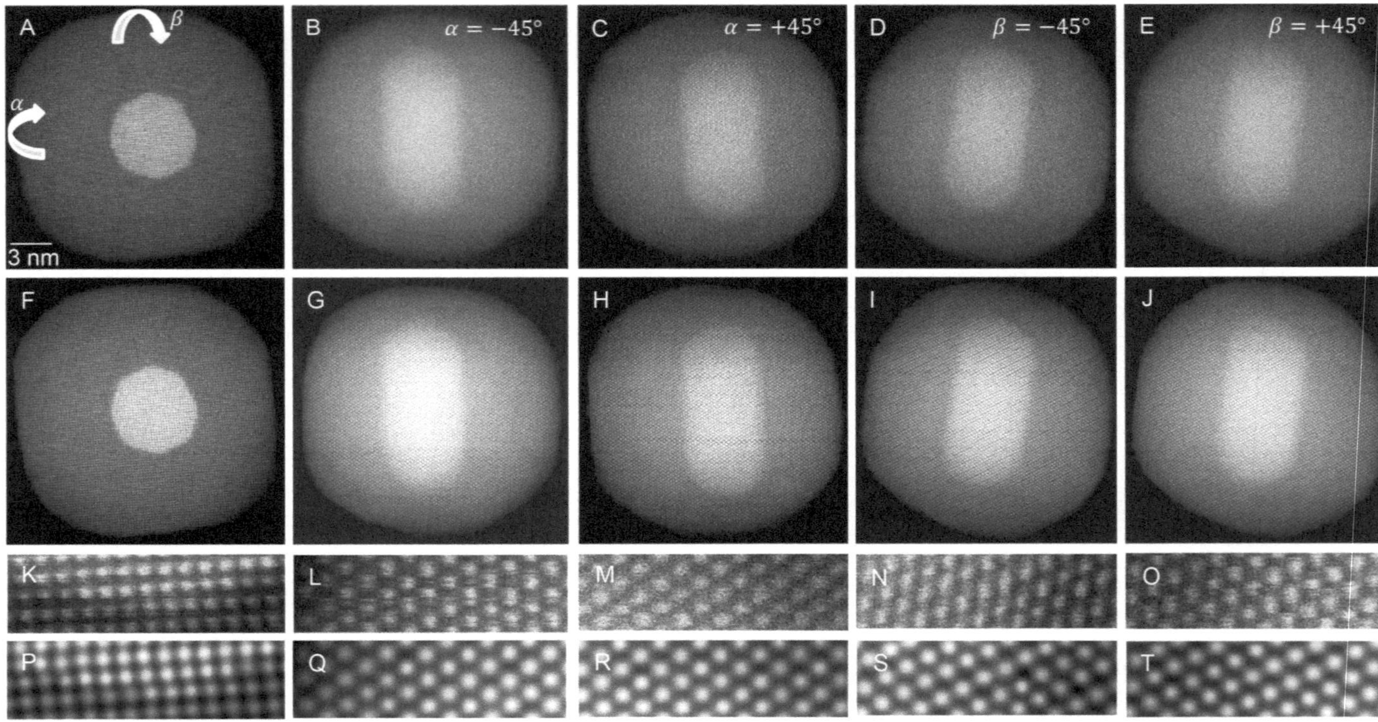

Figure 3.3 A-E: Five high-resolution HAADF STEM projection images of an Au@Ag nanorod obtained by tilting the nanorod along the different tilt axes that are mutually orthogonal. F-J: Refined models for experimental images A-E. K-T: More detailed view on the observations and corresponding models for each of the experimental images. Noise and scanning artifacts are reduced in the refined models.

between Au and Ag at the interface from the 2D HAADF STEM projection images. Therefore, model-estimation method A has been used, where each atomic column in the 2D projection images was estimated with a different width. The quality of the images was sufficient in order to use this model-estimation method. Figs. 3.3(F-J) show the corresponding estimated models of the originally recorded projection images shown in Figs. 3.3(A-E). A more detailed view on the observations and corresponding models for each of the experimental images is provided in Figs. 3.3(K-T). An excellent agreement between the refined models and the experimental projection images is observed, and indeed noise and scanning artifacts are reduced in the refined models. In this analysis, the model-based estimation acts as an image filter.

The refined models (F-J) of Fig. 3.3 are used as an input for 3D reconstruction. The result of this reconstruction and a detailed analysis of the position and type of the atoms is shown in Fig. 3.4. This analysis was done from the orthogonal slices through the reconstruction. Since HAADF STEM images were obtained from the underlying data for the 3D reconstruction, the intensity of each reconstructed atom is related to its atomic weight. Therefore, individual Ag and Au atoms can be distinguished, also at the interface by comparing their relative intensities. An intensity profile was acquired along the direction indicated by the white rectangle in Fig. 3.4(B). From this intensity profile, which is presented in Fig. 3.4(D), it is clear that Au and Ag atoms can indeed be identified from their intensities using a threshold value. In this manner, each atom in the slices presented in Figs. 3.4(B) and (C) was assigned to be Ag or Au. The result is shown in Figs. 3.4(E) and (F) and enables us to carry out a correct indexing of the type of interfacial planes. We thus conclude that the interface between the Au core and the Ag shell is sharp without intermixing. This result could only be obtained by a combination of statistical parameter estimation and a 3D reconstruction based on compressive sensing.

3.4. Displacement mapping

The analysis of a large field of view, with very precise and accurate measurements of the positions of the projected atomic columns allows one to construct displacement maps in the case of strained structures. As a first example, a displacement map is obtained for a five-fold-twinned decahedral Au nanoparticle (Goris et al., 2015). Secondly, the displacement of

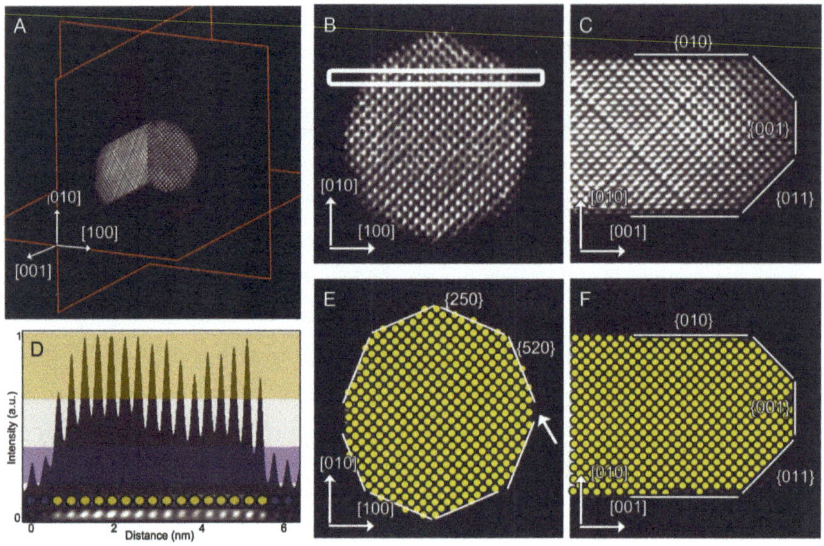

Figure 3.4 Orthogonal slices through the 3D reconstruction. (A) Three orthogonal slices through the reconstruction show the core-shell structure of the nanorod. The atomic lattice can be resolved in all three slices. (B, C) Detailed view of slices through the reconstruction perpendicular and parallel to the major axis of the nanorod. (D) An intensity profile is acquired along the direction indicated by the white rectangle in (B). (E, F) Slices corresponding to B and C, in which each Au atom is indicated by a yellow disk. In this manner, the correct interfacial planes can be determined based on intensity profiles through the reconstruction. Reprinted with permission from Goris et al. (2013). Copyright (2013) American Chemical Society.

Al atomic column positions within a $BiFeO_3/LaAlO_3$ heterostructure is discussed.

3.4.1 Five-fold-twinned decahedral Au nanoparticles

The first example is the mapping of the displacements within a five-fold-twinned decahedral Au nanoparticle (Goris et al., 2015). These complex decahedral particles exhibit highly localized surface plasmon resonances and are therefore of special interest because of their great potential for applications such as biosensing and cancer therapy (Jain et al., 2007; Willets & Van Duyne, 2007; Zhao et al., 2006). The exact optical properties are size and shape dependent. Current synthesis techniques enable a high degree of size and shape control resulting in decahedral nanoparticles with the same optical properties (Grzelczak et al., 2008; Pastoriza-Santos et al., 2007; Sánchez-Iglesias et al., 2006). Since the internal structure determines their growth, surface structures and stability, the characterization of the internal structure is required for further engineering their electronic and optical properties.

These types of Au particles are intrinsically strained as a result of their unique geometry as illustrated in Fig. 3.5(a). The decahedral particle consists of 5 tetrahedral subunits with an FCC crystal structure and {111} crystallographic facets. For perfect FCC tetrahedral subunits, the angle between adjacent {111} facets is 70.53° resulting in a solid-angle deficiency of 7.35°. As a result, these particles are strained or contain defects.

In several studies, different strain models have been proposed for these five-fold-twinned decahedral Au nanoparticles (Johnson et al., 2008). Often, strain maps and displacement fields are obtained by the geometrical phase technique (Hÿtch et al., 2003, 1998). In this work, though, statistical parameter estimation theory is used to create a displacement map for a five-fold-twinned decahedral Au nanoparticle in order to study its internal structure. A high-resolution HAADF STEM projection image of such a nanoparticle free of defects shown in Fig. 3.5(b) was acquired along the [110] zone axis, i.e. along the pentagonal axis. Next, the parameters of the model were estimated using model-estimation method B. Fig. 3.5(c) shows the estimated model which is in good agreement with the original image shown in Fig. 3.5(b). The estimated atomic column position coordinates were used to construct displacement maps for the atomic column positions. In order to measure the displacement for each atomic column, a reference lattice should be defined. For the five-fold-twinned decahedral nanoparticle, a reference lattice is defined for each tetrahedral subunit. It is assumed that the lattice is minimally distorted around the central projected atomic column. The lattice parameters of the reference lattice in a- and b-direction are then defined from the average interatomic distances in the two directions in the measured lattice around the central projected atomic column. Atomic columns up to two interatomic distances from the central column in a tetrahedral subunit are taken into account for the measurement of the reference lattice parameters. In this manner, the lattice parameters of the reference lattice are defined with a precision better than 1 pm. The angle under which the reference lattice should be expanded has also been measured from the estimated atomic column positions. The reference lattice is expanded from the central projected atomic column in each tetrahedral subunit following a perfect FCC structure. By comparing the atomic column positions from the reference lattice with the measured atomic column positions, displacements can be measured. The displacement map of the

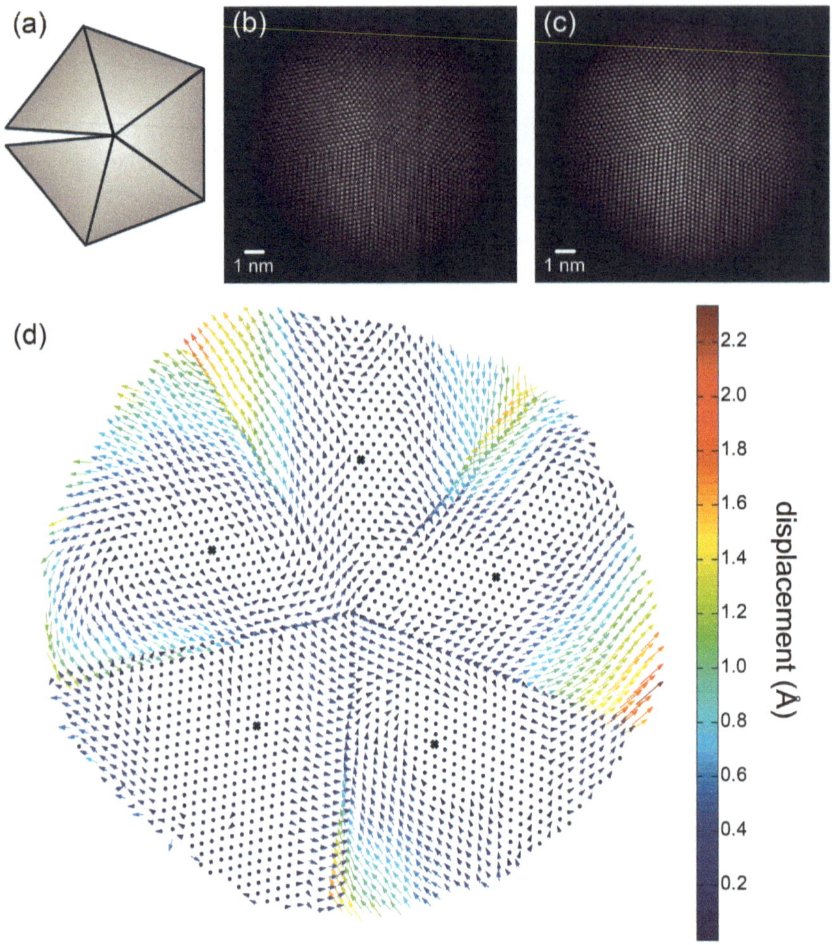

Figure 3.5 Defect-free five-fold-twinned decahedral Au nanoparticle. (a) For perfect FCC tetrahedral subunits a solid angle deficiency of 7.35° exists viewed along the pentagonal axis. (b) HAADF STEM image along the [110] zone axis. (c) Refined model. (d) Displacement map. The black crosses indicate from which atomic column in each tetrahedral subunit the reference lattice is extended.

five-fold-twinned decahedral nanoparticle is illustrated in Fig. 3.5(d). The black crosses indicate the atomic column position from which the reference lattice is expanded. For clarity, displacements of less than 0.25 Å are not plotted by arrows; in this case the reference lattice is plotted by dark blue dots. From this map, it can be seen that the displacements are not homogeneously distributed. It is expected that this information in combination with a 3D reconstruction at the atomic scale can give unique insights in the connection between structure and properties at the atomic scale of these nanoparticles.

3.4.2 $BiFeO_3/LaAlO_3$ heterostructure

The second application where a displacement map is created for a large field of view is for the Al atomic column positions within $BiFeO_3/LaAlO_3$, a functional oxide material. These types of materials have been studied for decades because of the abundance in interesting physical properties depending on the composites and more recently in interface engineering (MacLaren & Ramasse, 2014). Particularly, the contribution of perovskite compounds cannot be ignored.

Perovskite oxides comprise a large family of compounds with a general formula ABO_3, named after a mineral perovskite $CaTiO_3$. In a regular perovskite structure, small B cations (typically transition metal cations) are coordinated by octahedra of 6 oxygen atoms. The octahedra are connected into a corner-sharing 3D framework. Cuboctahedral cavities of the framework are occupied by relatively large A cations (e.g. alkali, alkali-earth or rare-earth cations). Due to the inherent flexibility of the octahedral framework, the perovskite structure can accommodate a large number of various cations both in the A and B position. Depending on the composition, structure distortions and specific interatomic interactions, perovskites can demonstrate a plethora of physical properties, such as electronic and ionic conductivity, ferroelectricity, magnetism, catalytic activity and many others (Abakumov et al., 2013; Mitchell, 2003).

Specifically, the ferroelectricity in perovskites is typically associated with the off-center displacements of the B cations away from the center of the octahedra. One of the well-known perovskite ferroelectrics is $PbZr_xTi_{1-x}O_3$, a so-called PZT ceramic (Hao et al., 2014). Above the Curie temperature (T_c), the ceramic has a cubic centrosymmetric structure with no electric dipole moment within its unit cell. Nevertheless, below T_c, the positively charged B-site cations of Ti/Zr shift away from the center along certain crystallographic directions and lower the crystal symmetry by distorting the lattice to noncentrosymmetric. Since positively charged cations are deviated from the central anions octahedron, i.e. the oxygen octahedron, an electric dipole is therefore present.

It is therefore obvious that the study of the deviation of B cations from the central oxygen octahedron is one of the key elements in understanding the fundamentals of perovskite materials in order to further engineer these materials. Particularly in the recent ten years, the discovery of multiferroics, where ferroelectricity and ferromagnetism coexist and interplay with each other, stimulates numerous studies in developing a variety of materials both experimentally and theoretically. One active field is the heterostructure design through thin film fabrication (Heron et al., 2014; Zeches et al., 2009). By epitaxially growing a thin film on a selected substrate, strain is introduced into the heterostructure due to the lattice mismatch. Consequently, local displacements of the atoms contribute enormously to novel behavior at the heterostructure interface.

The imaging mode of HAADF STEM is a powerful imaging technique in order to retrieve the information from the heterostructure interface where atomic columns of A/B cations could be imaged in a straightforward manner. Here, a quantitative measurement of the displacements of the B-site cations with regard to the A-site cations is demonstrated on a BFO/LAO ($BiFeO_3/LaAlO_3$) hetero-interface which is shown in Fig. 3.6(a). Moreover, a mapping of a large field of view is performed in order to investigate whether domains are present where polarization switches. In order to quantitatively measure the displacements, the atomic column positions are estimated using statistical parameter estimation theory. Therefore, the LAO region has been analyzed. The inset in Fig. 3.6(a) shows the refined model for a part of the LAO region. In order to construct a displacement map, the Al atomic column positions are compared with a reference position. The center of the 4 surrounding La atomic column positions defines the reference position as illustrated in Fig. 3.6(b). From the reference position and the estimated La atomic column position the displacement can be measured. The measured displacements are shown in Fig. 3.6(c) and range between 2 pm and 58 pm. From the displacement map, it can be observed that the polarization changes within the analyzed field of view. Since the thickness of $LaAlO_3$ is less than 10 nm, the displacements measured from the 2D projection image are representative. It is expected that displacement mapping like demonstrated here in combination with imaging of the oxygen atomic columns using ABF STEM imaging can enhance understanding the physical properties of functional oxide materials (Gao et al., 2014; Kim et al., 2014; Li et al., 2013; Lubk et al., 2013).

3.5. Conclusions

In this chapter, a new efficient algorithm for model estimation has been implemented in order to analyze large fields of view. This is impor-

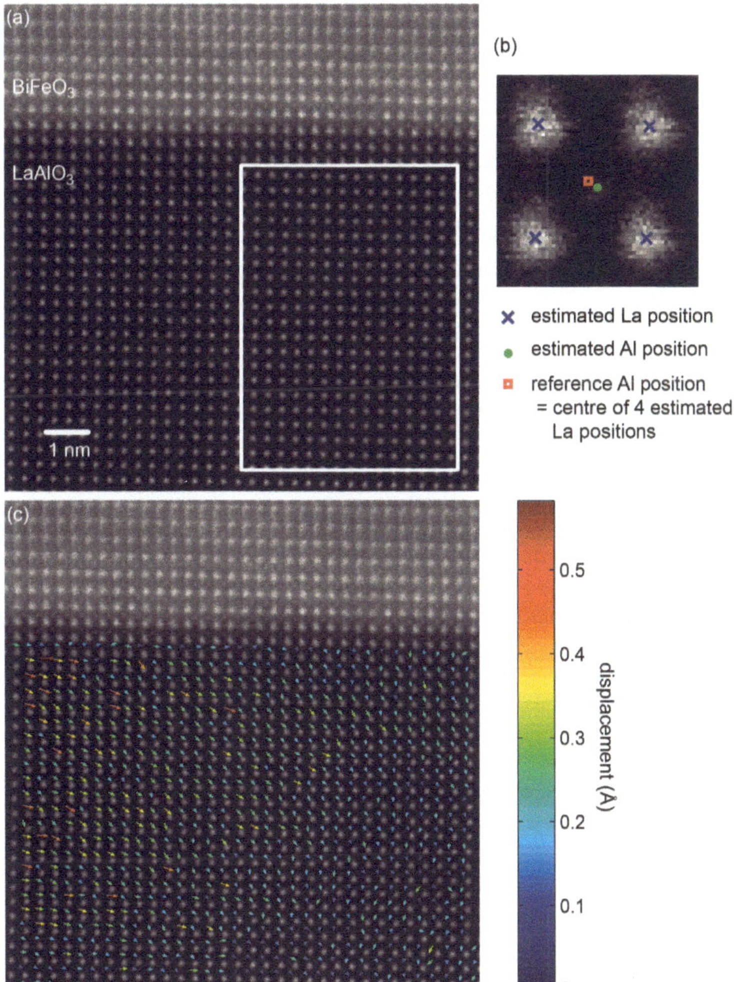

Figure 3.6 (a) HAADF STEM image of $BiFeO_3$/$LaAlO_3$. The inset in the white rectangle shows a part of the estimated model for the LAO region. (b) Illustration of the measurement of the displacement of the estimated Al atomic column from its reference position. (c) Displacement map.

tant since estimating all parameters of the parametric model at the same time is a computationally very intensive task. The basic idea of the new algorithm is the segmentation of the image into smaller sections containing individual atomic columns for the estimation of the nonlinear parameters of the model. The ideas presented here are combined in a free user-friendly software program, StatSTEM, to stimulate the use of model-based quantification (De Backer et al., 2016). The availability of this new algorithm opens up new possibilities for quantitative analyzes of large fields of view.

As a first application, model estimation has been combined with 3D reconstruction techniques allowing to reconstruct the 3D position and type of the atoms in an Au@Ag nanorod for the first time (Goris et al., 2013). Secondly, it has been illustrated how displacement mapping can be performed in a large field of view. This has been demonstrated for the atomic column positions in a five-fold-twinned decahedral Au nanoparticle (Goris et al., 2015) and for the displacement of the Al column within $LaAlO_3$. From these examples, it is expected that the improved algorithm will offer good prospects for quantitative analyzes of nanomaterials at the atomic scale with high precision.

Acknowledgments

This project has received funding from the European Research Council (ERC) under the European Union's Horizon 2020 research and innovation programme (Grant Agreement No. 770887 and No. 823717 ESTEEM3). The authors acknowledge financial support from the Research Foundation Flanders (FWO, Belgium) through Projects No. G.0502.18N, G.0267.18N, G.0368.15N, and a postdoctoral grant to A. D. B. The authors acknowledge funding from the University of Antwerp through a TOP project.

CHAPTER FOUR

Atom counting

Annick De Backer[a,b], Jarmo Fatermans[a,b,c], Arnold J. den Dekker[c], and Sandra Van Aert[a,b,*]
[a]Electron Microscopy for Materials Science (EMAT), University of Antwerp, Antwerp, Belgium
[b]NANOlab Center of Excellence, University of Antwerp, Antwerp, Belgium
[c]imec-Vision Lab, University of Antwerp, Wilrijk, Belgium
*Corresponding author. e-mail address: sandra.vanaert@uantwerpen.be

Contents

4.1. Statistics-based atom counting: methodology, possibilities, and inherent limitations

4.1.1 Introduction

In the past, different approaches have been proposed to count the number of atoms of a nanostructure in annular dark-field (ADF) scanning transmission electron microscopy (STEM). Erni et al. attempted to count the number of atoms by measuring the absolute differences in intensity of neighboring atomic columns (Erni et al., 2003). LeBeau et al. proposed to solve the counting problem by comparing simulated atomic column

Advances in Imaging and Electron Physics, Volume 217
ISSN 1076-5670
https://doi.org/10.1016/bs.aiep.2021.01.004

intensities with normalized experimentally measured atomic column intensities as illustrated in the introduction in Fig. 1.4 (LeBeau et al., 2010). Van Aert et al. introduced a statistical model-based approach to count the number of atoms from ADF STEM images of a structure viewed along a zone axis (Van Aert et al., 2011). This method has also been illustrated in the introduction in Figs. 1.3 and 1.5 and may lead to promising results for the determination of the three-dimensional (3D) arrangement of atoms in crystalline nanoparticles (Bals et al., 2011, 2012; Van Aert et al., 2011). By repeating the counting method for different viewing directions, the 3D atomic structure can be reconstructed using discrete tomography (Batenburg, 2007; Jinschek et al., 2008). Moreover, this statistics-based atom-counting method promises counting results with single atom sensitivity for each atomic column. In this chapter, this method to count the number of atoms of mono-type crystalline nanostructures from high-resolution ADF STEM images is discussed in detail. In the first part of this chapter, the possibilities and inherent limitations imposed by unavoidable statistical uncertainties will be investigated (De Backer et al., 2013). In the second part of this chapter, the proposed atom-counting method will be applied to different nanostructures.

Although ADF STEM images are highly sensitive to the number of projected atoms, it is impossible to directly count the atoms by means of visual interpretation only. Hence, the starting point of the method will be the use of the statistical model-based estimation procedures described in Chapter 2 to extract quantitative information from the ADF STEM images. In order to count the number of atoms, it is assumed that the total intensity of scattered electrons per atomic column in ADF STEM images, the so-called scattering cross-section, scales with the number of atoms in the column. The statistics-based approach proceeds in two steps.

In the first step of the counting procedure, the ADF STEM image will be considered as a data plane from which structure parameters have to be determined with the highest possible precision and accuracy (den Dekker et al., 2005). In Section 2.4 of Chapter 2 it is shown that statistical parameter estimation theory can be used to obtain accurate measurements with the highest possible precision. A parametric model is then fitted to the experimental data using a criterion of goodness of fit, quantifying the similarity between the experimental images and the model. Estimates for the scattering cross-sections per atomic column can be obtained from the estimates of the parameters of the parametric expectation model, as described by Eq. (2.43).

These estimates can be used as an input for the second step in the procedure to count the number of atoms per atomic column. In order to solve this problem, estimated scattering cross-sections will be regarded as a statistical draw from an unknown probability distribution consisting of a superposition of Gaussian components, i.e. a Gaussian mixture model. On the basis of the measured scattering cross-sections, the number of components of this distribution and their locations will be estimated. This will require a combination of an estimation procedure to determine the locations, width, and proportions of the components and the use of order selection criteria to determine the number of statistically significant components. Once the distribution is known, the number of atoms in a particular projected atomic column can be identified at the atomic scale by assigning its scattering cross-section to the component of the estimated probability distribution with the largest probability to generate this scattering cross-section.

By means of simulations, the possibilities and inherent limitations of the method to estimate the number of atoms will be studied. For this purpose, the performance of different order selection criteria will be compared and the interpretation of the order selection criterion that outperforms the others will be studied in more detail. Furthermore, the precision with which the number of atoms in a structure can be determined using this statistics-based method will be investigated.

The first part of this chapter will be organized as follows. In Section 4.1.2, the methodology to estimate the number of atoms of a nanostructure using statistics-based methods is described in detail. This is followed by an example in Section 4.1.3 illustrating the successive steps to count the number of atoms. In Section 4.1.4, the possibilities and inherent limitations of the method are investigated. Finally, in Section 4.1.5, conclusions are drawn. In the second part of this chapter, i.e. in Section 4.2, practical applications will be discussed.

4.1.2 Methodology

4.1.2.1 Model-based parameter estimation

The first step of the proposed method is to obtain precise and accurate estimates of the scattering cross-sections for each atomic column from the ADF STEM image. Therefore, use is made of statistical parameter estimation theory (den Dekker et al., 2005; Van Aert, Chang, et al., 2009; van den Bos, 2007), which is discussed in detail in Chapter 2. In Subsection 2.4 of that chapter, it is shown that accurate estimates with the highest possible precision are obtained with this approach.

The ADF STEM image is considered as a data plane from which the unknown parameters have to be estimated in a statistical manner and the expectation at the pixel (k, l) at the position (x_k, y_l) of the ADF STEM image for mono-type crystalline nanostructures can be described as:

$$f_{kl}(\boldsymbol{\theta}) = \zeta + \sum_{n=1}^{N} \eta_n \exp\left(-\frac{\left(x_k - \beta_{x_n}\right)^2 + \left(y_l - \beta_{y_n}\right)^2}{2\rho^2}\right). \tag{4.1}$$

The indices in this model are simplified here as compared to e.g. Eq. (2.9) since only one column type is present for mono-type crystalline nanoparticles. The unknown parameters of this model, defined by Eq. (2.10), are estimated in the least squares sense. From the obtained estimated parameters, the estimated scattering cross-sections can be calculated (Van Aert, Chang, et al., 2009):

$$\hat{V}_n = 2\pi \hat{\eta}_n \hat{\rho}^2. \tag{4.2}$$

These numbers are the volumes under the Gaussian peaks above the background and correspond to the total intensity of electrons scattered toward the ADF detector. In the next subsection, it will be described how these estimates can be analyzed to quantify the number of atoms in each projected atomic column.

4.1.2.2 Probability distribution of the estimated scattering cross-sections

The estimated scattering cross-sections calculated using statistical parameter estimation can be visualized in a histogram. Ideally, this histogram would consist of isolated components as shown in Fig. 4.1(a), where each component is generated by a set of atomic columns having the same number of atoms projected on top of each other. In practice, however, the components are smeared out such that it is impossible to visually determine their locations as shown in Fig. 4.1(b). This is due to a combination of different effects, such as experimental detection noise, instabilities of the material under the electron beam, different vertical onset of columns of the same height, vacancies, relaxation at the boundaries, contamination, intensity transfer between columns, and the influence of neighboring columns of different heights. Therefore, instead of visually interpreting histograms, the location and the number of components of the unknown distribution underlying the histogram are estimated on the basis of the measured scattering cross-sections. An algorithm to estimate the locations μ_g, the width

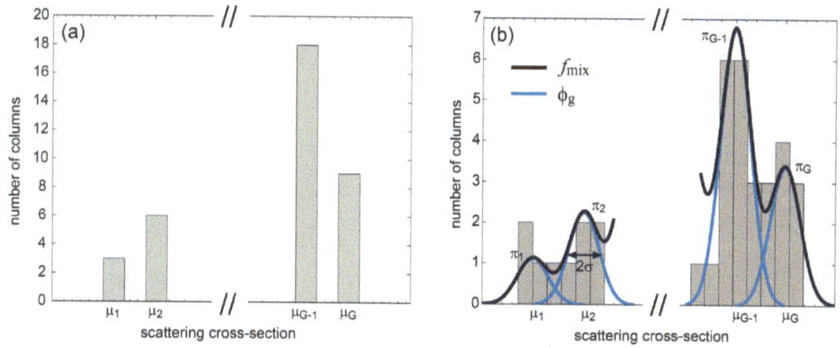

Figure 4.1 Distribution of the scattering cross-sections. (a) Example of a histogram of the scattering cross-sections in the ideal case. (b) Example of a histogram in the presence of noise. The parameters μ_g, π_g, and σ and the individual components ϕ_g of the mixture model f_{mix} are indicated.

σ, and the proportions π_g of the components in combination with the use of order selection criteria to evaluate the number of significant components is proposed (McLachlan & Peel, 2000; Van Aert et al., 2011, 2013). This procedure, of which the result only depends on the values of the estimated scattering cross-sections, leads to an accurate and precise estimate of the probability distribution as will be discussed in Section 4.1.4. Another advantage is that the result is independent of the subjective choice of bins in the histogram.

The estimated scattering cross-sections are regarded as independent statistical draws from a so-called Gaussian mixture model. In a sense, the assumption of independent statistical draws implies that cross-talk between neighboring atomic columns is not significantly contributing (Allen et al., 2003; Fertig & Rose, 1981; Nellist & Pennycook, 1999). The model is defined as a superposition of Gaussian components and describes the probability that a specific scattering cross-section value is observed. The probability density function of a mixture model with G components can parametrically be written as:

$$f_{\text{mix}}\left(V_n; \boldsymbol{\Psi}_G\right) = \sum_{g=1}^{G} \pi_g \phi_g \left(V_n; \mu_g, \sigma\right). \tag{4.3}$$

This parametric probability density function is illustrated in Fig. 4.1(b). The symbol V_n represents the stochastic variable for the nth scattering cross-section and the symbol $\boldsymbol{\Psi}_G$ represents the vector containing all unknown parameters in the mixture model with G components:

$$\boldsymbol{\Psi}_G = (\pi_1, \ldots, \pi_{G-1}, \mu_1, \ldots, \mu_G, \sigma)^T. \tag{4.4}$$

The parameters π_g, μ_g, and σ denote the mixing proportion of the gth component, the location of the gth component, and the width of the components, respectively. The mixing proportions π_g define the probability that a column contains g atoms with g ranging between 1 and a maximum number of components G, which is generally not known beforehand. The mixing proportions π_g sum up to unity. Therefore, the Gth mixing proportion is omitted in the parameter vector $\boldsymbol{\Psi}_G$. Each component $\phi_g\left(V_n; \mu_g, \sigma\right)$ of the mixture model is generated by a set of columns having the same number of atoms. Generally, if there is an increase of one atom between subsequent components, component g corresponds to atomic columns with g atoms, and G is the number of atoms found in the thickest atomic column. Depending on the nanostructure under study, it is possible that the first component does not correspond to one atom, but corresponds to a higher numbers of atoms. In that case, an offset should be taken into account when assigning number of atoms to atomic columns of the different components.

The components $\phi_g\left(V_n; \mu_g, \sigma\right)$ are modeled as homoscedastic normal components, i.e. with common variance σ^2:

$$\phi_g\left(V_n; \mu_g, \sigma\right) = \frac{1}{\sqrt{2\pi}\sigma} \exp\left(-\frac{\left(V_n - \mu_g\right)^2}{2\sigma^2}\right). \tag{4.5}$$

The parameters $\boldsymbol{\Psi}_G$ of the Gaussian mixture model, Eq. (4.3), can be computed using the maximum likelihood estimator (Day, 1969; Hasselblad, 1969; McLachlan & Peel, 2000; Wolfe, 1970) for which the estimates are found by maximizing the log-likelihood function of the unknown parameter vector $\boldsymbol{\Psi}_G$ for a given set of observed scattering cross-sections $\hat{V}$, which is defined as:

$$\log L(\boldsymbol{\Psi}_G) = \sum_{n=1}^{N} \log\left(\sum_{g=1}^{G} \pi_g \phi_g\left(\hat{V}_n; \mu_g, \sigma\right)\right). \tag{4.6}$$

The maximum likelihood estimates of $\boldsymbol{\Psi}_G$ are computed by solving the likelihood equation:

$$\frac{\partial \log L(\boldsymbol{\Psi}_G)}{\partial \boldsymbol{\Psi}_G} = 0. \tag{4.7}$$

This equation can only be solved iteratively. A very useful method is the so-called expectation maximization (EM) algorithm (Dempster et al., 1977). Using this algorithm, a solution for the parameters $\boldsymbol{\Psi}_G$ can be obtained for a fixed number of G components. The algorithm proceeds in two steps: the E-step for expectation and the M-step for maximization. In practice, the posterior probability that the nth scattering cross-section with observed value $\hat{V}_n$ belongs to the gth component of the mixture is calculated in the E-step, i.e.:

$$\tau_g\left(\hat{V}_n; \boldsymbol{\Psi}_G^{(k)}\right) = \frac{\pi_g^{(k)}\phi_g\left(\hat{V}_n; \mu_g^{(k)}, \sigma^{(k)}\right)}{\sum_{h=1}^{G} \pi_h^{(k)}\phi_h\left(\hat{V}_n; \mu_h^{(k)}, \sigma^{(k)}\right)} \tag{4.8}$$

for $g = 1, \ldots, G$ and $n = 1, \ldots, N$. The iteration step is denoted by k.

The M-step calculates the updated estimate $\boldsymbol{\Psi}_G^{(k+1)}$. The updated estimates for the mixing proportions, $\pi_g^{(k+1)}$, can be calculated as follows:

$$\pi_g^{(k+1)} = \sum_{n=1}^{N} \tau_g\left(\hat{V}_n; \boldsymbol{\Psi}_G^{(k)}\right)/N \tag{4.9}$$

for $g = 1, \ldots, G$. Updates for the unknown parameters $\mu_g^{(k+1)}$ and $\sigma^{(k+1)}$ of the component densities are given by:

$$\mu_g^{(k+1)} = \frac{\sum_{n=1}^{N} \tau_g\left(\hat{V}_n; \boldsymbol{\Psi}_G^{(k)}\right)\hat{V}_n}{\sum_{n=1}^{N} \tau_g\left(\hat{V}_n; \boldsymbol{\Psi}_G^{(k)}\right)} \tag{4.10}$$

$$\sigma^{(k+1)} = \sqrt{\frac{\sum_{g=1}^{G}\sum_{n=1}^{N} \tau_g\left(\hat{V}_n; \boldsymbol{\Psi}_G^{(k)}\right)\left(\hat{V}_n - \mu_g^{(k+1)}\right)^2}{N}}. \tag{4.11}$$

By means of simulations, which are discussed in Subsection 4.1.4.3, it has been found that unbiased estimates for the parameters of the mixture model are obtained with the following input values for the algorithm. Reliable starting values for the mixing proportions π_g are given by equal probabilities:

$$\pi_g^{(0)} = \frac{1}{G}. \tag{4.12}$$

The starting value for the variance σ of the components can be chosen as:

$$\sigma^{(0)} = \frac{\max\left(\hat{\mathbf{V}}\right) - \min\left(\hat{\mathbf{V}}\right)}{2G}. \tag{4.13}$$

For the starting values of the locations, different sets are used with the locations ranging between the minimum and maximum of the data set. The E- and M-step are iterated until convergence is attained and estimates are obtained for each set of starting parameters. The estimates $\hat{\mathbf{\Psi}}_G$ of the parameters in the mixture model are then given by the set of estimated parameters having the maximal log-likelihood value. For the true number of components G, the obtained locations μ_g of the components correspond to the average scattering cross-sections for sets of atom columns having the same number of atoms.

However, so far, only the estimation of the probability distribution of the scattering cross-sections is considered presuming a specific number of components. Subsection 4.1.2.3 deals with determining the number of components G in a mixture model using order selection criteria.

4.1.2.3 Assessing the number of components in the mixture model using order selection criteria

In Subsection 4.1.2.2, it was explained how we can obtain the probability density for a mixture model presuming a certain number of components in the mixture model. However, determining the order of the mixture model, i.e. the number of components in the distribution of the scattering cross-sections using so-called order selection criteria, is a difficult task. Different approaches are available to assess the number of components in a mixture model. A brief overview is given here. For other approaches and references, we refer to McLachlan and Peel (2000).

The information criteria for model selection calculate a measure for the goodness of fit taking the complexity of the model into account. Therefore, they include a criterion-specific penalty term which increases with the number of parameters in the Gaussian mixture model, i.e. with the number of components in the distribution. As such, a good trade-off between the goodness of fit and the complexity of the model arises. The general expression for information criteria is twice the negative penalized log-likelihood criterion:

$$-2\log L(\hat{\mathbf{\Psi}}_G) + 2C, \tag{4.14}$$

where the first term accounts for the lack of fit and the second term for the complexity of the model. The model that minimizes the negative penalized log-likelihood is selected as the probability density describing the observed scattering cross-sections. In practice, the values of the criterion are plotted as a function of the number of components. In the ideal case, the values of the criterion will show a sharp minimum for the true number of components of the underlying probability density function.

Different criteria which all account for the complexity of the model in a different manner have been proposed in the literature.

In Akaike's Information Criterion (AIC) (Akaike, 1974), the penalty term simply equals the total number of parameters of the model:

$$\mathrm{AIC}(G) = -2\log L(\hat{\boldsymbol{\Psi}}_G) + 2d \tag{4.15}$$

with d the number of parameters in the mixture model where $d = 2G$ assuming Eq. (4.3). A very similar criterion is the Generalized Information Criterion (GIC) (Broersen & Wensink, 1993), where the contribution of the penalty term can be modified by the parameter r:

$$\mathrm{GIC}(G) = -2\log L(\hat{\boldsymbol{\Psi}}_G) + r \cdot d. \tag{4.16}$$

Throughout this chapter, the parameter r of the GIC is chosen equal to 3 (Broersen & Wensink, 1996), penalizing more heavily for the complexity of the model as compared to AIC.

Another criterion considered here is the Bayesian Information Criterion (BIC) (Schwarz, 1978):

$$\mathrm{BIC}(G) = -2\log(\hat{\boldsymbol{\Psi}}_G) + d\log(N). \tag{4.17}$$

Here, the penalty term depends on the sample size N. If the sample size $N > 8$, then BIC accounts more for the complexity of the model as compared to AIC.

In the Classification Likelihood information Criterion (CLC) (Biernacki & Govaert, 1997), the entropy $\mathrm{EN}(\hat{\tau})$ associated to the model with G components is used as the term that penalizes the complexity of the model:

$$\mathrm{CLC}(G) = -2\log L(\hat{\boldsymbol{\Psi}}_G) + \mathrm{EN}(\hat{\tau}). \tag{4.18}$$

The entropy associated to the model with G components is given by:

$$\mathrm{EN}(\hat{\tau}) = -\sum_{g=1}^{G}\sum_{n=1}^{N} \tau_g\left(\hat{V}_n; \hat{\boldsymbol{\Psi}}_G\right) \log \tau_g\left(\hat{V}_n; \hat{\boldsymbol{\Psi}}_G\right) \tag{4.19}$$

where $\tau_g\left(\hat{V}_n; \hat{\boldsymbol{\Psi}}_G\right)$ is given by Eq. (4.8). The entropy of the model expresses how well the components of the mixture model are separated, i.e. how much overlap there is between neighboring components. If the components are well separated, then $\mathrm{EN}(\hat{\tau})$ is close to zero. In other words, the separation of the components in the mixture model defines how drastically the log-likelihood is penalized.

The Approximate Weight of Evidence (AWE) criterion depends on this entropy term, the number of parameters and the sample size N (Banfield & Raftery, 1993):

$$\mathrm{AWE}(G) = -2\log L(\hat{\boldsymbol{\Psi}}_G) + 2\mathrm{EN}(\hat{\tau}) + 2d(3/2 + \log N). \tag{4.20}$$

If the components are well separated, the performance of the AWE criterion resembles BIC.

Since BIC and CLC tend to underestimate and overestimate the number of components of the probability distribution, respectively (Biernacki et al., 1999; Celeux & Soromenho, 1996), the Integrated Classification Likelihood (ICL) criterion is introduced to overcome these shortcomings (Biernacki et al., 1998):

$$\mathrm{ICL}(G) = -2\log L(\hat{\boldsymbol{\Psi}}_G) + 2\mathrm{EN}(\hat{\tau}) + d\log N. \tag{4.21}$$

In Section 4.1.4.1, the performance of these criteria is compared by means of simulations.

4.1.3 Counting atoms based on a simulated gold wedge

An example illustrating the procedure to count the number of atoms is worked out in this section. The STEMsim software (Rosenauer & Schowalter, 2008) has been used to simulate an HAADF STEM image of an Au wedge. The parameters of the multislice simulation are listed in Table 4.1. The value for the Debye-Waller factor is taken from Gao and Peng (1999). The input structure for the STEM simulation is shown in Fig. 4.2(a). The noise in the simulated image, shown in Fig. 4.2(b), is Poisson distributed. First, estimates for the unknown parameters are obtained from the simulated image in the least squares sense using the expectation model given by

Table 4.1 Parameters for the multislice simulation of an HAADF STEM image of an Au wedge using the STEMsim software.

Parameter	Symbol	Value
Acceleration voltage	V_0 (kV)	300
Defocus	ϵ (Å)	−83.01
Spherical aberration	C_s (mm)	0.035
Spherical aberration of 5th order	C_5 (mm)	0
Objective aperture angle	α (mrad)	21.7881
Spatial incoherence of source	$FWHM_S$ (Å)	0.7
HAADF inner collection angle	β_1 (mrad)	50
HAADF outer collection angle	β_2 (mrad)	50
Slice thickness	z_{slice} (Å)	2.04
Debye-Waller factor	B (Å^2)	0.6373
Pixel size	$\Delta x = \Delta y$ (Å)	0.09955
Number of pixels x-direction	K	320
Number of pixels y-direction	L	170
Incident electron dose	N_d (e$^-$Å^{-2})	$6{\cdot}10^4$

(a) input structure

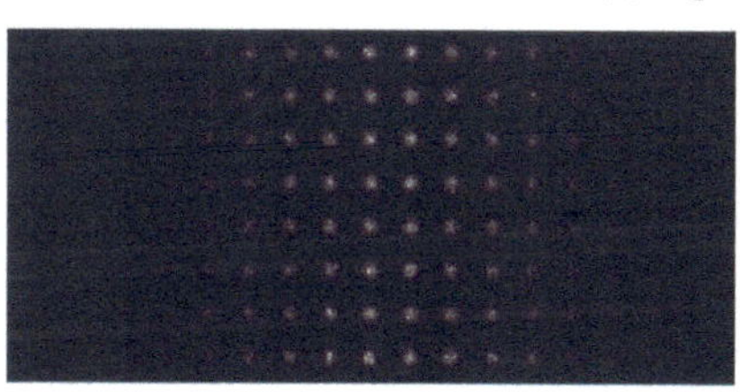

(b) simulation

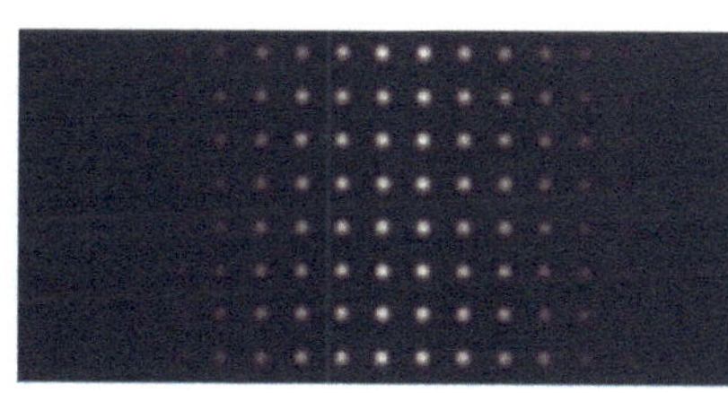

(c) model

Figure 4.2 (a) Input structure, (b) simulated HAADF STEM image, and (c) corresponding model of an Au wedge. Reprinted from De Backer et al. (2013), Copyright (2013), with permission from Elsevier.

Eq. (4.1). The thus obtained model for the HAADF STEM image evaluated at the estimated parameters is shown in Fig. 4.2(c). The volume under the peak above the background $\hat{\mathbf{V}}$, i.e. the scattering cross-section,

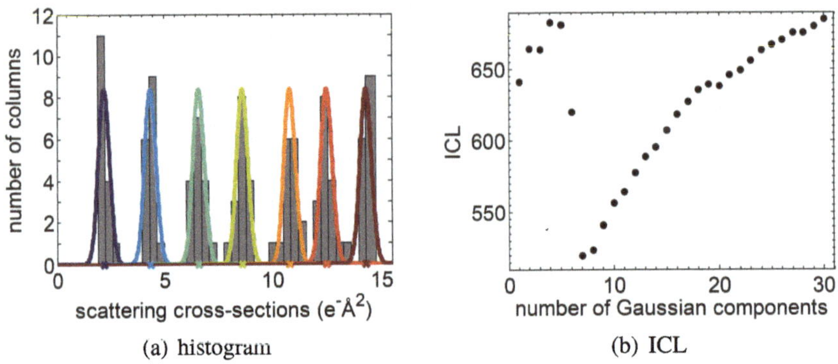

Figure 4.3 Analysis of the estimated scattering cross-sections. (a) Histogram of the estimated scattering cross-sections $\hat{\mathbf{V}}$ with the curve indicating the estimated mixture model and the points indicating the estimated locations of the components, (b) ICL criterion evaluated as a function of the number of Gaussian components in the Gaussian mixture model.

is calculated from the estimated parameters for each atomic column. These estimated scattering cross-sections $\hat{\mathbf{V}}$ are used to determine the number of atoms in each atomic column. The estimated scattering cross-sections $\hat{\mathbf{V}}$ are analyzed through the combination of the expectation maximization algorithm to determine the locations of the Gaussian mixture model and the ICL criterion to assess the number of significant components in the mixture model. Fig. 4.3(b) shows the ICL evaluation. The criterion shows a sharp minimum at 7, corresponding to the estimated number of components in the probability distribution. Fig. 4.3(a) shows the histogram of estimated scattering cross-sections together with the estimated mixture model. From Fig. 4.3(a), it is clear that the Gaussian components are approximately equidistantly separated indicating that the estimated scattering cross-sections increase monotonically with the number of atoms in an atomic column. This knowledge can be used to determine the number of atoms in the atomic column assigned to the first component. The mean of the first components equals 2.2 $e^{-}Å^{2}$. The average increment between the successive scattering cross-sections equals 2.0 $e^{-}Å^{2}$. From this ratio between both values, it can be concluded that the number of atoms in the atomic columns assigned to the first component equals one. Finally, based on the analysis shown in Fig. 4.3(a) and 4.3(b), a figure can be constructed that shows the number of atoms per atomic column by assigning each scattering cross-section to the component of the estimated probability distribution with the largest probability for this scattering cross-section.

This counting result equals the original input used for the STEMsim simulation shown in Fig. 4.2(a). The total number of atoms in the Au wedge equals 448.

4.1.4 Possibilities and limitations

In Sections 4.1.2 and 4.1.3, it was described and shown how values for the scattering cross-sections per atomic column can be obtained from ADF STEM images and how the distribution of these scattering cross-sections is analyzed to determine the number of atoms per atomic column using statistical model-based methods. In this section, we will discuss the possibilities and inherent limitations of this statistics-based atom-counting method.

The ultimate goal of an atom-counting procedure is to determine the number of atoms in an atomic column with the highest possible accuracy and precision. However, on the one hand, modeling errors in the Gaussian mixture model for the probability distribution limit the accuracy resulting in systematic errors for the number of atoms. In Subsections 4.1.4.1 and 4.1.4.2, it is studied under which conditions unbiased measurements can be obtained for the number of atoms in an atomic column by using information criteria to assess the order of the mixture model. On the other hand, the precision on the estimated number of atoms is limited due to the unavoidable presence of noise in the estimated scattering cross-sections, as discussed in Subsection 2.4 of Chapter 2. In Sections 4.1.4.3 and 4.1.4.4, the variance on the parameters of the Gaussian mixture model and, subsequently, the precision with which the number of atoms can be determined is analyzed.

4.1.4.1 Performance of the order selection criteria

In Subsection 4.1.2.3, different order selection criteria which are present in the literature are briefly discussed. In this subsection, the performance of these criteria is studied by simulating random draws from probability distributions as a function of the relative width of the Gaussian components. This relative width expresses the ratio between the width σ of the Gaussian components and the increment δ between successive components. The locations of the different components are equidistantly chosen as it is assumed that the scattering cross-sections almost linearly increase with the number of atoms in a column for thin specimens. Statistical draws from a probability distribution with 16 components located at $\mu = 1, 2, \ldots, 16$ are simulated using the probability distribution of the Gaussian mixture model given by Eq. (4.3). The relative width σ/δ of these components is varied

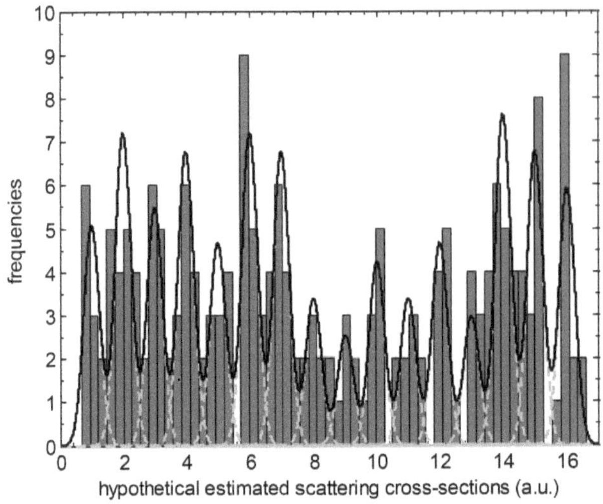

Figure 4.4 Example of a histogram (random draw) together with the underlying simulated probability distribution with 16 components and relative width $\sigma/\delta = 0.25$.

between $0.01 \leq \sigma/\delta \leq 0.30$ and describes how well the different components in the histogram are separated. The proportions π_g for the different components are multinomially distributed with equal probability of 1/16. The number of observations N, i.e. the number of atomic columns, equals 200. For each value of the relative width σ/δ, 100 random draws from the Gaussian mixture model are simulated. An example of such a random draw of a probability distribution is shown in Fig. 4.4 for $\sigma/\delta = 0.25$. The information criteria, which are discussed in Subsection 4.1.2.3, are evaluated for these simulated random draws from the distributions assuming that the number of components may vary between 5 and a maximum value of 30. The minimum of 5 is set based on the characteristic shape of the criterion discussed in more detail in Subsection 4.1.4.2. In Table 4.2, the performance of different order selection criteria is shown in terms of the percentage of correctly determined number of components as a function of the relative width of the Gaussian components. The number of components is chosen based on the minimum of the criterion without visual inspection of the criterion. From the results in Table 4.2, it can be concluded that the ICL criterion outperforms the other information criteria for assessing the number of components in a Gaussian mixture model for atom-counting applications. In combination with visual inspection of the criterion and a higher number of scattering cross-sections N, it is expected

Table 4.2 Performance of different order selection criteria in terms of the percentage of correctly determined number of components as a function of the relative width σ/δ of the components. The average number of data points per component N/G equals 12.5. The largest value σ/δ for which a percentage above 50 is still observed, is highlighted for each criterion. Reprinted from De Backer et al. (2013), Copyright (2013), with permission from Elsevier.

σ/δ	CLC	AIC	GIC	AWE	BIC	ICL
0.01	77	81	91	100	100	100
0.05	45	45	80	100	97	98
0.10	14	21	60	100	93	98
0.15	19	27	77	92	99	100
0.20	27	38	74	0	68	99
0.25	8	45	44	0	0	78
0.30	1	30	1	0	0	8

that the percentage of correctly determined number of components will increase. Consequently also more reliable results will be possible for higher relative widths σ/δ, as will be shown in Subsection 4.1.4.2.

4.1.4.2 Interpretation of the integration classification likelihood criterion

The purpose of this subsection is to get more insight in the ICL criterion since it turns out that this criterion has optimal properties. When evaluating ICL as a function of the number of components of the Gaussian mixture model, it has been observed that the true number of components often corresponds to a local optimum. This knowledge may help interpreting ICL in a correct manner and may improve the reliability of the selected number of components. Fig. 4.5 shows the shape of ICL for an underlying distribution consisting of 8 and 16 components as a function of an increasing number of observations corresponding to the number of atomic columns N. The random draws from the distributions are simulated under the same conditions as described in Subsection 4.1.4.1. The relative width of the components equals $\sigma/\delta = 0.25$. Two main conclusions can be drawn from Fig. 4.5. First, it is observed that the shape of the criterion is comparable for the different cases. It appears that ICL values are relatively small for a small number of components and that an extra minimum is shown at the correct number of components, especially when the number of components is larger. This behavior hampers finding the correct number of components

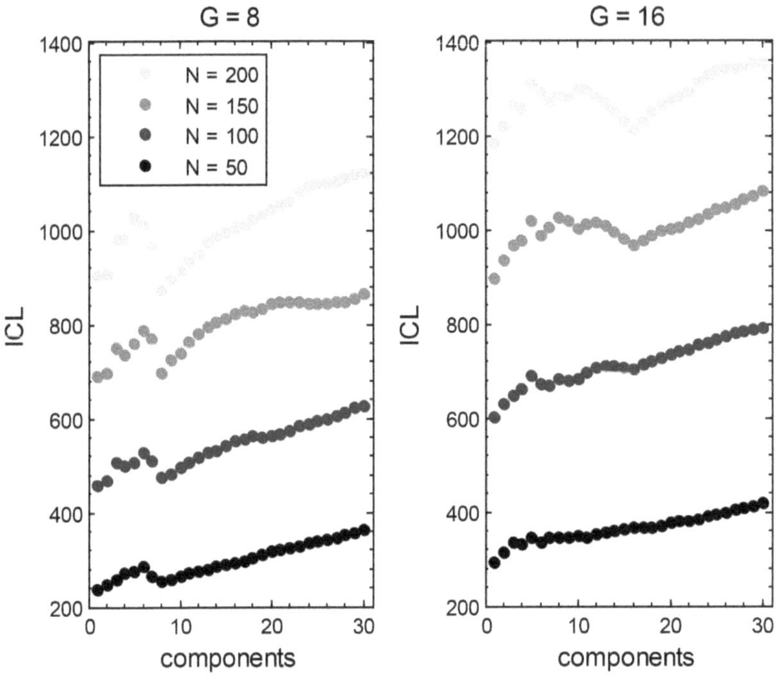

Figure 4.5 The ICL criterion as a function of an increasing number of atomic columns N for $G = 8$ components at the left-hand side and $G = 16$ components at the right-hand side with $\sigma/\delta = 0.25$. Reprinted from De Backer et al. (2013), Copyright (2013), with permission from Elsevier.

of the underlying distribution by using ICL without visual inspection and explains the sudden drop down in the performance of the ICL criterion in Table 4.2. Therefore, small values corresponding to a small number of components are neglected during visual inspection as it can be assumed that in the atom-counting analysis at least a certain number of components will be available in the distribution of the estimated scattering cross-sections. This number, related to the thickness, can often be determined based on some prior knowledge about the size of the particle under study. It should be noticed that if the number of components of the underlying distribution is small, no local minima are found for a higher number of components in ICL.

Secondly, it has been observed from Fig. 4.5 that the minimum of interest in the ICL criterion corresponding to the correct number of components becomes more pronounced if the number of observations N increases. In addition, if the underlying distribution consists of many components, the number of observations N should be larger than for a distribution with a small number of components to determine the correct order of the mixture model. This conclusion follows from Fig. 4.5 where a minimum in ICL for $G = 16$ components for $N = 50$ and $N = 100$ is rather unclear whereas a more pronounced minimum for $G = 8$ components is already observed for $N = 50$. This means that the quality of the ICL increases for an increasing ratio of N/G. While the number of observations corresponds to the number of atomic columns in an experimental image when a single image is recorded, this number can be increased by analyzing larger fields of view or multiple images, if experimental conditions are favorable.

Based on these conclusions, the number of observations, i.e. atomic columns, needed in order to obtain reliable results from the evaluation of ICL can be derived as a function of the number of components in the probability distribution. Therefore, simulations similar to those discussed in Subsection 4.1.4.1 are used. The relative width of the components equals $\sigma/\delta = 0.25$. The percentage of correctly determined number of components is calculated for random draws from probability distributions consisting of a certain number of atomic columns and components. This calculation is repeated while increasing the number of atomic columns till this percentage equals 95%. The results of this analysis are shown in Fig. 4.6. From this figure, it can be concluded that the necessary number of atomic columns increases almost linearly with the number of components of the underlying Gaussian mixture model. It should be mentioned that the results in Fig. 4.6 are obtained without any visual inspection of ICL. This means that the necessary number of observations needed to obtain reliable results can slightly be reduced when accompanying this analysis with a visual inspection of the ICL criterion.

Since the necessary number of atomic columns increases almost linearly with the number of components, the number of atomic columns per component N/G can be considered as a measure to extend this analysis for higher relative widths. Then, the reliability of the ICL can be derived more generally as a function of the relative width σ/δ. The results of these simulations are shown in Fig. 4.7 for relative widths σ/δ between 0.2 and 0.4. For these simulations, probability distributions with 10 components were used under the same simulation conditions as those of Section 4.1.4.1. The ICL is evaluated from 8 to 12 components. From the figure, it can be concluded that the number of atomic columns per component N/G should be increased for higher relative widths σ/δ in order to retain the high reliabil-

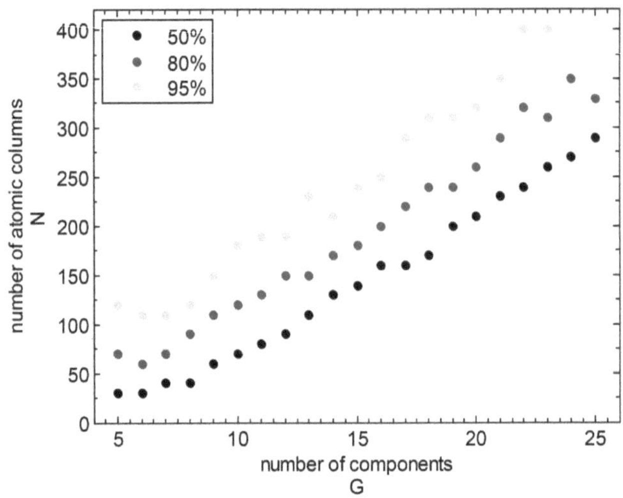

Figure 4.6 The number of atomic columns needed to obtain reliable results from the ICL criterion for a given number of components with $\sigma/\delta = 0.25$. Different levels have been chosen to express the reliability in terms of the percentage of correctly assigned number of components. The black, dark gray and light gray dots correspond to 50%, 80%, and 95% respectively. Reprinted from De Backer et al. (2013), Copyright (2013), with permission from Elsevier.

ity of the ICL. A reliability of 50% is still possible for high overlap between neighboring components.

In general, it can be concluded that the use of the ICL criterion leads to unbiased measurements for the number of components in a mixture model provided that enough atomic columns are available. Since the number of components determines the number of atoms that will be assigned to each atomic column, the analysis also reflects the accuracy of the method to count the number of atoms. In Subsection 4.1.4.3, it will be mentioned that unbiased estimates for the parameters of the Gaussian mixture model are obtained which are also required for accurate atom-counting results. The precision of the atom-counting method will be discussed in Subsections 4.1.4.3 and 4.1.4.4.

4.1.4.3 Precision of the estimated parameters of the Gaussian mixture model

The precision of the number of atoms is mainly determined by the width σ of the Gaussian components in the mixture model, and the precision of the locations μ_g of the components in the Gaussian mixture model. The

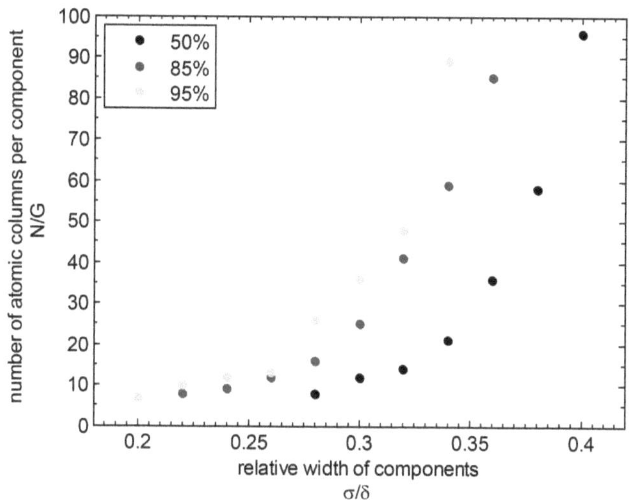

Figure 4.7 The number of atomic columns per component needed to obtain reliable results from the ICL criterion for a certain relative width. Different levels have been chosen to express the reliability in terms of the percentage of correctly assigned number of components. The black, dark gray and light gray dots correspond to 50%, 80%, and 95% respectively. Reprinted from De Backer et al. (2013), Copyright (2013), with permission from Elsevier.

parameter μ_g can be estimated by maximizing the likelihood function using the EM algorithm discussed in Subsection 4.1.2.2. As extensively discussed in Chapter 2, statistical parameter estimation theory provides an expression for the highest precision, or in other words, the minimum variance, with which the parameters of the Gaussian mixture model can be determined, i.e. the so-called Cramér-Rao lower bound (CRLB) (den Dekker et al., 2005; Van Aert, den Dekker, Van Dyck, et al., 2002b; van den Bos, 2007), which is the theoretical lower bound on the variance of any unbiased estimator for unknown model parameters. One of the properties of the maximum likelihood estimator is that it achieves the CRLB asymptotically, i.e. for an infinite number of observations. Consequently, if the CRLB is attained using the EM algorithm, the CRLB provides a useful measure for the error bar on the estimated parameters of the Gaussian mixture model. In Appendix 4.A on page 142, it is summarized how the CRLB is computed for the estimated parameters of the Gaussian mixture model.

In this subsection, it is investigated by means of simulations if the lower bound on the variance of the estimated parameters is attained. Therefore, 100 different noise realizations of a Gaussian mixture model with

Table 4.3 Attainability of the CRLB on the estimated parameters of a Gaussian mixture model with 5 components. The sample variances are computed from parameter estimates obtained from 100 noise realizations of the Gaussian mixture model and compared with the CRLB on the variance of the model parameters. Reprinted from De Backer et al. (2013), Copyright (2013), with permission from Elsevier.

	CRLB	sample variance	95% confidence interval
π_1	0.0017	0.0022	[0.0017; 0.0030]
π_2	0.0018	0.0026	[0.0020; 0.0035]
π_3	0.0018	0.0019	[0.0014; 0.0025]
π_4	0.0018	0.0017	[0.0013; 0.0029]
π_5	0.0017	0.0017	[0.0013; 0.0024]
μ_1	0.0038	0.0041	[0.0032; 0.0056]
μ_2	0.0048	0.0048	[0.0037; 0.0065]
μ_3	0.0048	0.0055	[0.0042; 0.0074]
μ_4	0.0048	0.0045	[0.0035; 0.0061]
μ_5	0.0038	0.0031	[0.0024; 0.0041]
σ	0.00044	0.00043	[0.00033; 0.00057]

5 components are simulated. The values for the true parameters in the simulation are set as follows: the proportions are the same for each component, i.e. $\pi_g = 0.2$, and the relative width equals $\sigma/\delta = 0.25$. The number of atomic columns N equals 100. In this analysis unbiased estimates have been obtained for the parameters of the Gaussian mixture model, which is fundamental for obtaining an accurate estimate for the number of atoms. The results for the attainability of the CRLB on the variance of the estimated parameters of the mixture model are presented in Table 4.3. This table shows the computed value for the lower bound on the variance of a particular parameter in the second column. The third column shows the sample variance obtained from the parameter estimates from the 100 noise realizations using the EM algorithm. In the last column the corresponding 95% confidence intervals on the sample variances are given. Since the 95% confidence interval of the sample variance encloses the CRLB, it can be concluded from Table 4.3 that the CRLB on the variance of the parameters of the Gaussian mixture model is attained using the expectation maximization algorithm. This means that the CRLB can be used as a measure for the variance on the estimated parameters of the Gaussian mixture model. This variance will be used to determine the precision with which the number

of atoms in a particular atomic column can be estimated and this will be subject of the following subsection.

4.1.4.4 Single atom sensitivity

The number of atoms in a particular atomic column is determined by assigning each estimated scattering cross-section to the component of the estimated probability distribution with the largest probability to generate this scattering cross-section. In this manner, the precision of the number of atoms is mainly defined by the overlap between neighboring components in the Gaussian mixture model. Single atom sensitivity is defined as an error bar of ± 0 atoms. If there is no overlap between neighboring components in the Gaussian mixture model, single atom sensitivity is attained as in this case an estimated scattering cross-section can unambiguously be assigned to a certain location μ_g. However, overlap between neighboring components in the Gaussian mixture model is unavoidable and is given by a combination of the precision with which the location parameter μ_g in the Gaussian mixture model can be determined, the width σ and the proportions π_g of the components in the Gaussian mixture model. Nevertheless, the probability to assign the correct number of atoms, i.e. the fraction of the analyzed atomic columns without error, can be computed.

The precision of the location parameter μ_g can be quantified using the Cramér-Rao lower bound on the variance of the estimated parameters of the Gaussian mixture model as discussed in Subsection 4.1.4.3, and will be denoted as $CRLB_\mu$. The width σ of the components is nonzero due to a combination of experimental detection noise and instabilities. These two parts define an effective width σ_{eff} for the normal components:

$$\sigma_{\mathrm{eff}} = \sqrt{CRLB_\mu + \sigma^2}. \tag{4.22}$$

The amount of overlap between neighboring components is furthermore influenced by the proportions π_g and is also used to define the probability to assign the correct number of atoms to a specific component. This procedure on how to compute these probabilities is illustrated in Fig. 4.8. The component shown in color (light and dark gray) will be used for this purpose and will be referred to as component ϕ_g. This distribution ϕ_g will be generated by a set of atomic columns having the same number of atoms in it. Atomic columns having scattering cross-sections belonging to interval I_g of the component ϕ_g indicated in Fig. 4.8, will be assigned to component ϕ_g. This central part I_g is bounded by the intersections of component

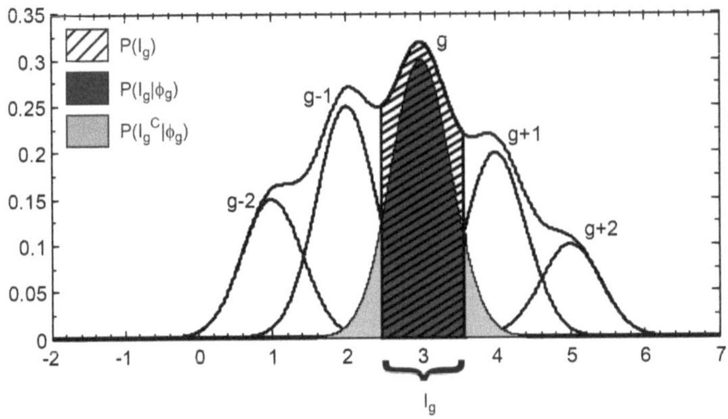

Figure 4.8 Overlap between neighboring components in the Gaussian mixture model to illustrate the concept of single atom sensitivity. Scattering cross-sections belonging to interval I_g will be assigned to component ϕ_g. The probability $P(I_g)$ that a scattering cross-section resides in I_g is indicated by the hatched area. The posterior probability $P(I_g|\phi_g)$ with which a scattering cross-section belonging to ϕ_g resides in I_g is indicated by the dark gray area. The posterior probability $P(I_g^C|\phi_g)$ with which a scattering cross-section belonging to ϕ_g resides not in I_g is indicated by the light gray area. Reprinted from De Backer et al. (2013), Copyright (2013), with permission from Elsevier.

ϕ_g with ϕ_{g-1} and ϕ_g with ϕ_{g+1}. The location of the intersection $S_{g(g+1)}$ between component ϕ_g and ϕ_{g+1}, can be expressed as a function of σ_{eff}, π_g, and μ_g:

$$S_{g(g+1)} = \frac{2\sigma_{\text{eff}}^2 \log\left(\frac{\pi_g}{\pi_{g+1}}\right) - \mu_g^2 + \mu_{g+1}^2}{2\left(\mu_{g+1} - \mu_g\right)}. \tag{4.23}$$

Using these locations of the intersections which define the interval I_g, the posterior probability with which a scattering cross-section belonging to component ϕ_g resides in I_g can be computed as follows:

$$P(I_g|\phi_g) = \Phi\left(\frac{S_{g(g+1)} - \mu_g}{\sigma_{\text{eff}}}\right) - \Phi\left(\frac{S_{(g-1)g} - \mu_g}{\sigma_{\text{eff}}}\right) \tag{4.24}$$

where $\Phi\left(\frac{S_{g(g+1)}-\mu_g}{\sigma_{\text{eff}}}\right)$ corresponds to the cumulative distribution function of the standard normal distribution evaluated at $\frac{S_{g(g+1)}-\mu_g}{\sigma_{\text{eff}}}$. The posterior probability that a scattering cross-section belonging to the central dark gray part I_g stems from component ϕ_g is then given by Bayes' rule:

$$P(\phi_g|I_g) = P(I_g|\phi_g) \cdot \frac{P(\phi_g)}{P(I_g)} \tag{4.25}$$

where $P(\phi_g)$ equals the mixing proportion π_g of the gth component and $P(I_g)$ corresponds to the area below the Gaussian mixture model between $S_{(g-1)g}$ and $S_{g(g+1)}$, hatched in Fig. 4.8:

$$P(I_g) = \int_{S_{(g-1)g}}^{S_{g(g+1)}} f_{\text{mix}}(V; \mathbf{\Psi}_G) dV. \tag{4.26}$$

Eq. (4.25) expresses the probability that a scattering cross-section of interval I_g originates from the correct component ϕ_g and thus identifies the percentage of atomic columns that has been determined with single atom sensitivity. For the whole structure of interest, the probability to count the number of atoms in an atomic column without error can then be obtained by averaging the probabilities of Eq. (4.25) for all the components in the Gaussian mixture model.

Similarly, the probability to miscount the number of atoms with ± 1 can be computed. This probability is related to the outermost light gray parts in Fig. 4.8. Atomic column scattering cross-sections whose scattering cross-sections belong to these parts will be assigned to component $g-1$ or $g+1$. In practice, the probability to miscount the number of atoms with ± 2 will generally equal zero, but can be computed in the same manner taking the overlap of distribution g with the second neighboring components into account.

If equal proportions are assumed for the components in the Gaussian mixture model, the intersection between neighboring components is located in the middle between the locations μ_g of the considered components. The probability of counting the number of atoms correctly, i.e. with an error of ± 0 atoms, is then defined only by σ and $CRLB_\mu$. In order to illustrate how this probability relates to the latter parameters, tolerance levels for single atom sensitivity are drawn on a plot of $CRLB_\mu$ as a function of the relative width σ/δ in Fig. 4.9. For this figure, the Cramér-Rao lower bound on the variance of the location parameter μ_g has been calculated for different values of the parameters π_g, μ_g, and σ of the Gaussian mixture model based on the formulae in Appendix 4.A. In addition, the number of components G and the number of atomic columns N have been varied. Based on the computed values, it was observed that the value of $CRLB_\mu$ is

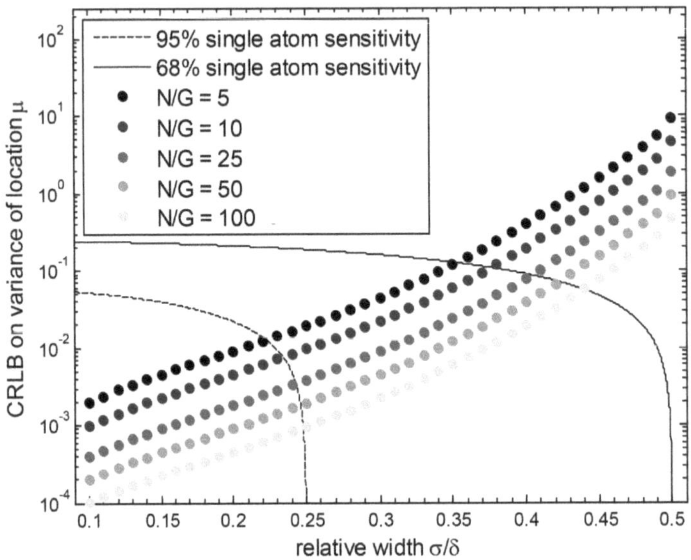

Figure 4.9 Precision on the location μ for a given relative width σ/δ and number of atomic columns per component N/G. The dashed line indicates the 95% tolerance level for single atom sensitivity. The full line indicates the 68% tolerance level for single atom sensitivity. Reprinted from De Backer et al. (2013), Copyright (2013), with permission from Elsevier.

determined by the relative width σ/δ of the components and the number of atomic columns per component N/G. Therefore, $CRLB_\mu$ has been calculated for the central component of a probability distribution with equal proportions for different values of σ/δ and N/G. The values of $CRLB_\mu$ increase for a decreasing ratio of N/G and an increasing relative width σ/δ. Based on Eq. (4.25), lines are drawn in Fig. 4.9 for the 68% and 95% tolerance levels. These lines show that the precision with which the number of atoms can be counted in a particular atomic column is limited by the relative width of the components. Indeed, if the relative width goes beyond a certain value, then neighboring components will strongly overlap. Fig. 4.9 can be used to give an indication of the probability to count the correct number of atoms for a specific case.

The precision with which the number of atoms is determined in the example of the Au wedge described in Section 4.1.3 can be derived using Eq. (4.25). For this example, it can be concluded that single atom sensitivity is attained for 99.9% of the analyzed atomic columns. The fact that this probability is larger than 95% can also be derived from Fig. 4.9, since the number of atomic columns per component N/G equals 16 and the average relative width σ/δ equals 0.15.

4.1.5 Conclusions

In the first part of this chapter, a statistics-based method to count the number of atoms in a nanostructure containing one atom type from ADF STEM images has been explained in detail together with a thorough study on the possibilities and inherent limitations imposed by unavoidable statistical uncertainties (De Backer et al., 2013). It has been shown that the method largely proceeds in two steps. First, a parametric model is fitted to the experimental ADF STEM images in order to extract quantitative values for the scattering cross-sections. These values can be used to determine the number of atoms per atomic column, if it is assumed that the scattering cross-section scales with the number of atoms. In the second step, the number of atoms per atomic column is determined by fitting a Gaussian finite mixture model to the probability distribution of the estimated scattering cross-sections using the expectation maximization algorithm. This probability distribution consists of components of sets of atomic columns having the same number of atoms. The number of components in the probability distribution needs to be determined by the use of an order selection criterion.

By means of simulations, the possibilities and inherent limitations of this statistics-based atom-counting method have been investigated. The performance of different order selection criteria has been evaluated. It turns out that the Integrated Classification Likelihood, or shortly ICL, criterion outperforms other selection criteria when one is evaluating probability distributions for counting the number of atoms in a nanostructure. In addition, it has been shown that different aspects need to be taken into account for a correct interpretation of the ICL criterion. The number of atomic columns N, the number of components G in the probability distribution, the width of the components σ, the average increment between the components δ, and the typical shape of the criterion directly affect the accuracy with which the number of atoms can be determined. Also the precision with which the number of atoms in a particular atomic column can be estimated has been studied. It has been shown that the variance of the expectation maximization estimates for the parameters of the Gaussian mixture model attains the Cramér-Rao lower bound, i.e. the highest precision with which the parameters of the model can be determined. Moreover, it has been shown that single atom sensitivity is feasible.

In conclusion, the statistics-based method to count the number of atoms in mono-type crystalline nanoparticles can be used without prior knowledge of the atomic structure and is independent of image simulations. This powerful method enables one to identify the percentage of atomic columns that has been determined with single atom sensitivity. In the second part of this chapter, the practical use of this statistics-based atom-counting method will be tested on experimental images and applied to different nanostructures.

4.2. Practical applications: nanoparticle atom counting from experimental ADF STEM images

4.2.1 Introduction

In Van Aert et al. (2011), it has been proven that the statistics-based atom-counting method described in the first part of this chapter, Section 4.1, is successful for the determination of the number of atoms in a metal nanocluster embedded into a stabilizing matrix with the same crystal structure. The atom-counting results were consistent for two different viewing directions. Furthermore, the first 3D reconstruction at the atomic scale could be realized by combining these counting results using discrete tomography as illustrated in Chapter 1. In this part of this chapter, the statistics-based method for atom counting will be applied to other challenging structures: Ge clusters, larger Au nanorods, and Pt/Ir bimetallic catalysts.

As a first application in Section 4.2.2, an attempt will be made to count the number of atoms in a Ge cluster. Since this Ge cluster is not embedded in a stabilizing matrix, it is moving easily under illumination of the electron beam. This makes the atom counting much more challenging as compared to atom counting of a cluster embedded in a stabilizing matrix with the same crystal structure as in Van Aert et al. (2011). The atom-counting method described in Section 4.1, will be recapitulated briefly for this application. Although our method is a straightforward procedure, this exploratory example will show that the interpretation of the output is absolutely not straightforward. Therefore, a combination with other complementary methods will be required in order to obtain reliable atom counts with single atom sensitivity.

As a second application, atom counting for an Au nanorod will be discussed in Section 4.2.3. The counting results will be validated using a comparison with detailed ADF STEM image simulations (Van Aert et al., 2013). This requires the intensity variations to be placed on an absolute scale by normalizing the measured image intensities to the incident beam. Furthermore, atom-counting results from an Au nanorod will be compared with the number of atoms that can be counted from a slice through the 3D atomic scale reconstruction from electron tomography. Additionally, it will be illustrated how the atom-counting results can be used to create 3D atomistic models.

In a third, more elaborate analysis in Section 4.2.4, the number of atoms will be counted for a Pt/Ir nanoparticle, i.e. an industrial bimetallic catalyst (De Backer, Martinez, et al., 2015). The goal here is to elucidate the experimental constraints on the atom-counting reliability which become extremely important when looking at beam-sensitive systems. The inherent limitations of the atom-counting method discussed in the first part of this chapter, Section 4.1, are clearly illustrated by the analysis of the experimental data. Finally, the conclusions of the second part of this chapter will be summarized in Section 4.2.5.

4.2.2 Ge cluster: an exploratory example

In this section, atom counting on a Ge cluster using the statistics-based method will be illustrated. For this purpose, the statistics-based atom-counting method will be briefly summarized. Next, it will be shown that the interpretation of the output of the atom-counting method not always leads to a unique trustworthy solution.

4.2.2.1 Experiment

HAADF STEM images of a nanometer-sized Ge cluster were taken at the Qu-Ant-EM, a double-corrected FEI Titan^3 working at 300 kV with a 21.4 mrad semiconvergence angle. It is assumed that defocus ϵ, spherical aberration C_s, and spherical aberration of the fifth order C_5 equal 0 for this experiment. A series of images was taken while the Ge cluster was rotating under the electron beam. The images were taken under the same conditions and the pixel size Δx equals 0.25 Å. The structure along the [110] direction has a dumbbell structure, i.e. pairs of atoms with a distance of 140 pm. The images along the [110] direction are shown in Figs. 4.10(A-E). Because of instabilities of the material under the electron beam and slight mistilt, overlap of neighboring atomic columns in the Ge dumbbells cannot be avoided, and therefore the dumbbells are not resolved in the high-resolution HAADF STEM images.

4.2.2.2 Methodology

The statistics-based atom-counting method, explained in detail in the first part of this chapter, Section 4.1, has been employed to obtain atom counts for the 5 high-resolution HAADF STEM images shown in Figs. 4.10(A-E). As a first step, the scattering cross-sections are quantified atomic column-by-atomic column by fitting the empirical imaging model to the experimental images, as described in detail in Chapter 2 and Subsection 4.1.2.1 of this chapter. Strong correlation effects hinder precise measurements of the individual scattering cross-sections of each column in the Ge dumbbell. Therefore, the scattering cross-sections along each column of Ge dumbbells have been estimated rather than the individual column scattering cross-sections. The refined models of the observations are shown in Figs. 4.10(F-J) and are in excellent agreement with the recorded images in Figs. 4.10(A-E) demonstrating the quality of the parametric model used. The estimated scattering cross-sections are expected to increase monotonically with the number of atoms. The estimated scattering cross-sections of the images of Figs. 4.10(A-E) are visualized in a histogram as shown in Figs. 4.11(F-J), respectively. A visual interpretation of experimental histograms may lead to deceptive results. Instead, the statistical approach introduced in (Sub)sections 4.1.2.2, 4.1.2.3, and 4.1.3 of this chapter is used allowing one to determine the locations of all statistically meaningful components. The number of significant components, i.e. the model order G can be retrieved by evaluating the so-called Integrated Classification Likelihood (ICL) criterion, which is shown in Figs. 4.11(A-E) for the images of Figs. 4.10(A-E). The estimated model order is given by the number of components for which ICL reaches a local minimum, which is arrowed in Figs. 4.11(A-E). The ICL criterion indeed shows the typical shape described in Subsection 4.1.4.2 of this chapter. Figs. 4.11(F-J) show the corresponding estimated Gaussian mixture models for the number of components arrowed in the ICL criterion. Based on these estimated probability distributions, the number of atoms along the third dimension in each Ge atomic dumbbell can be identified by assigning each scattering cross-section to the components of the mixture model with the largest probability to generate this scattering cross-section. As discussed in Section 4.1.3 of this chapter, the location of the first component with respect to the average increment δ between the successive components can be used to determine the number of atoms in the atomic columns assigned to this first component. This has been done for the construction of the scatter plots in Figs. 4.11(K-O). Here, the locations of the components of the Gaussian mixture model

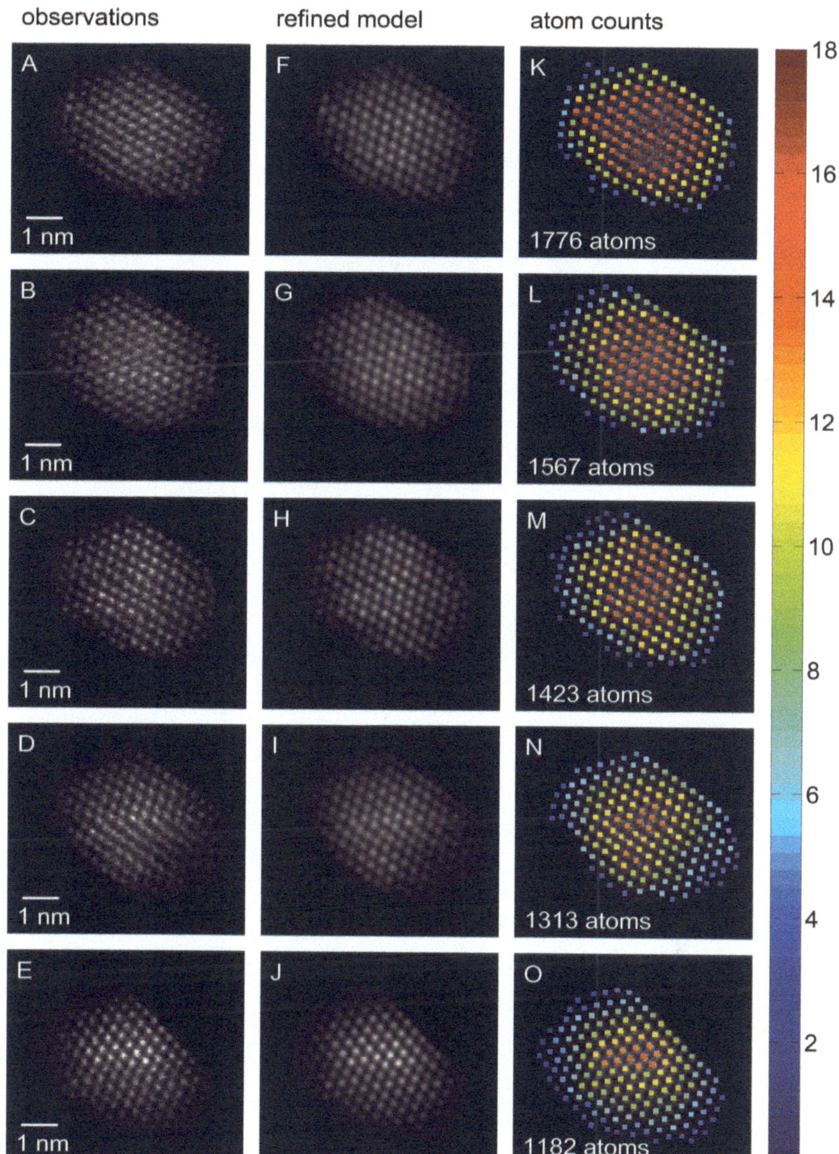

Figure 4.10 Analysis of a Ge cluster. (A-E) Experimental HAADF STEM images of the cluster along the [110] direction. (F-J) Corresponding refined models for the experimental observations. (K-O) Quantification of the number of Ge atoms in each atomic dumbbell based on the analysis presented in Fig. 4.11.

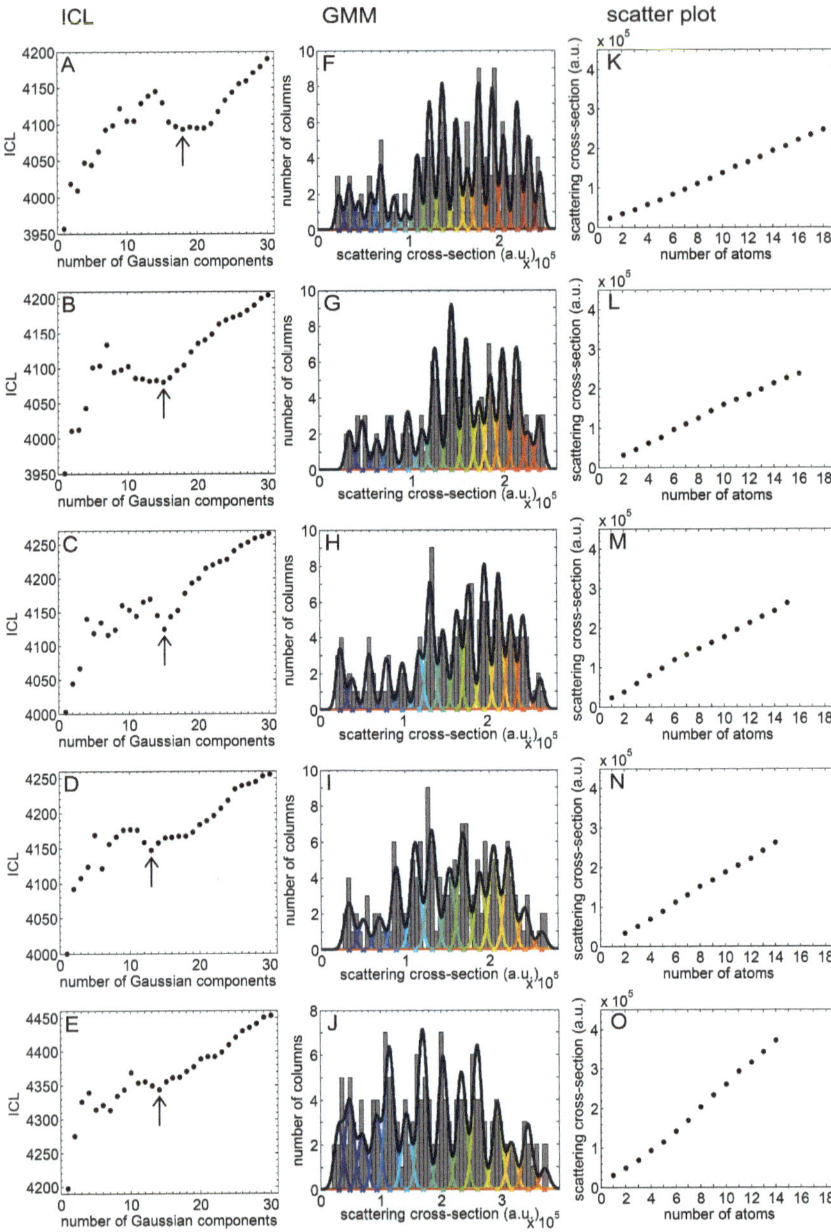

Figure 4.11 Atom-counting analysis of a Ge cluster. (A-E) Integrated Classification Likelihood criterion evaluated as a function of the number of Gaussian components in the Gaussian mixture model (GMM). (F-J) Histograms of the scattering cross-sections of the Ge dumbbells together with the estimated mixture model (black curve) and its individual components (colored curves). (K-O) Scatter plot of the estimated locations of the components in the Gaussian mixture model as a function of the number of atoms in each atomic dumbbell corresponding to this component.

are plotted as a function of the number of atoms corresponding to the component. Finally, atom counts along the third dimension in each Ge atomic dumbbell of the nanoparticle are quantified in Figs. 4.10(K-O) for the different observed images together with the estimated total number of atoms in the Ge cluster. These results will be discussed in the next subsection.

4.2.2.3 Results and discussion

In the previous subsection, the statistics-based atom-counting method described in detail in the first part of this chapter, Section 4.1, has been summarized briefly and has been applied to the experimental set of HAADF STEM images of a Ge cluster. The results are shown in detail in Figs. 4.10 and 4.11. In this subsection, these results will be discussed.

First, it is noted that the total number of atoms is reduced by more than 30% from the first to the last frame as can be seen from Figs. 4.10(K-O). This can be caused by beam damage, sputtering atoms from the nanoparticle. Then, the mean scattering cross-section value of the last component, i.e. the thickest atomic column, should decrease. Therefore, second, the scatter plots of Figs. 4.11(K-O), in which the locations of the components are plotted versus the corresponding number of atoms are studied into more detail. For these scatter plots, it is expected that the slope of the curve is the same for each recorded frame, since a certain number of atoms corresponds to a specific value for the scattering cross-section which is determined from the Gaussian mixture model estimation. However, the slope in the scatter plots here changes through the different frames. Furthermore, if the total number of atoms is decreasing throughout the experiments, it is expected that the largest scattering cross-section value is observed for the first frame whereas the opposite is observed for these data.

From these observations, it is clear that for these experimental data the interpretation of the output of the atom-counting method is not straightforward and cannot provide a unique solution. In addition, the available data do not allow to diagnose whether the atom-counting method results in inaccurate atom counts for this experiment or whether the experimental conditions have been changed during the recording of the HAADF STEM images. Therefore, a combination with other techniques will be required in order to obtain trustful atom-counting results with single atom sensitivity. The combination of the statistics-based atom-counting method with detailed STEM image simulations and electron tomography will be used for the analysis of Au nanorods and Pt/Ir particles, which will be the subject of the next sections.

4.2.3 Atom counting in Au nanorods

In the previous section, it has been illustrated that, although the application of the statistical methods is relatively straightforward, the interpretation of the output of these methods is certainly not straightforward. If no unique solution can be found from the output of the statistics-based atom-counting method, this statistics-based method should be complemented with another approach. In the literature, also a simulation-based approach for atom counting is available as illustrated in Fig. 1.4 in the introduction of this work (LeBeau et al., 2010). This method is completely independent of the statistics-based method. In contrast to the statistics-based method, the simulation-based method does not provide the precision with which the number of atoms can be counted, but can validate the atom-counting results obtained with the statistics-based atom-counting method. In Subsection 4.2.3.1, it will first be shown how atom-counting results can be obtained with single atom sensitivity for an Au nanorod through a combination of the simulation- and statistics-based approach (Van Aert et al., 2013). Second, in Subsection 4.2.3.2, an alternative for validation of the atom-counting results is illustrated by comparing the obtained atom counts with the result from atomic-resolution tomography (De Backer et al., 2017). After that, it will be illustrated in Subsection 4.2.3.3, how 3D atomistic models can be created from reliable atom-counting results (De Backer et al., 2017; Jones et al., 2014).

4.2.3.1 Validation using detailed image simulations

4.2.3.1.1 Experiment and simulation

In order to evolve toward accurate atom-counting results, the statistics-based atom-counting method described in the first part of this chapter, Section 4.1, is combined with detailed image simulations (Van Aert et al., 2013). This is done for an Au nanorod, shown in Fig. 4.12(a), which has been imaged along the [100] zone axis at the Qu-Ant-EM, a double aberration-corrected Titan^3 operated at 300 kV with a 21.8 mrad convergence angle. The detailed experimental settings are listed in Table 4.4. The value for the Debye-Waller factor is taken from Gao and Peng (1999). When the experimental data are normalized with respect to the incoming electron beam (LeBeau & Stemmer, 2008; Rosenauer et al., 2009), the scattering cross-sections resulting from the experiment can be compared with the scattering cross-sections determined from a set of simulations with increasing thickness.

Table 4.4 Experiment and frozen lattice simulations settings for Au in [100] zone axis.

Parameter	Symbol	Value
Acceleration voltage	V_0 (kV)	300
Defocus	ϵ (nm)	−8.301
Spherical aberration	C_s (mm)	0.035
Objective aperture angle	α (mrad)	21.8
Spatial incoherence of source	FWHM (Å)	0.7
HAADF inner collection angle	β_1 (mrad)	62
HAADF outer collection angle	β_2 (mrad)	190
Number of unit cells per supercell	$N_a \times N_b$	11 × 11
Supercell size	$N_a \times N_b$ (Å^2)	44.88 × 44.88
Maximum specimen thickness	$z_{[100]}$ (nm)	40.8
Debye-Waller factor	B (Å^2)	0.637
Pixel size	$\Delta x = \Delta y$ (Å)	0.102
Pixel size to sample atomic potential	$\Delta x_V = \Delta y_V$ (Å)	0.0261
Number of phonon configurations		20

Following the method described in Krause et al. (2016); LeBeau and Stemmer (2008); Mehrtens, Müller, et al. (2013); Rosenauer et al. (2009), the experimental image intensities I_{exp} are converted to a fraction of the incident probe using the following equation:

$$I_{\text{norm}} = \frac{I_{\text{exp}} - I_{\text{vac}}}{I_{\text{det}} - I_{\text{vac}}} \tag{4.27}$$

where I_{norm} is the normalized intensity, which can directly be compared with simulated images; I_{vac} is the mean number of counts in the vacuum due to the detector's amplifier offset and I_{det} is the mean number of counts generated by the probe hitting the detector area. In other words, this normalization places STEM experiments on the same absolute intensity scale as used in theoretical simulations. As such, the accuracy of the counting procedure can be tested.

Nevertheless, this analysis, allowing for a comparison between experimental and simulated scattering cross-sections, should be carried out very carefully. The accurate measurement of inner and outer ADF detector angle is of critical importance since the variation of the scattering cross-section value highly depends on it, as shown in Martinez et al. (2015). Moreover, the importance of the detector response is often overlooked leading to a discrepancy between experiment and simulation. A full characterization of the ADF detector should be carried out if the highest quality for a match between simulations and experiment is to be achieved (LeBeau & Stemmer, 2008; Rosenauer et al., 2009). This procedure requires to scan the STEM probe over the ADF detector to obtain a response map, which has been found to be both inhomogeneous and asymmetric (Findlay & LeBeau, 2013; Krause et al., 2016; MacArthur et al., 2014). The detector response at small scattering angles varies significantly from that at higher angles, and therefore, this should be taken into account while performing image simulations. Here, this effect is incorporated by including a detector sensitivity profile in the image simulation, which azimuthally weights the contribution of the scattered electron with respect to its scattering angle (Grieb et al., 2012; Mehrtens, Schowalter, et al., 2013). Including this detector sensitivity profile allows a significantly improved agreement between experiment and simulated images.

Detailed image simulations were carried out using the STEMsim program (Rosenauer & Schowalter, 2008) under the frozen lattice approach taking into account the microscope settings and detector sensitivity. Table 4.4 summarizes the settings used throughout the simulations. The frozen lattice approximation considers uncorrelated phonon distributions of atomic displacements due to the thermal vibrations of the crystal and therefore describes the so-called thermal diffuse scattering (TDS).

4.2.3.1.2 Atom-counting results and single atom sensitivity

The experimental image of the Au nanorod is shown in Fig. 4.12(a). This image can be analyzed along the same lines as the Ge clusters discussed in the previous subsection. The refined model of the experimental image is shown in Fig. 4.12(b). The ICL criterion, shown in Fig. 4.12(d). For this type of nanorods, one might expect a symmetrical structure. The ICL criterion reveals a local minimum at 47 components. Using the selection of 47 components in the Gaussian mixture model, the thickness can be derived from this atom-counting analysis along the perpendicular direction. This thickness matches the diameter obtained from the projection image in Fig. 4.12(a). The histogram of scattered intensities is shown in Fig. 4.12(e) together with the estimated probability distribution. The number of Au atoms for every column of the nanorod is shown in Fig. 4.12(c). Since the intensities in the HAADF STEM image have been normalized with respect to the incident beam as discussed in the previous paragraph, the accuracy of the counting procedure can be tested. This validation is required since more local minima are present in the ICL criterion as shown in Fig. 4.12(d).

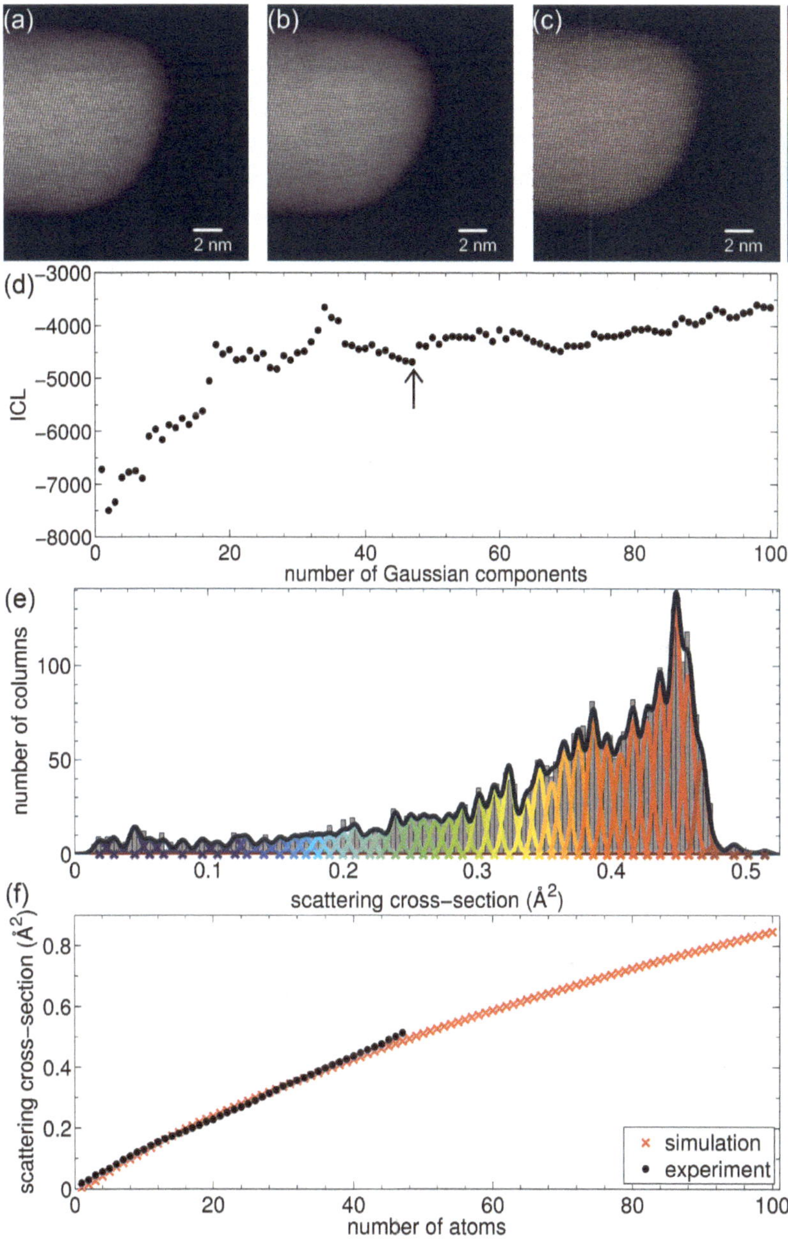

Figure 4.12 (a) HAADF STEM image of an Au nanorod. (b) Refined model. (c) Number of Au atoms per column. (d) The Integrated Classification Likelihood criterion evaluated as a function of the number of Gaussian components in the mixture model. (e) Histogram of scattering cross-sections of the Au columns together with the estimated mixture model and its individual components. (f) Comparison of experimental and simulated scattering cross-sections. Reprinted figure with permission from Van Aert et al. (2013) Copyright (2013) by the American Physical Society.

Fig. 4.12(f) shows the experimental mean scattering cross-sections - corresponding to the component locations in Fig. 4.12(e) - together with the scattering cross-sections estimated from the frozen phonon calculations using the STEMsim program under the same experimental conditions which are listed in Table 4.4 (Rosenauer & Schowalter, 2008). The excellent match of the experimental and simulated scattering cross-sections within the expected 5%–10% error range validates the accuracy of the obtained atom counts (LeBeau et al., 2008; Rosenauer et al., 2011). Minor deviations can indeed not be avoided because of remaining uncertainties in the microscope settings, the effect of amorphous layers, and practical limitations to fully take into account the complex dependence of the Debye-Waller factor on both the particle's size and its different behavior for surface atoms as compared to bulk-like atoms (Aveyard et al., 2014; Buffat, 1977; Clark et al., 1965; Mkhoyan et al., 2008). However, from an alternative point of view, the experimental mean scattering cross-sections can be used as a reference to further improve the accuracy of imaging models taking all important physical effects into account. The steady increase of the simulated scattering cross-sections, even above 47 atoms as shown in Fig. 4.12(f), proves the high sensitivity of the scattering cross-sections for the number of atoms. The precision, on the other hand, is limited by the unavoidable presence of noise in the experimental images resulting into fluctuations of the scattering cross-sections $\hat{V}$ around the mean values μ_g. The precision is then set by the overlap of the Gaussian components shown in Fig. 4.12(e). When the overlap increases, the probability that a wrong number of atoms is assigned to a particular column indeed increases. The procedure on how to calculate the percentage of atomic columns for which the number of atoms is counted with single atom sensitivity has been discussed in Subsection 4.1.4.4 of this chapter. First, the posterior probability $P(I_g|\phi_g)$ given by Eq. (4.24) that a scattering cross-section belonging to component ϕ_g resides in the central part of a component I_g, i.e. interval I_g as illustrated in Fig. 4.8 of this chapter, is computed. On average, that is, by averaging fractions for the components 1 up top 47, this probability corresponds to 79.89%. Next, the posterior probability $P(\phi_g|I_g)$ given by Eq. (4.25) that a scattering cross-section belonging to the central part I_g stems from component ϕ_g is calculated for each component. On average, this percentage equals 80.55%. This means that for 80.55% of all atomic columns the number of atoms is measured without error. Similarly, the probability to miscount the number of atoms with ± 1 can be computed. For this purpose, the posterior probability that the scattering cross-section belonging to the component ϕ_{g-1}

or ϕ_{g+1} resides in the interval I_g equals $P(I_g|\phi_{g-1}) + P(I_g|\phi_{g+1}) = 19.91\%$, and the probability to miscount the number of atoms with ± 1 equals $P(\phi_{g-1}|I_g) + P(\phi_{g+1}|I_g) = 19.43\%$, which is the posterior probability that a scattering cross-sections belonging to I_g stems from component ϕ_{g-1} or ϕ_{g+1}. Consequently, the probability to have an error of ± 2 atoms is almost zero. In contrast to the results obtained when using simulations alone, the amount of overlap with neighboring components can be determined when using this statistics-based procedure, resulting in an estimation of the precision that can be achieved. However, since both approaches, a simulations-based and statistics-based atom-counting method, are independent of each other, the combination of the two methods allows for reliable atom-counting results with single atom sensitivity.

4.2.3.2 Comparison with atomic-resolution electron tomography

In this section, an alternative validation for the atom-counting results using electron tomography is briefly illustrated.

Three projection ADF STEM images of an Au nanorod have been acquired along different major zone axes using the X-Ant-EM, a probe corrected Titan[3] operated at 200 kV with a 24 mrad convergence angle. The inner and outer detector collection angles ranged from 45–160 mrad. The experimental images are shown in Figs. 4.13(a-c). These images were used as an input for a 3D reconstruction using a compressive sensing based tomography method as explained in Goris, Bals, et al. (2012). For a 3D reconstruction at the atomic scale, the sparsity of the object can be exploited in the reconstruction algorithm, as only a limited number of voxels contain an atom and most voxels correspond to vacuum. For each of the projection images, also the number of atoms in the projected atomic columns have been counted using the statistics-based method described in the first part of this chapter. The refined model and the atom-counting results for the projection image along $-45°$ are illustrated in Fig. 4.13(d-e). Since different viewing directions of the nanorod are available, the choice of the local minimum in the ICL criterion was guided by the prior knowledge of a symmetric nanorod. More details can be found in (De Backer et al., 2017).

In the previous subsection, the atom-counting results have been validated through a comparison with detailed image simulations. Here, the thus obtained atom counts are compared with the number of atoms that can be derived from an orthogonal slice from the atomic-resolution reconstruction from electron tomography (De Backer et al., 2017). This comparison is illustrated in Fig. 4.14. Slight deviations can be expected, since the 3D

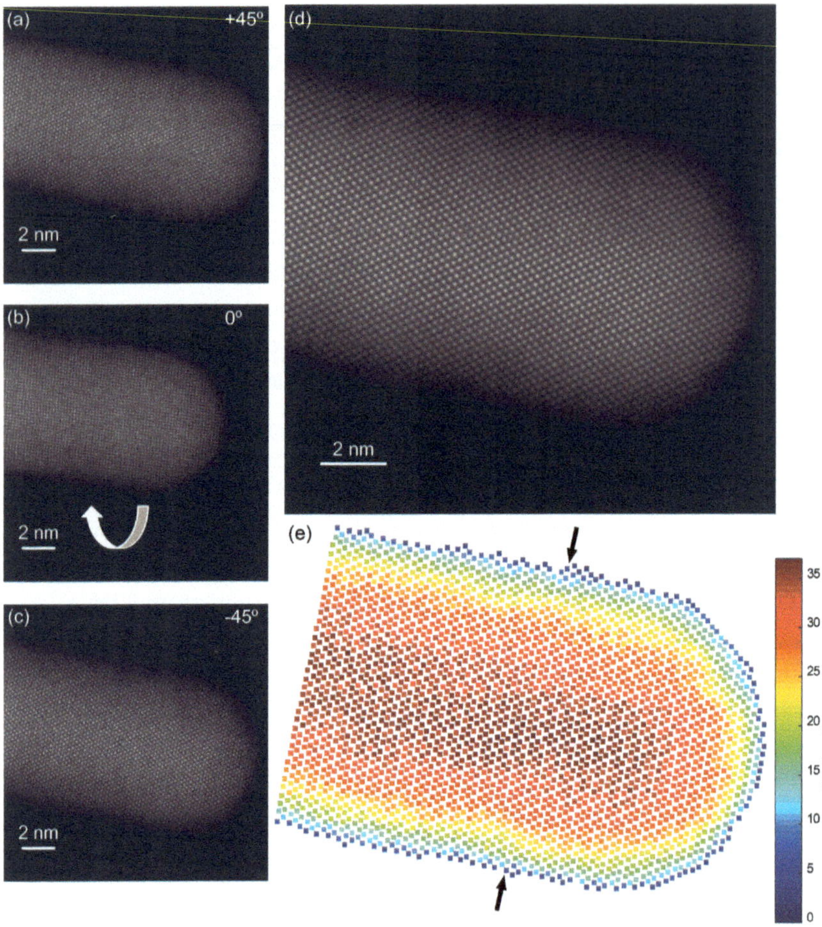

Figure 4.13 High resolution ADF STEM projection images of an Au nanorod and statistical atom counting. (a, b, c) Three projection images obtained by tilting the nanorod along the long axis. (d) Refined model of projection image (c). (e) Atom-counts for this projection image. Reproduced from De Backer et al. (2017) with permission from The Royal Society of Chemistry.

reconstruction using tomography represents an average structure based on the three high-resolution projection images, whereas the atom counts result from each single image. The number of atoms has been counted in the 40th atomic plane from the tip of the Au nanorod. Good agreement between the counts derived from the 3D reconstruction and from the statistics-based atom-counting method is obtained. The width of the nanorod appears to be underestimated for the tomography reconstruction, which is possibly

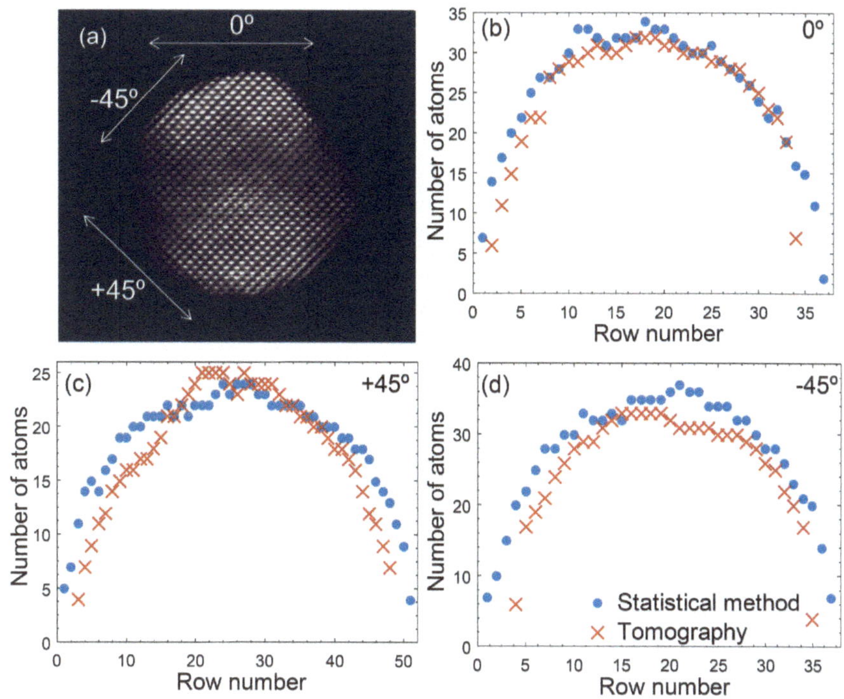

Figure 4.14 Comparison between atom counts and atomic-resolution electron tomography. (a) Orthogonal slice from the 3D reconstruction of the 40th atomic plane from the tip. (b-d) Comparison of the counts along the different orientations. Adapted from De Backer et al. (2017) with permission from The Royal Society of Chemistry.

caused by artifacts introduced by the missing wedge. The missing wedge causes the intensities to be smeared out. As such, these intensities will be lower than the applied threshold to visualize the atoms in the orthogonal slice.

4.2.3.3 Toward 3D atomistic models using atom counting

Reliable atom-counting results can be used as an input for creating 3D atomistic models (De Backer et al., 2017; Jones et al., 2014). This approach provides an alternative to electron tomography for a 3D reconstruction at the atomic scale. Electron tomography requires multiple exposures which is not always feasible in practice. For example, in situ experiments or the characterization of beam-sensitive materials are nearly impossible using electron tomography. In this subsection, 3D atomistic models are created for the Au nanorod of Subsection 4.2.3.2. Therefore, atom counts along with

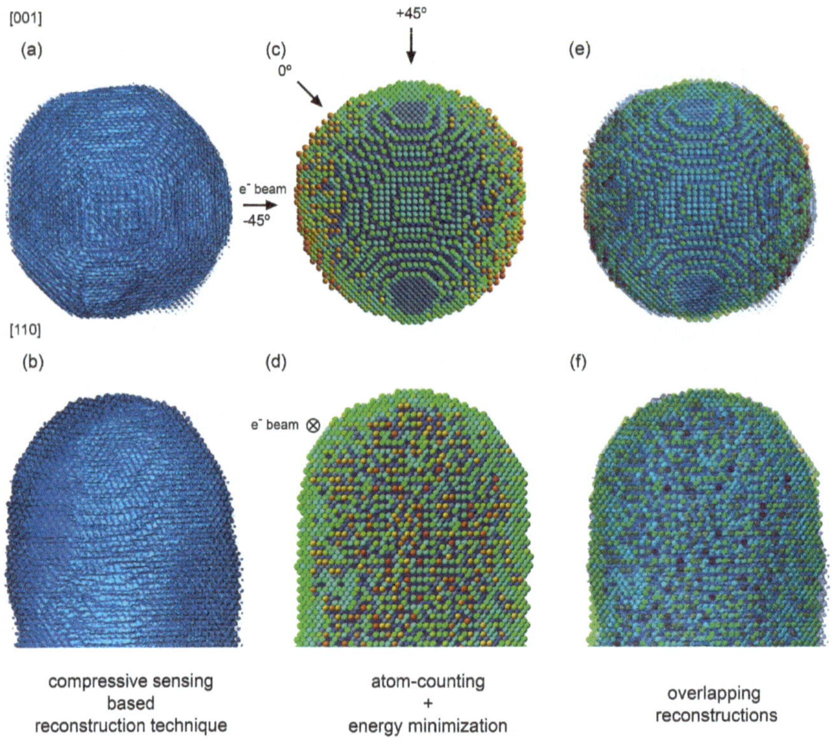

Figure 4.15 Comparison of the reconstructions based on atom counting with energy minimization and those obtained by atomic resolution tomography. (a, b) Compressive sensing based reconstruction of Au nanorod viewed along [001] and [110] direction. (c, d) Reconstruction based on atom counting and energy minimization using a single projection image (Fig. 4.13(a)) viewed along [001] and [110] orientation (the coloring of the Au atoms indicates the nearest-neighbor coordination, from 1 in red to 12 in dark blue). (e, f) Overlap of the reconstructions shown in (a, c) and (b, d). Adapted from De Backer et al. (2017) with permission from The Royal Society of Chemistry.

the prior knowledge of a face-centered cubic geometry are used to create an atomistic model. An energy minimization using a Monte Carlo approach has then been used to relax the nanoparticle's 3D structure (Jones et al., 2014). The atomistic model built from the atom-counting results in Fig. 4.13(o) is shown in Fig. 4.15(c-d). The beam direction is arrowed in this figure. As expected, the surface roughness is higher along this direction. Nevertheless, different surface facets of the Au nanorod can be observed from this 3D reconstruction. These surface facets are of special interest for materials scientists since they determine the physical properties of nanostructures (Chang et al., 2010; Goris, Bals, et al., 2012; Katz-Boon et al., 2011, 2015; Pecharromán et al., 2008). Furthermore, Fig. 4.15(a-b) shows the 3D reconstruction using the compressive sensing based tomography method. Fig. 4.15(e-f) compare both reconstructions in 3D. An excellent visual match of the overall morphology of the nanorod has also been found. This agreement is only made possible because of the high accuracy of the statistically determined atom-counts. More details on the reconstruction of 3D atomic models from a single projection using atom counting and Z-contrast imaging can be found in De Backer et al. (2017). This methodology opens up the possibility for the study of beam-sensitive materials, 2D self-assembled structures (Geuchies et al., 2016), and where *in situ* hardware makes tomography impossible (Altantzis et al., 2019).

4.2.4 Dose-limited reliability for atom counting in Pt/Ir nanoparticles

In the previous section, atom counting has been demonstrated on Au nanorods which are relatively stable under the incoming electron beam. Here, ADF STEM experiments for an atom-counting analysis are conducted on an industrial bimetallic catalyst providing a severe characterization challenge because of its beam sensitivity and limited size (De Backer, Martinez, et al., 2015). As discussed in the first part of this chapter, the reliability of the statistics-based atom-counting method depends on different parameters which can be linked with the quality of the recorded images, such as total number of atomic columns in the experimental STEM image, the number of columns having a different number of atoms, and the amount of noise. In the previous section, it is shown that through a combination of the simulation- and statistics-based method, being completely independent of each other, one is able to count atoms with trustworthy single atom sensitivity. In this section, it will be shown that this combined method allows us to diagnose limits of atom counting and to investigate the effect of dose, sampling, background behind the particle of interest, tilt, and sample geometry. The latter parameters become important when studying industrially catalyst particles such as the Pt/Ir mixed-alloy nanoparticles supported on carbon black discussed in this section.

In Subsection 4.2.4.1, the material under study as well as the experimental and simulation details are presented. The methodology for counting the number of atoms for these catalyst particles is illustrated in Subsection 4.2.4.2. The findings for atom counting conducted on catalytic particles are discussed in Subsection 4.2.4.3.

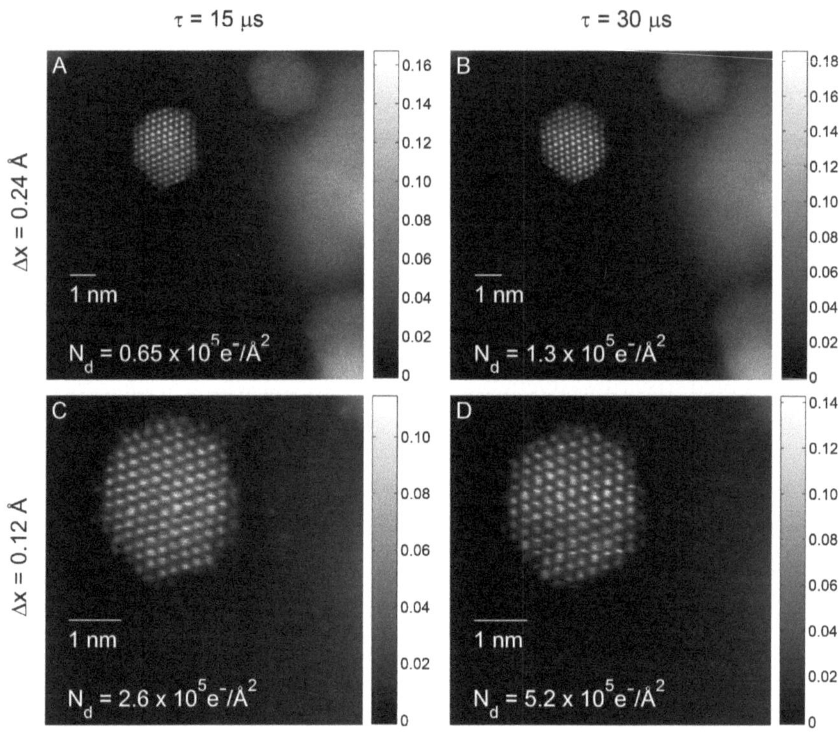

Figure 4.16 Normalized ADF STEM images of a Pt/Ir particle together with the experimental conditions (dwell time τ, pixel size *dx*, and incoming electron dose N_d). Reprinted from De Backer, Martinez, et al. (2015), Copyright (2015), with permission from Elsevier.

4.2.4.1 Experiment and simulations

Bimetallic nanoparticles have shown promise for hydrogen fuel-cell applications. Not only do they provide a reduction in the amount of platinum used, but they have also demonstrated a significantly higher catalytic activity. Pt/Ir particles, in particular, show improved resistance to CO poisoning, a by-product in the reforming of hydrocarbons to H_2 gas (Ralph & Hogarth, 2002). The Pt/Ir particles were supported on a 3D carbon black support and received in powder form dusted onto a carbon coated copper grid. Images were taken at the Qu-Ant-EM, a double corrected FEI Titan[3] working at 300 kV with a 20.2 mrad semiconvergence angle under different dose conditions as shown in Fig. 4.16. The ADF detector ranges from 35 to 190 mrad corresponding to a camera length of 145 mm. For unique atom-counting results, we require a monotonic increase for the

Table 4.5 Parameters for the frozen lattice simulation of Pt in [110] zone axis. Reprinted from De Backer, Martinez, et al. (2015), Copyright (2015), with permission from Elsevier.

Parameter	Symbol	Value
Acceleration voltage	V_0 (kV)	300
Defocus	ϵ (nm)	0
Spherical aberration	C_s (nm)	406
Objective aperture angle	α (mrad)	20.2
Spatial incoherence of source	FWHM (Å)	0.7
Inner detector angle	β_1 (mrad)	35
Outer detector angle	β_2 (mrad)	190
Debye-Waller factor	B (Å^2)	0.384
Pixel size	Δx (Å)	0.1154
Number of phonon configurations		20

scattering cross-sections with thickness. As will be shown in Chapter 5, for small enough particles such as in this study, a lower ADF detector inner angle is preferred for atom counting because of the higher detected signal for a given incident electron dose. Although coherent contributions are present in the detected signal, a monotonic increase in scattering cross-section is observed for this ADF detector collection range. Furthermore, the scattering cross-sections are robust to tilt when using a lower ADF detector inner angle.

The experimental images were normalized with respect to the incident beam (Krause et al., 2016; LeBeau & Stemmer, 2008; Rosenauer et al., 2009) in order to be comparable to image simulations. Image simulations were performed using the STEMsim software (Rosenauer & Schowalter, 2008) under the frozen lattice approach. The simulation settings are listed in Table 4.5. The value for the Debye-Waller factor is taken from Gao and Peng (1999). Since the difference in atomic number only equals 1 for the two elements in the bimetallic nanoparticle, the simulated scattering cross-sections of Ir ($Z = 77$) and Pt ($Z = 78$) differ less than 3% up to 15 atoms in a projected atomic column and, therefore, for purposes of the analysis presented here, they can be treated as being monometallic.

4.2.4.2 Methodology

If a spatially varying background originating from e.g. the carbon support is present in the image, this background has to be subtracted since ignoring it may lead to errors in the quantitative atom-counting results. To account for

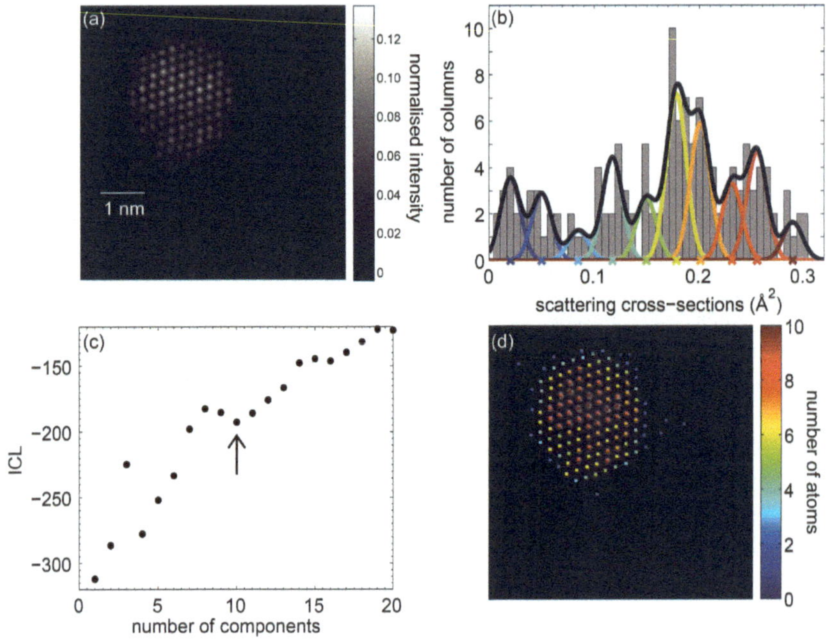

Figure 4.17 (a) Fitted model of the background-subtracted image of Fig. 4.16(D). (b) Histogram of the scattering cross-sections of the atomic columns. The black curve shows the estimated mixture model. The individual components are shown as colored curves which correspond to the colors for the number of atoms in (d). (c) The ICL criterion evaluated as a function of the number of Gaussian components in a mixture model. (d) Number of atoms per column. Reprinted from De Backer, Martinez, et al. (2015), Copyright (2015), with permission from Elsevier.

the image intensity contribution of such a carbon support under the sample, a mask can be generated which contains the region of interest for quantification, i.e. the nanoparticle. The intensity outside this mask represents the contribution from the support only and will be used to estimate the values within the mask using a technique known as in-painting. For small areas, and where the support does not contain abrupt changes, the simplest form of in-painting, known as iterative Gaussian-blurring or isotropic diffusion can be used (Oliveira et al., 2001).

Next, the scattering cross-sections can be quantified atomic column-by-atomic column by fitting the empirical imaging model to the background-subtracted experimental images. The fitted model for the background-subtracted version of Fig. 4.16(D) is shown in Fig. 4.17(a). The scattering cross-sections are presented in Fig. 4.17(b) in the form of a histogram.

By evaluating the Integrated Classification Likelihood criterion, as shown and arrowed in Fig. 4.17(c), the presence of 10 significant components has been found whose locations are indicated by means of colored crosses in Fig. 4.17(b). From the estimated component locations, the number of atoms in each projected atomic column can be quantified leading to the result in Fig. 4.17(d).

4.2.4.3 Results and discussions: single atom sensitivity for challenging nanoparticles?

The ultimate goal is to obtain atom-counting results for nanoparticles of arbitrary shape, size, and atom type with high precision and high accuracy which are both necessary conditions to conclude single atom sensitivity. High precision can be observed from the Gaussian mixture model when there is nearly no overlap between neighboring components, such as in Fig. 4.17(b). High accuracy can be confirmed when comparing the results of the statistics-based atom-counting method with simulated values from detailed image simulations as discussed in Section 4.2.3. So far, this method has successfully been demonstrated on Au nanorods, which are relatively stable under the incoming electron beam, and this procedure promises single atom sensitivity. In this subsection, the accuracy of the atom-counting results is investigated for the catalyst of Fig. 4.16 for which the characterization is more challenging because of the beam sensitivity and the limited size of the particles. Different parameters which affect the reliability of the atom-counting results are studied.

4.2.4.3.1 Effect of background subtraction

From Fig. 4.16 it can be seen that a spatially varying background is present in the images. This background has been subtracted following the method described in Subsection 4.2.4.2. In Fig. 4.18, the effect of the background subtraction on the atom counts is illustrated. Fig. 4.18(a) shows the difference map of the atom counts per atomic column with and without background subtraction and Fig. 4.18(b) shows the vertically average mean error in atom counts. Ignoring the spatially varying background clearly leads to a counting error of ± 1 atom. This type of error thus results from bias of the estimated scattering cross-sections caused by model misspecification (den Dekker et al., 2005). The subtracted background represents the carbon support and is shown in Fig. 4.18(c). As expected, the reconstructed carbon ramp corresponds to the mean error in atom counts of Fig. 4.18(b).

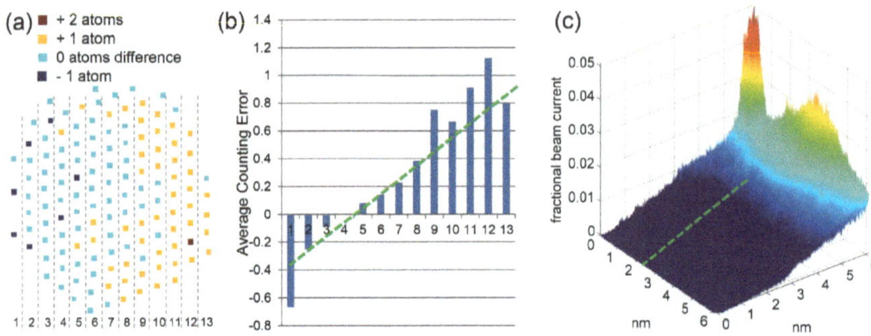

Figure 4.18 (a) The error observed when the carbon intensity is not considered. (b) Vertically averaged error. (c) The observed carbon contribution in the background-only image. Reprinted from De Backer, Martinez, et al. (2015), Copyright (2015), with permission from Elsevier.

4.2.4.3.2 Comparison with simulations

In order to check the accuracy, the estimated mean scattering cross-sections of the components are compared with the library values from the detailed image simulations for this experiment. The library values are shown for Pt, the values for Ir are not plotted on this graph since they cannot be distinguished from the scattering cross-sections of Pt up to 15 atoms. From Fig. 4.19, where the simulated values are shown for Pt, it can be seen that for the image shown in Fig. 4.16(D), shortly referred to in the following as image D, an excellent match has been found between the estimated mean scattering cross-sections from the experiment and the simulated values. The observed agreement is within the expected 5–10% error range proving that the number of components when using the ICL criterion as well as the locations of these components in Fig. 4.17(b) are accurately determined. Slight deviations cannot be avoided because of remaining uncertainties as discussed in Subsection 4.2.3.1.

Similar analyzes were performed for the remaining images in Fig. 4.16, being referred to as images A, B, and C. Their results are also shown in Fig. 4.19. A mismatch between the estimated and simulated scattering cross-sections can be observed for image C, but is apparent and most pronounced for images A and B. For these images, the number of components in the Gaussian mixture model is considerably underestimated. Based on a comparison with the simulated values, 13 and 15 components are expected for images A and B, respectively. In the following subsections, the origin of the mismatch of images A, B, and C with the simulated values is discussed.

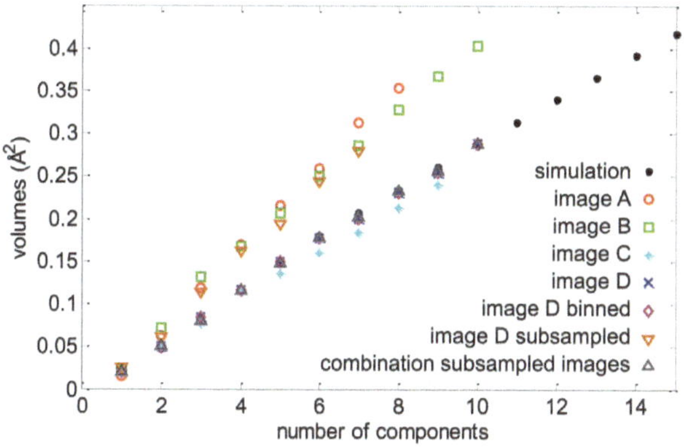

Figure 4.19 Overview of the analyzed data in comparison with the simulations. Reprinted from De Backer, Martinez, et al. (2015), Copyright (2015), with permission from Elsevier.

4.2.4.3.3 Effect of sample mis-tilt

The slight mismatch of image C from the simulated curve in Fig. 4.19 stems from a small sample mis-tilt (E et al., 2013). This can be understood from electron channeling. This phenomenon, where an aligned column of atoms parallel to the incident beam direction exhibits a small lensing effect on the beam, causes a column of atoms to have a larger scattering cross-section than the sum of the individual scattering cross-sections of its constituent atoms (E et al., 2013). This results in a percentage loss in scattering cross-section for small sample mis-tilts away from a low-order zone axis. The mis-tilt measured based on a geometry calculation and the ellipticity of the column equals 31.6 mrad for image C. Based on image simulations, this corresponds to 7% loss of scattering cross-section which is in agreement with the results shown in Fig. 4.19. One should be particularly aware of this effect for beam-sensitive samples which are liable to reconstruct and rotate under the beam.

4.2.4.3.4 Effect of sampling

The effect of sampling on the atom-counting results is considered since the pixel size of images A and B is double the pixel size (half the magnification) of image D, as can be seen from Fig. 4.16. Consequently, the number of sampling points in images A and B is a quarter of the number of sampling points in image D. In E et al. (2013), it is shown that pixel size has no effect on the scattering cross-sections in the absence of noise. Here, the effect of sampling on the atom-counting analysis is studied by analyzing a new image generated by binning 2 × 2 pixels of image D. In that case, the estimated mean scattering cross-sections are perfectly in accordance with the simulated values as can be concluded from Fig. 4.19. This means that sampling itself has no effect on the accuracy of the atom counts so long as atomic resolution is preserved in the ADF STEM image. The underestimation of the model order for images A and B can thus not be explained as a sampling effect.

4.2.4.3.5 Effect of dose

In the previous subsection, the effect of sampling was studied as a possible reason for the underestimation of the number of components in images A and B. It should be noticed that while binning 2 × 2 pixels, the total recorded electron dose is preserved such that only the effect of sampling is studied. However, the electron dose is 4 times lower in image B as compared with image D, and even 8 times lower for image A. This suggests that a reduction of incoming electron dose causes an underestimation of the model order when evaluating the ICL criterion. The lower magnification and electron dose of image B with respect to image D can be mimicked by subsampling image D, i.e. by taking every second pixel in x and y direction (one quarter overall). The result of the analysis of the subsampled image D is also shown in Fig. 4.19. This result resembles the analysis of images A and B thus explaining the mismatch with the simulation for these images. The reduced dose leads to less precise measurements of the scattering cross-sections resulting in insufficient statistics for the determination of the number of components by the evaluation of the ICL criterion. The effect of dose on the accuracy of the atom-counting results will be discussed in more detail in the following subsection.

4.2.4.3.6 Discussion of the accuracy of the atom-counting results

The most pronounced mismatch with the simulations in Fig. 4.19 can be explained as an effect of reduced dose, resulting in less accurate measurements for the number of atoms in an atomic column. This effect of the reduced dose can be studied in more detail from the Gaussian mixture model. If the dose is reduced, the neighboring components of the Gaussian mixture model start overlapping more, because of the less precise measurements of the scattering cross-sections. This significant overlap between neighboring components in the Gaussian mixture model can be measured in terms of the relative width of the components. This relative width expresses the ratio of the width of the components of the Gaussian mixture

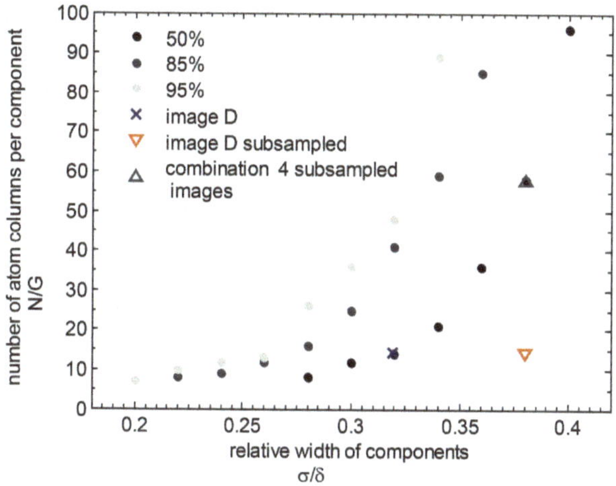

Figure 4.20 Probability of choosing the correct number of components based on the ICL criterion as a function of σ/δ and N/G. The corresponding values for image D, subsampled image D, and a combination of 4 subsampled images are also shown. Reprinted from De Backer, Martinez, et al. (2015), Copyright (2015), with permission from Elsevier.

model and the average increment in scattering cross-section for 1 extra atom, i.e. the σ/δ-ratio discussed in Subsection 4.1.4.2 of this chapter. In this subsection, it has been shown that this σ/δ-ratio in combination with the N/G-ratio is important when it comes to accurately determining the minimum of the ICL criterion, where N is the number of analyzed atomic columns in the ADF STEM image and G the true model order. This was summarized in Fig. 4.7. Here, the probability of choosing the correct number of components based on the minimum of the ICL criterion is resumed in Fig. 4.20 as a function of the σ/δ-ratio and N/G-ratio. The corresponding values for image D and subsampled image D are indicated in Fig. 4.20 to illustrate the effect of the reduced dose. It is clear that the reduced dose widens the components hence limiting the precision. As a consequence, the probability for choosing the correct minimum of ICL significantly decreases. Nevertheless, Fig. 4.20 suggests that this loss in accuracy can be compensated by increasing the N/G-ratio. In practice, this can be realized by analyzing collectively the scattering cross-sections originating from images recorded under the same conditions. From Fig. 4.20, it can be seen that a combination of 4 images should be sufficient. The results of such a combined analysis of the collective scattering cross-sections originating from 4 different subsampled versions of image D are shown in Fig. 4.19.

The estimated mean scattering cross-sections again coincidence with the simulated values. It is important to mention that this collective analysis only pays off in terms of accuracy but leaves the precision unaffected. Indeed, the precision of the atom-counting results is determined by the overlap in the Gaussian mixture model and is set by the precision with which the scattering cross-sections are known. This precision is characterized by the electron dose and the value of the scattering cross-sections, but is independent of the number of analyzed scattering cross-sections. For low-dose acquisitions, the precision could in principle be compensated when averaging 'identical' images prior to determination of scattering cross-sections. In practice, however, the assumption of multiple images of exactly the same structure is very unrealistic, especially for beam-sensitive samples.

4.2.5 Conclusions

In the second part of this chapter, it has been shown how precise and accurate atom-counting results can be obtained for different types of nanoparticles from ADF STEM images. First, it has been illustrated that the statistics-based atom-counting method should be combined with another independent, complementary method. Here, a combination with a simulations-based approach (Van Aert et al., 2013) or electron tomography (De Backer et al., 2017) is presented. The independence of both methods allows us to test the reliability of the atom counts obtained with the statistics-based approach.

The scattering cross-sections have been shown to be a very successful measure to count the number of atoms in an atomic column from a single STEM image, since they monotonically increase with the number of atoms in an atomic column. This is in contrast to the peak intensities which saturate already at very low thicknesses. Using the scattering cross-sections, it has been shown that for an Au nanorod, 80% of the analyzed atomic columns has been measured without error and 20% with an error of 1 atom (Van Aert et al., 2013). Furthermore, it has been shown that the atom-counting results are consistent with the number of atoms counted from a slice through a 3D reconstruction obtained using atomic-resolution electron tomography. In addition, it has been illustrated how atom-counting results can be used for the creation of 3D atomistic models (De Backer et al., 2017).

In the atom-counting analysis of an industrial bimetallic catalyst, it has been shown that the interplay of several effects can greatly impact the reliability of atom counting in quantitative ADF STEM (De Backer, Martinez, et al., 2015). This includes not only parameters such as electron dose and lateral sampling, but also aspects often beyond the control of the experimentalist, including particle tilt and contrast contributions from the carbon support. Here, a study of the minimum dose and sampling requirements to count atoms with single atom sensitivity could be reported using a hybrid statistical and library method to quantify experimental images. It is discovered that a varying sample support should be taken into account in order to avoid atom-counting errors and that care must be taken in the interpretation of column intensities for slightly tilted nanoparticles. For samples that undergo transformation under the beam, poor image quality cannot be countered by simple image averaging, rather the frames must be analyzed collectively. Finally, this study suggests that while dose is often lowered to reduce sample damage, some minimum level is required to retain single atom precision. This trade-off between damage and reliability then guides experiment design suggesting that optimum dose and sampling conditions exist for each instrument/sample such that the highest possible precision can be obtained from the hardware available.

In general, based on the analysis of the different experimental images, it can be concluded that single atom sensitivity can be obtained for a sufficiently high incident electron dose, whereas less precise but still accurate counting results can be obtained when imaging with a lower incident electron dose. The availability of accurate and precise atom-counting results opens up new possibilities for the characterization of materials where electron tomography cannot be applied. This includes beam-sensitive materials or in-situ experiments, where the 3D atomic structure of the object under study changes (Altantzis et al., 2019; De Backer et al., 2017; Geuchies et al., 2016). Furthermore, from an alternative point of view, the atom-counting results can be used as a reference to further improve the accuracy of imaging models and electron tomography reconstruction algorithms. On the one hand, the atom-counting results can help taking important physical effects into account in the development of new imaging models. On the other hand, a synergistic combination of atom counting and compressive sensing reconstruction algorithms for electron tomography will enable to include quantitative prior information on the number of atoms during the reconstruction. Since the atomic positions are not fixed during this reconstruction, this enables one to detect deviations from the perfect crystal structure, which is not possible when using only the atom-counting results. Moreover, this additional information in the reconstruction algorithm might reduce the number of 2D projection images that is needed and will allow for a reliable 3D reconstruction of larger nanoparticles at the atomic scale.

Acknowledgments

This project has received funding from the European Research Council (ERC) under the European Union's Horizon 2020 research and innovation programme (Grant Agreement No. 770887 and No. 823717 ESTEEM3). The authors acknowledge financial support from the Research Foundation Flanders (FWO, Belgium) through Projects No. G.0502.18N, G.0267.18N, G.0368.15N, and a postdoctoral grant to A. D. B. The authors acknowledge funding from the University of Antwerp through a TOP project.

Appendix 4.A. CRLB for the parameters of the Gaussian mixture model: Fisher information and attainable precision

The concept of Fisher information and attainable precision is discussed in detail in Chapter 2. Here, these principles will be applied for the parameters of the Gaussian mixture model. For the Gaussian mixture model, the observations, i.e. the estimated scattering cross-sections, are assumed to be statistically independent having the following joint probability distribution:

$$p(\mathbf{V}, \mathbf{\Psi}_G) = \prod_{n=1}^{N} f_{\text{mix}}(V_n; \mathbf{\Psi}_G) \tag{4.A.1}$$

where $f_{\text{mix}}(V_n; \mathbf{\Psi}_G)$ is the probability density function of a mixture model given by Eq. (4.3) and $\mathbf{V}$. The dependence of the joint probability density function on the unknown parameters $\mathbf{\Psi}_G$ defines the Fisher information matrix and the Cramér-Rao lower bound. The Fisher information matrix is defined as:

$$F_{\mathbf{\Psi}} = -\mathbb{E}\left[\frac{\partial^2 \log p(\mathbf{V}; \mathbf{\Psi}_G)}{\partial \mathbf{\Psi}_G \partial \mathbf{\Psi}_G^T}\right]. \tag{4.A.2}$$

In practice, the following integral is numerically integrated:

$$F_{\mathbf{\Psi}_G} = -N \int_{-\infty}^{\infty} \frac{\partial^2 \log f_{\text{mix}}(V; \mathbf{\Psi}_G)}{\partial \mathbf{\Psi}_G^2} f_{\text{mix}}(V; \mathbf{\Psi}_G) dV. \tag{4.A.3}$$

The calculation of the Fisher information matrix of Eq. (4.A.3) requires the calculation of the second order derivatives of the logarithm of the joint probability distribution with respect to the parameters of the Gaussian mixture model. In order to simplify the notation for the derivatives, we will use the following notation:

$$\phi_g(V_n; \mu_g, \sigma) = \phi_g, \tag{4.A.4}$$

$$f_{\text{mix}}(V_n; \mathbf{\Psi}_G) = f_{\text{mix}}, \tag{4.A.5}$$

$$p(\mathbf{V}, \mathbf{\Psi}_G) = p. \tag{4.A.6}$$

The first order derivatives are given by the following expressions:

$$\frac{\partial \log p}{\partial \pi_i} = \sum_{n=1}^{N} \frac{(\phi_i - \phi_G)}{f_{\text{mix}}},$$

$$\frac{\partial \log p}{\partial \mu_i} = \sum_{n=1}^{N} \frac{\pi_i \phi_i}{f_{\text{mix}}} \frac{V_n - \mu_i}{\sigma^2},$$

$$\frac{\partial \log p}{\partial \sigma} = \sum_{n=1}^{N} \frac{1}{f_{\text{mix}}} \sum_{g=1}^{G} \pi_g \phi_g \left(\frac{(V_n - \mu_g)^2}{\sigma^3} - \frac{1}{\sigma} \right).$$

The different second order derivatives are then given by:

$$\frac{\partial^2 \log p}{\partial \pi_i^2} = -\sum_{n=1}^{N} \frac{(\phi_i - \phi_G)^2}{f_{\text{mix}}^2},$$

$$\frac{\partial^2 \log p}{\partial \pi_i \partial \pi_k} = -\sum_{n=1}^{N} \frac{(\phi_i - \phi_G)(\phi_k - \phi_G)}{f_{\text{mix}}^2},$$

$$\frac{\partial^2 \log p}{\partial \pi_i \partial \mu_i} = \sum_{n=1}^{N} \frac{\phi_i (V_n - \mu_i)}{f_{\text{mix}} \sigma^2} \left(1 - \frac{\pi_i (\phi_i - \phi_G)}{f_{\text{mix}}} \right),$$

$$\frac{\partial^2 \log p}{\partial \pi_i \partial \mu_k} = -\sum_{n=1}^{N} \frac{\pi_k \phi_k}{f_{\text{mix}}^2} \frac{V_n - \mu_k}{\sigma^2} (\phi_i - \phi_G),$$

$$\frac{\partial^2 \log p}{\partial \pi_i \partial \mu_G} = -\sum_{n=1}^{N} \frac{\phi_G (V_n - \mu_G)}{f_{\text{mix}} \sigma^2} \left(1 + \frac{\pi_G (\phi_i - \phi_G)}{f_{\text{mix}}} \right),$$

$$\frac{\partial^2 \log p}{\partial \pi_i \partial \sigma} = \sum_{n=1}^{N} \frac{1}{f_{\text{mix}}} \left(\left(\frac{(V_n - \mu_i)^2}{\sigma^3} - \frac{1}{\sigma} \right) \phi_i - \left(\frac{(V_n - \mu_G)^2}{\sigma^3} - \frac{1}{\sigma} \right) \phi_G \right.$$
$$\left. - (\phi_i - \phi_G) \frac{\sum_{g=1}^{G} \pi_i \phi_i \left(\frac{(V_n - \mu_i)^2}{\sigma^3} - \frac{1}{\sigma} \right)}{f_{\text{mix}}} \right),$$

$$
\begin{aligned}
\frac{\partial^2 \log p}{\partial \mu_i^2} &= \sum_{n=1}^{N} \frac{\pi_i \phi_i}{f_{\text{mix}} \sigma^2} \left(\frac{(V_n - \mu_i)^2}{\sigma^2} - 1 - \frac{\pi_i \phi_i (V_n - \mu_i)^2}{f_{\text{mix}} \sigma^2} \right), \\
\frac{\partial^2 \log p}{\partial \mu_i \partial \mu_k} &= -\sum_{n=1}^{N} \frac{f_{\text{mix}}^2}{\pi_k \phi_k \pi_i \phi_i} \frac{(V_n - \mu_k)}{\sigma^2} \frac{(V_n - \mu_i)}{\sigma^2}, \\
\frac{\partial^2 \log p}{\partial \mu_i \partial \sigma} &= \sum_{n=1}^{N} \frac{\phi_i (V_n - \mu_i) \pi_i}{f_{\text{mix}} \sigma^2} \left(\frac{(V_n - \mu_i)^2}{\sigma^3} - \frac{3}{\sigma} \right. \\
&\quad \left. - \frac{\sum_{g=1}^{G} \pi_i \phi_i \left(\frac{(V_n - \mu_i)^2}{\sigma^3} - \frac{1}{\sigma} \right)}{f_{\text{mix}}} \right), \\
\frac{\partial^2 \log p}{\partial \sigma^2} &= \sum_{n=1}^{N} \frac{1}{f_{\text{mix}}} \left(\sum_{g=1}^{G} \frac{\pi_i \phi_i}{\sigma^2} \left(\left(\frac{(V_n - \mu_i)^2}{\sigma^2} - 1 \right)^2 + 1 - \frac{3(V_n - \mu_i)^2}{\sigma^2} \right) \right. \\
&\quad \left. - \frac{1}{f_{\text{mix}}} \left(\sum_{g=1}^{G} \pi_i \phi_i \left(\frac{(V_n - \mu_i)^2}{\sigma^3} - \frac{1}{\sigma} \right) \right)^2 \right).
\end{aligned}
$$

These second order derivatives are used for the calculation of the Fisher information matrix of Eq. (4.A.3).

CHAPTER FIVE

Optimal experiment design for nanoparticle atom counting from ADF STEM images

Annick De Backer[a,b], Jarmo Fatermans[a,b,c], Arnold J. den Dekker[c], and Sandra Van Aert[a,b,*]

[a]Electron Microscopy for Materials Science (EMAT), University of Antwerp, Antwerp, Belgium
[b]NANOlab Center of Excellence, University of Antwerp, Antwerp, Belgium
[c]imec-Vision Lab, University of Antwerp, Wilrijk, Belgium
*Corresponding author. e-mail address: sandra.vanaert@uantwerpen.be

Contents

5.1. Introduction

In Chapter 4, different practical applications are worked out using the statistics-based atom-counting method. From the last application presented in that chapter, it is clear that for more challenging nanoparticles, it is a very tedious task to determine the number of atoms from an annular dark-field (ADF) scanning transmission electron microscopy (STEM)

Advances in Imaging and Electron Physics, Volume 217
ISSN 1076-5670
https://doi.org/10.1016/bs.aiep.2021.01.005

image. Therefore, it is important to optimize the experiment design for atom counting in order to increase the detected dose for atom counting and to determine the minimally required electron dose in order to obtain a pre-specified precision.

In this chapter, the theoretical limits to the precision with which the number of atoms in a projected atomic column can be estimated from high-resolution STEM images will be explored (De Backer, De wael, et al., 2015). Therefore, STEM images are interpreted quantitatively and the optimal experiment design to count the number of atoms in a projected atomic column is investigated. So far, the Cramér-Rao lower bound (CRLB) has been proved to be an optimal tool to determine a theoretical lower bound on the precision with which continuous parameters such as atomic column positions and intensities can be estimated (Bals et al., 2006; Bettens et al., 1999; De Backer et al., 2011; den Dekker et al., 2013, 1999, 2005, 2001; Van Aert, den Dekker, & Van Dyck, 2004; Van Aert, den Dekker, van den Bos, et al., 2002, 2004; Van Aert et al., 2005; Van Aert, den Dekker, Van Dyck, et al., 2002a; Van Aert et al., 2012; Van Aert & Van Dyck, 2001; Van den Broek et al., 2011; Wang, Van Aert, et al., 2012). However, this lower bound can only be derived for continuous parameters. Therefore, an alternative measure using the principles of detection theory will be introduced (den Dekker et al., 2013; Gonnissen et al., 2014; Kay, 2009) for problems concerning the estimation of the atomic number, which is a discrete parameter. This measure will here be extended to the atom-counting problem, where the number of atoms in a projected atomic column can also be considered as a discrete parameter.

Statistical detection theory allows us to derive a criterion to describe the performance of atom counting, provided that the expectations as well as statistical fluctuations of the experimental images can be accurately modeled. In this framework, the atom-counting problem is formulated as a statistical hypothesis test, where each hypothesis corresponds to a specific number of atoms in a projected atomic column. The probability to choose the wrong hypothesis needs to be minimized. This so-called probability of error can therefore be considered as optimality criterion. In order to compute this probability of error, use can be made of realistic simulations to describe the experimental images (Grieb et al., 2012; LeBeau et al., 2008; Rosenauer et al., 2009; Rosenauer & Schowalter, 2008; Van Aert et al., 2013), and knowledge about the statistics of the image pixel values. Ultimately, the image pixel values will be Poisson distributed because of the unavoidable presence of counting noise. This inherent presence of statistical fluctuations will therefore set fundamental limits to the precision with which the number of atoms can be determined.

Statistical detection theory enables us to decide between different image-performance measures that can be used for atom counting in an objective manner. Here, this comparison will be performed for peak intensities, scattering cross-sections and image intensities on a pixel-by-pixel basis. In addition, by minimizing the expression for the probability of error, the optimal experiment design can be derived. In this chapter, the optimal inner and outer angles of an annular STEM detector will be computed. So far, STEM imaging using a high-angle ADF (HAADF) detector is often used for atom counting (Bals et al., 2011, 2012; Erni et al., 2003; Hartel et al., 1996; LeBeau et al., 2010; Pennycook & Boatner, 1988; Singhal et al., 1997; Van Aert et al., 2011, 2013; Voyles et al., 2002). Here, it will be investigated if this imaging mode remains optimal in terms of atom-counting precision. The proposed procedure to derive the optimal STEM detector can then be applied in practical future research problems so as to further improve the precision of counting atoms in a projected atomic column.

The organization of this chapter is as follows. In Sections 5.2 and 5.3, detection theory is discussed and the probability of error is derived for both a binary and a multiple hypothesis test and the so-called sum of Kullback-Leibler divergences is introduced. In Section 5.4, the proposed method will first be applied to a simple linear imaging model, i.e. a simple Gaussian model for which the intensity linearly increases with the number of atoms in the projected atomic column, in order to study the dependence on the incident electron dose and number of atoms analytically. Secondly, the probability of error will be computed for a realistic simulation of a $SrTiO_3$-crystal in order to derive optimal inner and outer angles of the annular STEM detector. This will be done for a binary as well as for a multiple hypothesis test. In Section 5.5, the results found for the binary hypothesis test will be discussed and compared to those for the multiple hypothesis test. In Section 5.6, conclusions are drawn.

5.2. Detection theory: binary hypothesis testing

When one is considering the problem of counting the number of atoms in a projected atomic column of a mono-type crystalline structure from high-resolution STEM images, the goal is to obtain the number of atoms in a column as precisely as possible. In this section, an expression for the probability to miscount the number of atoms, the so-called probability of error, will be derived. Also, a tightly connected performance measure, the so-called sum of Kullback-Leibler divergences, will be introduced. To start simple, a binary hypothesis test is used in which only two possible successive numbers of atoms are considered as possible outcomes. In the next section, this theory will be extended toward a multiple hypothesis test. In that case, all possible hypotheses up to a certain thickness for the atomic column have to be included.

In the case where one wants to know if there are n or $n+1$ atoms in a projected atomic column, the hypotheses are given by

$$\begin{aligned} \mathcal{H}_0 &: n_{\mathcal{H}_0} = n \\ \mathcal{H}_1 &: n_{\mathcal{H}_1} = n+1 \end{aligned} \tag{5.1}$$

where $\mathcal{H}_0$ is referred to as the null hypothesis, $\mathcal{H}_1$ as the alternative hypothesis and $n_{\mathcal{H}_i}$ denotes the number of atoms under hypothesis $\mathcal{H}_i$. In order to express a prior belief in the likelihood of the hypotheses, it is assumed that the prior probabilities $P(\mathcal{H}_0)$ and $P(\mathcal{H}_1)$ associated with these hypotheses are known. It is reasonable to assign equal prior probabilities of $1/2$ if both hypotheses are equally likely. Throughout this chapter, it will always be assumed that both hypotheses are equally likely. By following the quantitative method proposed in den Dekker et al. (2013), the goal is now to quantify the probability of assigning the wrong hypothesis. In a so-called Bayesian approach, this probability of error P_e is defined as:

$$\begin{aligned} P_e &= \Pr\{\text{decide } \mathcal{H}_0, \mathcal{H}_1 \text{ true}\} + \Pr\{\text{decide } \mathcal{H}_1, \mathcal{H}_0 \text{ true}\} \\ &= P(\mathcal{H}_0|\mathcal{H}_1)\,P(\mathcal{H}_1) + P(\mathcal{H}_1|\mathcal{H}_0)\,P(\mathcal{H}_0) \end{aligned} \tag{5.2}$$

where $P(\mathcal{H}_i|\mathcal{H}_j)$ is the conditional probability of deciding $\mathcal{H}_i$ when $\mathcal{H}_j$ is true. Using the criterion of Eq. (5.2), the two possible errors are weighted appropriately to yield an overall error measure. Decision rules are now defined such that the probability of error is minimized. For this purpose, it is shown in Kay (2009) that one should then decide $\mathcal{H}_1$ if:

$$\frac{p(w;\mathcal{H}_1)}{p(w;\mathcal{H}_0)} > \frac{P(\mathcal{H}_0)}{P(\mathcal{H}_1)} = \gamma = 1, \tag{5.3}$$

otherwise $\mathcal{H}_0$ is decided. In this expression, $p(w;\mathcal{H}_i)$ is the conditional probability function assuming $\mathcal{H}_i$ to be true evaluated for the observation w. The exact expression for this probability function will be discussed for three different measures that can be used as a criterion to count the number of atoms in a projected atomic column: peak intensities (PI), scattering cross-sections (CS), and high-resolution STEM image pixel values of a projected atomic column. The observations are modeled as stochastic variables w. The stochastic variable w will take a different form for these three measures:

$$w^{\mathrm{PI}} \qquad \text{for peak intensities} \tag{5.4}$$

$$w^{\mathrm{CS}} = \sum_{k=1}^{K}\sum_{l=1}^{L} w_{kl} \cdot \Delta x^2 \qquad \text{for cross-sections} \tag{5.5}$$

$$\mathbf{w}^{\mathrm{Im}} = [w_{11}, w_{12}, \ldots, w_{KL}]^T \quad \text{for images} \tag{5.6}$$

where w_{kl} corresponds to the random variables describing the pixel intensities of the STEM image of the atomic column and w^{PI} corresponds to the random variable describing the pixel intensity at the position of the atomic column. The index kl denotes the probe position $(x_k, y_l)^T$ for a set of KL pixel observations, and Δx denotes the pixel size of the STEM image.

5.2.1 Peak intensities

When assuming that the pixels in a STEM image are statistically independent electron counting results, which are modeled as a Poisson distribution, the conditional probability function for the pixel intensity at the position of a projected atomic column is given by:

$$p\left(\omega^{\mathrm{PI}}; \mathcal{H}_i\right) = \frac{\left(\lambda_{\mathcal{H}_i}^{\mathrm{PI}}\right)^{\omega^{\mathrm{PI}}}}{\omega^{\mathrm{PI}}!} \exp -\lambda_{\mathcal{H}_i} \tag{5.7}$$

where the parameter $\lambda_{\mathcal{H}_i}^{\mathrm{PI}} = \mathbb{E}_{\mathcal{H}_i}\left[w^{\mathrm{PI}}\right]$ corresponds to the expectation value for the pixel intensity of the STEM image at the position of the projected atomic column under $\mathcal{H}_i$. Since this expectation value will depend on which hypothesis $\mathcal{H}_i$ is assumed to be true, also the probability function depends on $\mathcal{H}_i$. In general, this expectation value can be computed under each hypothesis using software that allows one to simulate a STEM image for a given input material's structure and a given set of microscope parameters (Rosenauer & Schowalter, 2008). Following the decision rule of Eq. (5.3), for equal prior probabilities $P(\mathcal{H}_0) = P(\mathcal{H}_1) = 1/2$, one should decide $\mathcal{H}_1$ for an observation w^{PI} if

$$p\left(w^{\mathrm{PI}}; \mathcal{H}_1\right) > p\left(w^{\mathrm{PI}}; \mathcal{H}_0\right) \tag{5.8}$$

otherwise $\mathcal{H}_0$ is decided. An illustration of the conditional probability functions $p\left(\omega^{\mathrm{PI}}; \mathcal{H}_i\right)$ of Eq. (5.7) of both hypotheses is given in Fig. 5.1(a) for

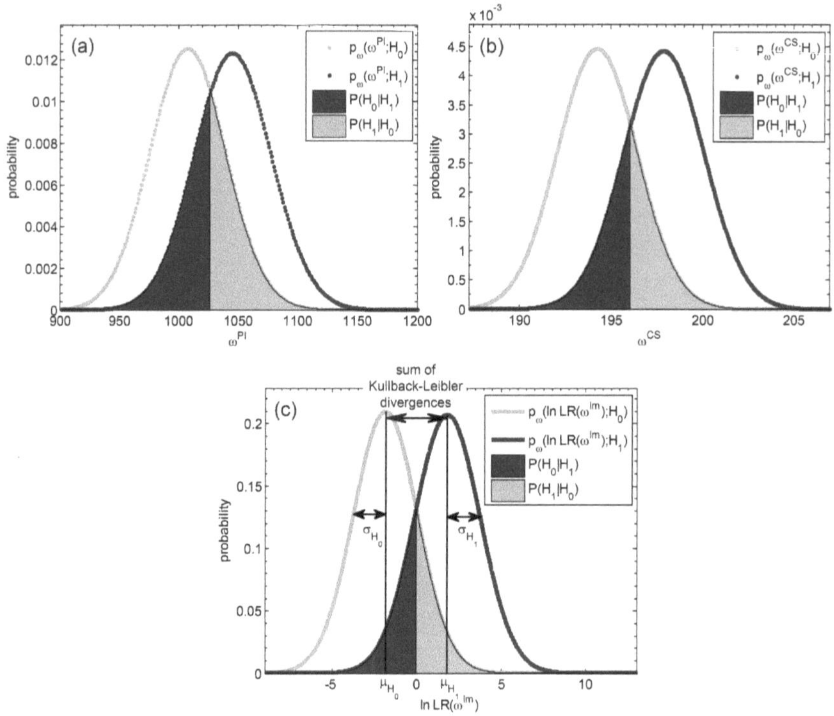

Figure 5.1 Calculation of the probability of error for a binary hypothesis test for (a) peak intensities ω^{PI} (electron dose $= 10^6$ e$^-$/Å^2 and $n_{\mathcal{H}_0} = 15$), (b) scattering cross-sections ω^{CS} (electron dose $= 10^5$ e$^-$/Å^2 and $n_{\mathcal{H}_0} = 30$), and (c) images ω^{Im} (electron dose $= 10^5$ e$^-$/Å^2 and $n_{\mathcal{H}_0} = 30$) of a Sr column with a detector collection range of 60–100 mrad and settings of Table 5.1 which are explained in Subsection 5.4.3. Reprinted from De Backer, De wael, et al. (2015), Copyright (2015), with permission from Elsevier.

the peak intensities of a Sr column consisting of either 15 or 16 atoms. The dark gray region denotes the error which is made if $\mathcal{H}_0$ is chosen while $\mathcal{H}_1$ is correct, and vice versa for the light gray region. It is clear from this figure that the probability of error corresponds to the overlapping area of the two conditional probability functions $p\left(\omega^{\mathrm{PI}}; \mathcal{H}_0\right)$ and $p\left(\omega^{\mathrm{PI}}; \mathcal{H}_1\right)$. This probability can analytically be computed using the cumulative distribution function of the Poisson distribution:

$$
\begin{aligned}
P_{\mathrm{e}} &= \frac{1}{2} P(\mathcal{H}_0|\mathcal{H}_1) + \frac{1}{2} P(\mathcal{H}_1|\mathcal{H}_0) \\
&= \frac{1}{2} F\left(x^{\mathrm{PI}}; \lambda^{\mathrm{PI}}_{\mathcal{H}_1}\right) + \frac{1}{2}\left[1 - F\left(x^{\mathrm{PI}}; \lambda^{\mathrm{PI}}_{\mathcal{H}_0}\right)\right]
\end{aligned}
\tag{5.9}
$$

where $F\left(x^{\mathrm{PI}}; \lambda_{\mathcal{H}_i}^{\mathrm{PI}}\right)$ equals the Poisson cumulative distribution function with parameter $\lambda_{\mathcal{H}_i}^{\mathrm{PI}}$ evaluated at the value x^{PI}, with x^{PI} the intersection between the two conditional distribution functions $p\left(\omega^{\mathrm{PI}}; \mathcal{H}_0\right)$ and $p\left(\omega^{\mathrm{PI}}; \mathcal{H}_1\right)$ given by:

$$x^{\mathrm{PI}} = \frac{\lambda_{\mathcal{H}_0}^{\mathrm{PI}} - \lambda_{\mathcal{H}_1}^{\mathrm{PI}}}{\log \frac{\lambda_{\mathcal{H}_0}^{\mathrm{PI}}}{\lambda_{\mathcal{H}_1}^{\mathrm{PI}}}}. \tag{5.10}$$

5.2.2 Scattering cross-sections

As a second measure, next to the peak intensities, one can also use scattering cross-sections under both hypotheses. The scattering cross-section values are here computed in the following manner (Eq. (5.5)) (E et al., 2013):

$$w^{\mathrm{CS}} = \sum_{k=1}^{K} \sum_{l=1}^{L} w_{kl} \cdot \Delta x^2. \tag{5.11}$$

Voronoi cells are used to define the integration area for the computation of the scattering cross-sections. A Voronoi cell of an atomic column is the cell formed by the perpendicular bisectors of the direct connections to the neighboring columns. This definition of the scattering cross-sections is used here for simplicity as the scattering cross-section values can directly be calculated from the simulated STEM images in this manner, since the positions of the atomic columns are known from the simulation. The computed values from the integration of the Voronoi cells are equivalent to volumes under the estimated Gaussian peaks at atomic column positions as used in the previous chapters. The use of the volumes provides some extra advantages, since the parametric model takes into account overlap between neighboring atomic columns. The use of scattering cross-sections from the integration of Voronoi cells is afforded here, since there are no thickness variations in the simulations.

The expected scattering cross-section value equals:

$$\mathbb{E}_{\mathcal{H}_0}\left[w^{\mathrm{CS}}\right] = \sum_{k=1}^{K} \sum_{l=1}^{L} \lambda_{\mathcal{H}_0;kl} \cdot \Delta x^2 = \mathrm{CS}_n \tag{5.12}$$

$$\mathbb{E}_{\mathcal{H}_1}\left[w^{\mathrm{CS}}\right] = \sum_{k=1}^{K} \sum_{l=1}^{L} \lambda_{\mathcal{H}_1;kl} \cdot \Delta x^2 = \mathrm{CS}_{n+1} \tag{5.13}$$

where $\lambda_{\mathcal{H}_i,kl} = \mathbb{E}_{\mathcal{H}_i}[w_{kl}]$ corresponds to the expectation values for the pixel intensities of the STEM image of the atomic column and CS_n is the expected scattering cross-section for n atoms. Since a sum of independent Poisson distributed variables is known to be Poisson distributed (Mood et al., 1974), the variable

$$\sum_{k=1}^{K}\sum_{l=1}^{L} w_{kl} \tag{5.14}$$

is therefore also Poisson distributed. The expected value $\lambda_{\mathcal{H}_i}$ then equals:

$$\lambda_{\mathcal{H}_i} = \mathbb{E}_{\mathcal{H}_i}\left[\sum_{k=1}^{K}\sum_{l=1}^{L} w_{kl}\right] = \sum_{k=1}^{K}\sum_{l=1}^{L} \lambda_{\mathcal{H}_i,kl}. \tag{5.15}$$

The conditional probability function for a scattering cross-section taking into account the constant factor Δx^2, is then given by:

$$p\left(\omega^{CS};\mathcal{H}_i\right) = \frac{\left(\lambda_{\mathcal{H}_i}\right)^{\omega^{CS}/\Delta x^2}}{\left(\frac{\omega^{CS}}{\Delta x^2}\right)!}\exp\left(-\lambda_{\mathcal{H}_i}\right). \tag{5.16}$$

Analogous to the decision rule for the peak intensities of Eq. (5.8), one should decide $\mathcal{H}_1$ for the observations of the scattering cross-section w^{CS} when assuming equal prior probabilities $P(\mathcal{H}_0) = P(\mathcal{H}_1) = 1/2$ if

$$p\left(w^{CS};\mathcal{H}_1\right) > p\left(w^{CS};\mathcal{H}_0\right), \tag{5.17}$$

otherwise $\mathcal{H}_0$ is decided. This decision rule and the calculation of the probability function given by Eq. (5.16) is illustrated in Fig. 5.1(b). The probability of error is then reformulated as follows for the scattering cross-sections:

$$\begin{aligned} P_e &= \frac{1}{2}P(\mathcal{H}_0|\mathcal{H}_1) + \frac{1}{2}P(\mathcal{H}_1|\mathcal{H}_0) \\ &= \frac{1}{2}F\left(\frac{x^{CS}}{\Delta x^2};\lambda_{\mathcal{H}_1}\right) + \frac{1}{2}\left[1 - F\left(\frac{x^{CS}}{\Delta x^2};\lambda_{\mathcal{H}_0}\right)\right] \end{aligned} \tag{5.18}$$

where $F\left(x^{CS}/\Delta x^2;\lambda_{\mathcal{H}_i}\right)$ equals the Poisson cumulative distribution function with parameter $\lambda_{\mathcal{H}_i}$ evaluated at $x^{CS}/\Delta x^2$ and x^{CS} is the intersection of the two probability functions $p\left(\omega^{\mathrm{CS}};\mathcal{H}_0\right)$ and $p\left(\omega^{\mathrm{CS}};\mathcal{H}_1\right)$ given by:

$$x^{\mathrm{CS}}=\frac{\left(\lambda_{\mathcal{H}_0}-\lambda_{\mathcal{H}_1}\right)\Delta x^2}{\log\frac{\lambda_{\mathcal{H}_0}}{\lambda_{\mathcal{H}_1}}}. \tag{5.19}$$

5.2.3 STEM images

As a third measure to count the number of atoms, the STEM images of the atomic columns are considered. When assuming that the pixels in a STEM image are statistically independent electron counting results, the conditional probability function for a STEM image of an atomic column is given by:

$$p\left(\boldsymbol{\omega}^{\mathrm{Im}};\mathcal{H}_i\right)=\prod_{k=1}^{K}\prod_{l=1}^{L}\frac{\left(\lambda_{\mathcal{H}_i,kl}\right)^{\omega_{kl}}}{\omega_{kl}!}\exp\left(-\lambda_{\mathcal{H}_i,kl}\right). \tag{5.20}$$

Since the stochastic variable $\mathbf{w}^{\mathrm{Im}}$ of the STEM image of an atomic column is characterized by a *joint* probability function, it is not possible to compute the probability of error directly from the overlapping areas of $p\left(\boldsymbol{\omega}^{\mathrm{Im}};\mathcal{H}_0\right)$ and $p\left(\boldsymbol{\omega}^{\mathrm{Im}};\mathcal{H}_1\right)$. However, as shown in Gonnissen et al. (2014), it is possible to calculate this probability of error analytically by reformulating the decision rule using the so-called log-likelihood ratio $\log\mathrm{LR}\left(\boldsymbol{\omega}^{\mathrm{Im}}\right)$. After reformulation of Eq. (5.3), one should then decide $\mathcal{H}_1$ for an observation $\mathbf{w}^{\mathrm{Im}}$, if:

$$\log\mathrm{LR}\left(\mathbf{w}^{\mathrm{Im}}\right)\equiv\log\left(\frac{p\left(\mathbf{w}^{\mathrm{Im}};\mathcal{H}_1\right)}{p\left(\mathbf{w}^{\mathrm{Im}};\mathcal{H}_0\right)}\right)>\log(1)=0, \tag{5.21}$$

otherwise $\mathcal{H}_0$ is decided, for equal prior probabilities. This corresponds to choosing the hypothesis for which the log-likelihood function is maximal. The function $\mathrm{LR}\left(\mathbf{w}^{\mathrm{Im}}\right)$ is called the likelihood function since it indicates for each set of observations of $\mathbf{w}^{\mathrm{Im}}$ the likelihood of $\mathcal{H}_1$ versus the likelihood of $\mathcal{H}_0$. Given the decision rule of Eq. (5.21), the expression for the probability of error P_{e} given by Eq. (5.2), can be rewritten as follows:

$$\begin{aligned}P_{\mathrm{e}}&=\frac{1}{2}P(\mathcal{H}_0|\mathcal{H}_1)+\frac{1}{2}P(\mathcal{H}_1|\mathcal{H}_0)\\&=\frac{1}{2}P\left(\log\mathrm{LR}\left(\mathbf{w}^{\mathrm{Im}}\right)<0|\mathcal{H}_1\right)+\frac{1}{2}P\left(\log\mathrm{LR}\left(\mathbf{w}^{\mathrm{Im}}\right)>0|\mathcal{H}_0\right).\end{aligned} \tag{5.22}$$

When using the conditional joint probability function for STEM images given by Eq. (5.20), the log-likelihood ratio defined by Eq. (5.21) can be rewritten as:

$$\begin{aligned}\log \mathrm{LR}\left(\mathbf{w}^{\mathrm{Im}}\right) &= \sum_{k=1}^{K}\sum_{l=1}^{L} \log\left(\frac{\lambda_{\mathcal{H}_1,kl}^{\omega_{kl}} \exp\left(-\lambda_{\mathcal{H}_1,kl}\right)}{\lambda_{\mathcal{H}_0,kl}^{\omega_{kl}} \exp\left(-\lambda_{\mathcal{H}_0,kl}\right)}\right) \\ &= \sum_{k=1}^{K}\sum_{l=1}^{L} \log\left(\lambda_{\mathcal{H}_1,kl}^{\omega_{kl}}\right) - \lambda_{\mathcal{H}_1,kl} - \log\left(\lambda_{\mathcal{H}_0,kl}^{\omega_{kl}}\right) + \lambda_{\mathcal{H}_0,kl} \\ &= \sum_{k=1}^{K}\sum_{l=1}^{L} \left(\omega_{kl} \log\left(\frac{\lambda_{\mathcal{H}_1,kl}}{\lambda_{\mathcal{H}_0,kl}}\right) - \lambda_{\mathcal{H}_1,kl} + \lambda_{\mathcal{H}_0,kl}\right). \end{aligned} \tag{5.23}$$

Following the central limit theorem, the log-likelihood ratio tends to be normally distributed:

$$p\left(\log \mathrm{LR}\left(\mathbf{w}^{\mathrm{Im}}\right); \mathcal{H}_i\right) = \frac{1}{\sigma_{\mathcal{H}_i}\sqrt{2\pi}} \exp\left(-\frac{\left(\log \mathrm{LR}\left(\mathbf{w}^{\mathrm{Im}}\right) - \mu_{\mathcal{H}_i}\right)^2}{2\sigma_{\mathcal{H}_i}^2}\right). \tag{5.24}$$

For STEM images the expected value and variance characterizing this normal distribution of $\log \mathrm{LR}\left(\mathbf{w}^{\mathrm{Im}}\right)$ can be computed from Eq. (5.24) when assuming $\mathcal{H}_i$ to be true giving the following results:

$$\begin{aligned}\mu_{\mathcal{H}_i} &= \mathbb{E}_{\mathcal{H}_i}\left[\log \mathrm{LR}\left(\mathbf{w}^{\mathrm{Im}}\right)\right] \\ &= \sum_{k=1}^{K}\sum_{l=1}^{L} \left(\lambda_{\mathcal{H}_i,kl} \log \frac{\lambda_{\mathcal{H}_1,kl}}{\lambda_{\mathcal{H}_0,kl}} - \lambda_{\mathcal{H}_1,kl} + \lambda_{\mathcal{H}_0,kl}\right), \end{aligned} \tag{5.25}$$

$$\sigma_{\mathcal{H}_i}^2 = \mathrm{var}_{\mathcal{H}_i}\left[\log \mathrm{LR}\left(\mathbf{w}^{\mathrm{Im}}\right)\right] = \sum_{k=1}^{K}\sum_{l=1}^{L} \lambda_{\mathcal{H}_i,kl} \left(\log \frac{\lambda_{\mathcal{H}_1,kl}}{\lambda_{\mathcal{H}_0,kl}}\right)^2. \tag{5.26}$$

In this derivation, use is made of the property that the variance of a Poisson distributed variable equals its expectation value, $\mathbb{E}_{\mathcal{H}_i}\left[w_{kl}\right] = \mathrm{var}_{\mathcal{H}_i}\left[w_{kl}\right] = \lambda_{\mathcal{H}_i,kl}$. The explicit description of the distribution of the log-likelihood ratio now enables us to unambiguously compute the probability of error given by Eq. (5.22), resulting in the following general expression:

$$P_{\mathrm{e}} = \frac{1}{2}\left[\Phi\left(\frac{-\mu_{\mathcal{H}_1}}{\sigma_{\mathcal{H}_1}}\right) + \Phi\left(\frac{\mu_{\mathcal{H}_0}}{\sigma_{\mathcal{H}_0}}\right)\right] \tag{5.27}$$

with $\Phi(\pm\mu/\sigma)$ the cumulative distribution function of the standard normal distribution evaluated at $\pm\mu/\sigma$. The meaning of the expected value $\mu_{\mathcal{H}_i}$ and variance $\sigma_{\mathcal{H}_i}^2$ characterizing the normal distributions of $\log \mathrm{LR}\left(\mathbf{w}^{\mathrm{Im}}\right)$ given by Eq. (5.24) under both hypotheses is illustrated in Fig. 5.1(c) based on STEM images for a Sr column consisting of either 30 or 31 atoms. The decision rule is also clarified: $\mathcal{H}_0$ is decided for $\log \mathrm{LR}\left(\mathbf{w}^{\mathrm{Im}}\right) < 0$, otherwise $\mathcal{H}_1$ is decided to be correct. The dark gray region denotes the error which is made if $\mathcal{H}_0$ is chosen while $\mathcal{H}_1$ is correct, and vice versa for the light gray region. It is clear from this figure that the probability of error also here corresponds to the overlap between both distributions.

5.2.4 Sum of Kullback-Leibler divergences

A tightly connected performance measure that can be investigated as an alternative for the probability of error is based on the so-called Kullback-Leibler divergence (Kullback, 1959; Kullback & Leibler, 1951). This measure quantifies the difference between two probability distributions and can be defined for peak intensities, scattering cross-sections and STEM images. Here, the Kullback-Leibler divergence will be discussed for a general stochastic variable w, which can be replaced by one of the corresponding stochastic variables for peak intensities, scattering cross-sections or STEM images, respectively. The Kullback-Leibler divergence from $p_{\mathcal{H}_1} = p(w; \mathcal{H}_1)$ to $p_{\mathcal{H}_0} = p(w; \mathcal{H}_0)$ is defined as:

$$D\left(p_{\mathcal{H}_1}, p_{\mathcal{H}_0}\right) \equiv \mathbb{E}_{p_{\mathcal{H}_1}}\left[\log \frac{p(w; \mathcal{H}_1)}{p(w; \mathcal{H}_0)}\right] = \mathbb{E}_{p_{\mathcal{H}_1}}\left[\log \mathrm{LR}(w)\right] \tag{5.28}$$

where $\log \mathrm{LR}(w)$ corresponds to the log-likelihood ratio as defined in Eq. (5.21) for the STEM images. This log-likelihood ratio can in the same manner be defined for the peak intensities and scattering cross-sections using their respective probability functions. The Kullback-Leibler divergence from $p_{\mathcal{H}_1}$ to $p_{\mathcal{H}_0}$ of Eq. (5.28) corresponds to the expected or mean log-likelihood ratio assuming $\mathcal{H}_1$ to be true. Similarly the Kullback-Leibler divergence from $p_{\mathcal{H}_0}$ to $p_{\mathcal{H}_1}$ equals:

$$\begin{aligned} D\left(p_{\mathcal{H}_0}, p_{\mathcal{H}_1}\right) &= \mathbb{E}_{p_{\mathcal{H}_0}}\left[\log \frac{p(w; \mathcal{H}_0)}{p(w; \mathcal{H}_1)}\right] \\ &= -\mathbb{E}_{p_{\mathcal{H}_0}}\left[\log \frac{p(w; \mathcal{H}_1)}{p(w; \mathcal{H}_0)}\right] = -\mathbb{E}_{p_{\mathcal{H}_0}}\left[\log \mathrm{LR}(w)\right]. \end{aligned} \tag{5.29}$$

From Eqs. (5.28) and (5.29), it follows that:

$$D\left(p_{\mathcal{H}_1}, p_{\mathcal{H}_0}\right) + D\left(p_{\mathcal{H}_0}, p_{\mathcal{H}_1}\right) = \mathbb{E}_{p_{\mathcal{H}_1}}\left[\log \mathrm{LR}(w)\right] - \mathbb{E}_{p_{\mathcal{H}_0}}\left[\log \mathrm{LR}(w)\right]. \tag{5.30}$$

The sum of Kullback-Leibler divergences thus corresponds to the difference in mean log-likelihood ratio under $\mathcal{H}_1$ and the corresponding value when assuming $\mathcal{H}_0$ to be true. An illustration for the sum of Kullback-Leibler divergences is also given by Fig. 5.1(c), where the distance between $\mu_{\mathcal{H}_0}$ and $\mu_{\mathcal{H}_1}$ corresponds to the sum of Kullback-Leibler divergences for STEM images. The sum of Kullback-Leibler divergences only takes into account the separations between the distributions of the log-likelihood ratio whereas the probability of error also takes into account the width of the distributions. Since the width and the separation between the distributions are related, as will be shown for the scattering cross-sections in Subsection 5.4.2, it is likely to assume that the probability to assign the wrong hypothesis will decrease when the sum of Kullback-Leibler divergences increases. A more explicit expression for Eq. (5.30) can be derived using Eq. (5.23):

$$\begin{aligned} &D\left(p_{\mathcal{H}_1}, p_{\mathcal{H}_0}\right) + D\left(p_{\mathcal{H}_0}, p_{\mathcal{H}_1}\right) \\ &\quad = \left(\mathbb{E}_{\mathcal{H}_1}[w] - \mathbb{E}_{\mathcal{H}_0}[w]\right)\left(\log \mathbb{E}_{\mathcal{H}_1}[w] - \log \mathbb{E}_{\mathcal{H}_0}[w]\right). \end{aligned} \tag{5.31}$$

5.3. Detection theory: multiple hypothesis testing

A binary hypothesis test is no longer adequate if different choices of the two hypotheses lead to different designs. Therefore, the approach of binary hypothesis testing for atom counting is extended toward multiple hypothesis testing. The multiple hypothesis test will be described for the scattering cross-sections, which are most optimal for atom-counting applications as will be shown in Section 5.4.4. The number of hypotheses in the multiple hypothesis test equals the maximum number of atoms in a projected atomic column which depends on the sample under study. In this case, a differentiation between $1, 2, \ldots, M$ atoms in a projected atomic column should be made and therefore one should decide among M possible hypotheses:

$$\{\mathcal{H}_0, \mathcal{H}_1, \ldots, \mathcal{H}_{M-1}\}. \tag{5.32}$$

In practice, the number of hypotheses that is included for the decision rule is chosen sufficiently large such that the observed scattering cross-sections for sure correspond to one of the considered hypotheses; M will typically be larger than the unknown thickness of the sample. Equivalent to Eq. (5.3), the decision rule is now defined such that the probability of

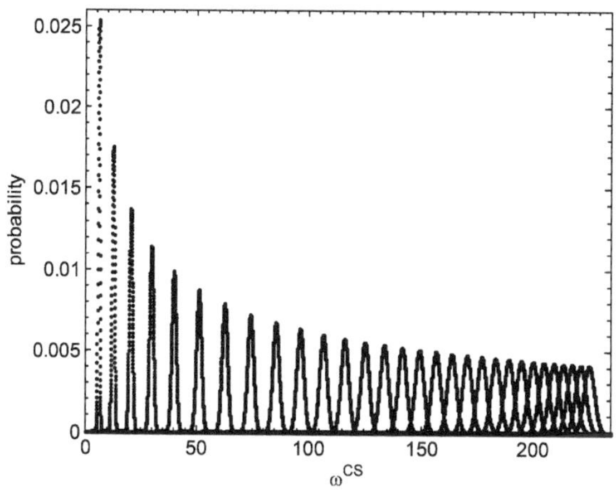

Figure 5.2 Conditional probability functions for a multiple hypothesis test of $M = 30$ for a Sr column with an incoming electron dose $= 10^5$ $e^-/Å^2$, for a detector collection range of 60–100 mrad and settings of Table 5.1 which are explained in Subsection 5.4.3. Reprinted from De Backer, De wael, et al. (2015), Copyright (2015), with permission from Elsevier.

error is minimized. The minimum probability of error decision rule is then to decide $\mathcal{H}_k$ if:

$$p\left(w^{\mathrm{CS}}; \mathcal{H}_k\right) P\left(\mathcal{H}_k\right) > p\left(w^{\mathrm{CS}}; \mathcal{H}_i\right) P\left(\mathcal{H}_i\right) \qquad \forall i \neq k. \tag{5.33}$$

For equal prior probabilities $P(\mathcal{H}_i) = 1/M$, which is assumed here, one should decide $\mathcal{H}_k$ if:

$$p\left(w^{\mathrm{CS}}; \mathcal{H}_k\right) > p\left(w^{\mathrm{CS}}; \mathcal{H}_i\right) \qquad \forall i \neq k. \tag{5.34}$$

An illustration of the conditional probability functions of a multiple hypothesis test consisting of $M = 30$ hypotheses is given in Fig. 5.2. Analogously to the binary test of Subsection 5.2.2, the probability of error for the scattering cross-sections of a multiple hypothesis test can be calculated from the overlapping areas of the conditional probability functions, weighted by their prior probability:

$$P_{\mathrm{e}} = \sum_{i=0}^{M-1} \sum_{j=0}^{M-1} C_{ij} P\left(\mathcal{H}_i | \mathcal{H}_j\right) P\left(\mathcal{H}_j\right), \tag{5.35}$$

where

$$C_{ij} = \begin{cases} 1 & i \neq j \\ 0 & i = j \end{cases}. \tag{5.36}$$

The number of terms for the calculation of the probability of error of a multiple hypothesis test used in Eq. (5.35) equals $M(M-1)$. Therefore, it is more efficient to calculate $P_c = 1 - P_e$, where P_c is the probability of a correct decision. In this case, the number of terms in the summation reduces to M:

$$P_c = \sum_{i=0}^{M-1} P(\mathcal{H}_i|\mathcal{H}_i)\, P(\mathcal{H}_i). \tag{5.37}$$

This expression can be calculated analytically for the scattering cross-sections, since their conditional probability functions are well-known from Eq. (5.16). The probability of a correct decision for equal prior probabilities $P(\mathcal{H}_i) = 1/M$ then corresponds to:

$$\begin{aligned} P_c = \frac{1}{M}\Bigg[& F\left(\frac{x^{CS_{0,1}}}{\Delta x^2}, \lambda_{\mathcal{H}_0^s}\right) \\ & + F\left(\frac{x^{CS_{1,2}}}{\Delta x^2}, \lambda_{\mathcal{H}_1^s}\right) - F\left(\frac{x^{CS_{1,0}}}{\Delta x^2}, \lambda_{\mathcal{H}_1^s}\right) \\ & + F\left(\frac{x^{CS_{2,3}}}{\Delta x^2}, \lambda_{\mathcal{H}_2^s}\right) - F\left(\frac{x^{CS_{2,1}}}{\Delta x^2}, \lambda_{\mathcal{H}_2^s}\right) \\ & + \ldots \\ & + \left(1 - F\left(\frac{x^{CS_{M-1,M-2}}}{\Delta x^2}, \lambda_{\mathcal{H}_{M-1}^s}\right)\right)\Bigg] \end{aligned} \tag{5.38}$$

where $\mathcal{H}_i^s$ denotes the sorted hypotheses according to the expected values of the scattering cross-sections, $x^{CS_{ij}} = x^{CS_{ji}}$ corresponds to the intersection between two neighboring probability distribution functions which is given by Eq. (5.19), $\lambda_{\mathcal{H}_i}$ corresponds to Eq. (5.15), and $F\left(x^{CS_{ij}}/\Delta x^2, \lambda_{\mathcal{H}_i^s}\right)$ equals the Poisson cumulative distribution function with parameter $\lambda_{\mathcal{H}_i^s}$ evaluated at $x^{CS_{ij}}/\Delta x^2$. This allows us to calculate analytically the probability of error P_e for atom counting. For a multiple hypothesis test with M different hypotheses, the maximum value for the probability of error equals $(M-1)/M$. This can be derived from the fact that when the decision for a certain hypothesis between M different hypotheses is completely random, the probability for a correct decision equals $1/M$ when assuming equal prior probabilities.

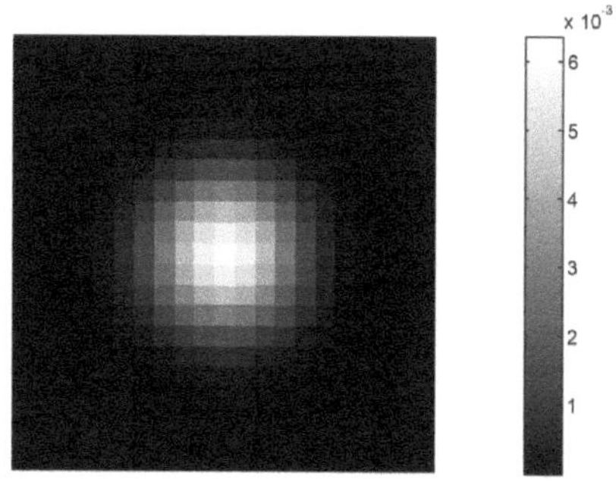

Figure 5.3 Building block of the linear imaging model. Normalized intensity for 1 atom in a projected atomic column ΔI_{kl}.

The probability of error and optimal experiment design can now be investigated for a simplified model and a realistic simulation experiment, using the analytical expressions for the probability of error for both the binary and the multiple hypothesis test. The obtained results for the proposed measures, i.e. peak intensities, scattering cross-sections, or STEM images, will be compared.

5.4. Optimal experiment design

5.4.1 Analytical results for a linear imaging model

In this subsection, the sum of Kullback-Leibler divergences and the probability of error for a binary hypothesis test will be studied for a linear imaging model, i.e. a simple Gaussian model for which the intensity linearly increases with the number of atoms in the projected atomic column. The analysis here will be done for the three different measures: STEM images, scattering cross-sections, and peak intensities. Distinguishing between n and $n+1$ atoms can be formulated as a binary hypothesis test as in Eq. (5.1). An isolated atomic column is considered for the linear imaging model. The parameters of the Gaussian model corresponding to a single atom equal $\rho = 0.5$ Å for the width and $\eta = 1$ for the intensity. This Gaussian shape has been simulated in the center of a two-dimensional grid consisting of $K \times L = 21 \times 21$ pixels having a pixel size $\Delta x = \Delta y = 0.2$ Å. This simulation then corresponds to the fraction of intensity with respect to the incoming electron beam of a single atom in a projected atomic column. This building block ΔI_{kl} for the linear imaging model is illustrated in Fig. 5.3. The model

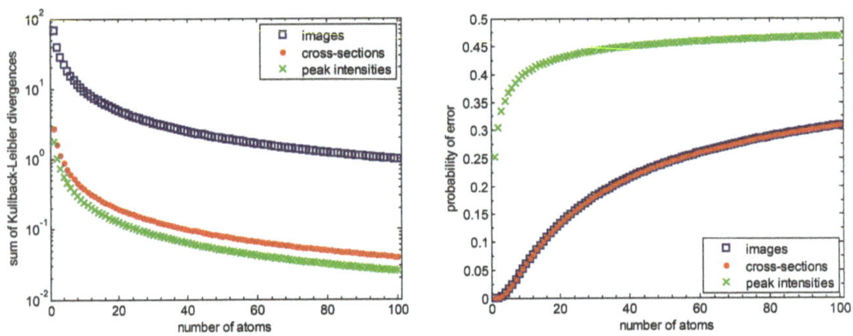

(a) sum of Kullback-Leibler divergences (b) probability of error

Figure 5.4 Sum of Kullback-Leibler divergences and probability of error as a function of the number of atoms $n_{\mathcal{H}_0}$ in a projected atomic column for STEM images (blue squares), scattering cross-sections (red dots), and peak intensities (green crosses) for an incident electron dose of 10^4 $e^-/\text{Å}^2$ using a linear imaging model.

for $n_{\mathcal{H}_i}$ atoms in the columns equals then:

$$N_p n_{\mathcal{H}_i} \Delta I_{kl}, \tag{5.39}$$

with N_p the incident electron dose per pixel. The sum of Kullback-Leibler divergences and probability of error can be calculated analytically for this model for the three different measures as a function of experimental parameters such as incident electron dose N_d and number of atoms $n_{\mathcal{H}_0}$ considered in the imaging model. Such expressions can be used to have more physical insight on how the probability of error depends on these parameters.

The analytical expressions of the sum of Kullback-Leibler divergences are derived in Appendix 5.A and are given by Eqs. (5.A.3), (5.A.8), and (5.A.12) and the probability of error for equal prior probabilities is given by Eqs. (5.9), (5.18), and (5.27) in which the expressions given by Eqs. (5.A.4), (5.A.9), and (5.A.13)-(5.A.14) are used, respectively, for the linear imaging model. They will be used to calculate these measures as a function of the number of atoms and the incident electron dose. The results of this analysis are shown in Figs. 5.4 and 5.5. Fig. 5.4 shows the sum of Kullback-Leibler divergences and the probability of error as a function of the number of atoms $n_{\mathcal{H}_0}$ for an incident electron dose of 10^4 $e^-/\text{Å}^2$. In Fig. 5.5 the sum of Kullback-Leibler divergences and the probability of error is shown as a function of the incident electron dose for $n_{\mathcal{H}_0} = 10$ atoms. From these figures, it is clear that when the sum of Kullback-Leibler divergences increases, the probability of error decreases. Although the an-

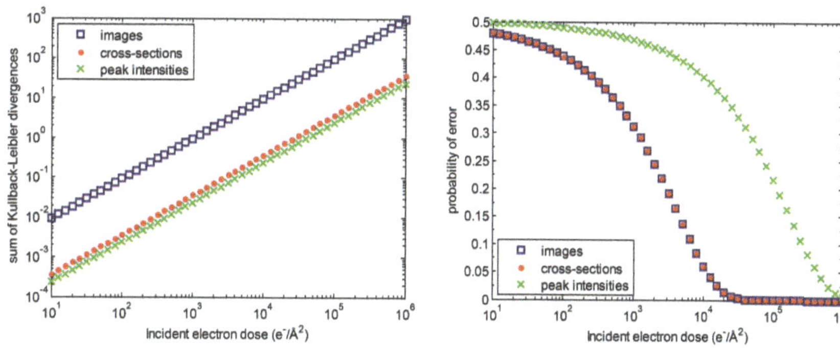

(a) sum of Kullback-Leibler divergences (b) probability of error

Figure 5.5 Sum of Kullback-Leibler divergences and probability of error as a function of the incident electron dose for STEM images (blue squares), scattering cross-sections (red dots), and peak intensities (green crosses) for $n_{\mathcal{H}_0} = 10$ atoms using a linear imaging model.

alytical expressions of the Kullback-Leibler divergences are more easily to interpret as compared to the expressions for the probability of error, it can be seen from Figs. 5.4 and 5.5 that it is necessary to include the overlap of the distributions to mutually compare the different measures, i.e. peak intensities, scattering cross-sections, and STEM images. From Figs. 5.4(b) and 5.5(b), it can be concluded that the scattering cross-sections perform equally well as the STEM images and prove more robust than peak intensities, whereas this conclusion could not be drawn from the sum of Kullback-Leibler divergences shown in Figs. 5.4(a) and 5.5(a). This can also be seen from Eqs. (5.A.8) and (5.A.12) for the sum of Kullback-Leibler divergences where an extra factor $1/\Delta x^2$ is present in the expression for the STEM images. This explains the higher value of the sum of Kullback-Leibler divergences for the STEM images. For the calculation of probability of error, the width of the probability distributions is also taken into account. The increased width of the probability distributions for the STEM images compensates the increased distance between the distributions. As such, the probability of error for the scattering cross-sections equals the probability of error for the STEM images.

The probability of error increases with the number of atoms as can be seen from Fig. 5.4(b), meaning that it is easier to differentiate e.g. between 1 and 2 atoms as compared to 20 and 21 atoms. As expected, the probability of error decreases for an increasing incident electron dose, as shown in Fig. 5.5(b). Figs. 5.4(b) and 5.5(b) can be combined into one summarizing graph, where the probability of error is shown for different levels as

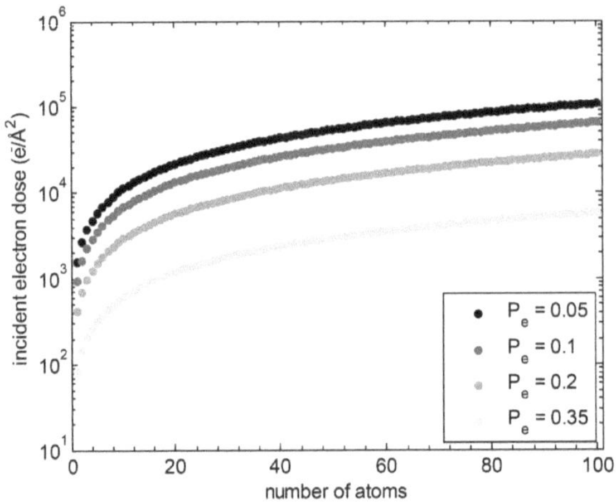

Figure 5.6 Probability of error as a function of the number of atoms $n_{\mathcal{H}_0}$ and the incident electron dose for the scattering cross-sections or STEM images using a linear imaging model.

a function of the number of atoms and the incident electron dose. This graph is shown in Fig. 5.6 for the scattering cross-sections or STEM images. From this figure, the minimally required incident electron dose for a certain probability of error can be deduced for a certain number of atoms.

In this subsection, a preliminary example for the probability error for atom counting for a linear imaging model has been discussed. In the following subsections, the probability of error will be calculated for realistic detailed STEM simulations. The sum of Kullback-Leibler divergences will no longer be considered since it has been shown for this preliminary example that the probability of error provides us a better insight into the precision that can be obtained for atom counting.

5.4.2 General analytical expression for scattering cross-sections

Similar to the analytical derivation for the linear imaging model, the probability of error can be evaluated analytically for a general model. This will be done in this subsection for the scattering cross-sections, since they seem to prove more robust than peak intensities and perform similar as STEM images based on the results of the linear imaging model. From Fig. 5.1(b), it can be seen that the probability of error decreases if the two distributions are more separated. A high difference between the expected values is therefore desired, as well as small values for the width of the distributions. The difference between the expected values of ω^{CS} for each hypothesis can be expressed as:

$$\mathbb{E}_{\mathcal{H}_1}\left[w^{\mathrm{CS}}\right] - \mathbb{E}_{\mathcal{H}_0}\left[w^{\mathrm{CS}}\right] = \left(\mathrm{CS}_n + \Delta\mathrm{CS}_{n,n+1}\right) - \mathrm{CS}_n = \Delta\mathrm{CS}_{n,n+1} \quad (5.40)$$

with $\Delta\mathrm{CS}_{n,n+1}$ the expected difference in scattering cross-section between two columns having n and $n+1$ atoms. The difference in expected values of w^{CS} for each hypothesis equals $\Delta\mathrm{CS}_{n,n+1}$. However, when $\Delta\mathrm{CS}_{n,n+1}$ increases, the variances also increase. The variances are calculated from the probability distribution given by Eq. (5.16) and can be expressed as:

$$\mathrm{var}_{\mathcal{H}_0}\left[w^{\mathrm{CS}}\right] = \mathrm{CS}_n \cdot \Delta x^2 \quad (5.41)$$
$$\mathrm{var}_{\mathcal{H}_1}\left[w^{\mathrm{CS}}\right] = \left(\mathrm{CS}_n + \Delta\mathrm{CS}_{n,n+1}\right) \cdot \Delta x^2. \quad (5.42)$$

Since the width of the distribution is proportional to $\sqrt{\mathrm{CS}_n}$ and $\sqrt{\mathrm{CS}_n + \Delta\mathrm{CS}_{n,n+1}}$ for $\mathcal{H}_0$ and $\mathcal{H}_1$ respectively, the probability of error will become smaller for increasing $\Delta\mathrm{CS}_{n,n+1}$, that is when the difference in scattering cross-section between two atomic columns increases.

5.4.3 Simulation experiments

In this and the next subsections, the results of a realistic simulation study will be presented. This study has been performed in order to investigate the probability of error defined in Section 5.2 to evaluate and optimize the inner and outer detector radii of an annular STEM detector in terms of quantitative atom counting. Simulations of STEM images are performed for a 30 nm thick $SrTiO_3$ crystal, i.e. a thickness of 75 atoms, using the STEMsim software (Rosenauer & Schowalter, 2008), for an aberration-corrected microscope at Scherzer defocus. The multislice approach is used in which thermal diffuse scattering is included as an absorptive potential. The simulation parameters which are used are listed in Table 5.1. The values for the Debye-Waller factors are taken from Abramov et al. (1995); Peng (1997).

5.4.4 Results from binary hypothesis test

In Fig. 5.7, the results for the probability of error from a binary hypothesis test with equal prior probabilities, defined by Eq. (5.2) are shown as a function of the number of atoms $n_{\mathcal{H}_0}$ in a Sr column for the three proposed measures and for two different incident electron doses. The dose only changes the value of the probability of error: for higher electron dose, the

Table 5.1 Parameter values used in the simulation software STEMsim (Rosenauer & Schowalter, 2008). Reprinted from De Backer, De wael, et al. (2015), Copyright (2015), with permission from Elsevier.

Parameter	Symbol	Value
Acceleration voltage	V (kV)	300
Defocus	ε (Å)	−14.03
Spherical aberration	C_s (mm)	0.001
Spherical aberration of fifth order	C_5 (mm)	0
Objective aperture angle	α (mrad)	20
Spatial incoherence of source	FWHM (Å)	0.7
Size of the supercell	$N_a \times N_b$ (nm^2)	4.3 × 4.3
Slice thickness	z_{slice} (Å)	1.95
Debye-Waller factor Sr	B (Å^2)	0.6214
Debye-Waller factor Ti	B (Å^2)	0.4398
Debye-Waller factor O	B (Å^2)	0.7323
Pixel size	Δx (Å)	0.1562
Total number of scanned pixels	$K \times L$	25 × 25

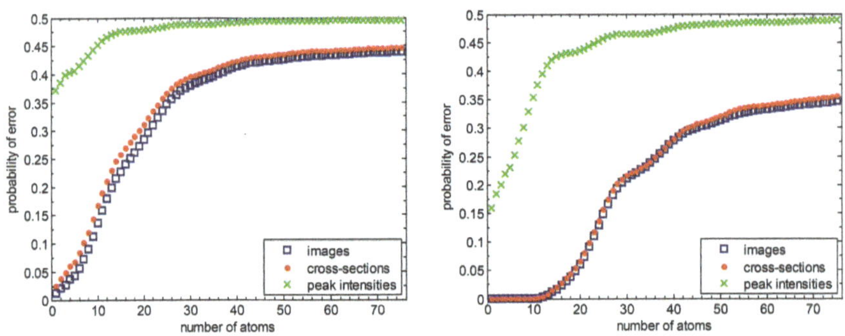

(a) $D = 10^4$ e$^-$/Å^2 (b) $D = 10^5$ e$^-$/Å^2

Figure 5.7 Probability of error as a function of the number of atoms in a Sr column with (a) incident electron dose $D = 10^4$ e$^-$/Å^2, and (b) $D = 10^5$ e$^-$/Å^2, for a detector collection range of 60–100 mrad using a binary hypothesis test. Reprinted from De Backer, De wael, et al. (2015), Copyright (2015), with permission from Elsevier.

probability of error becomes lower. It can be seen that the probability of error increases for an increasing number of atoms in a similar manner as for the linear imaging model of which the results are shown in Fig. 5.4(b). This means that it is easier to differentiate between 1 and 2 atoms than to differentiate between 75 and 76 atoms in a column. Another important conclusion that can be drawn from this figure for the realistic simulations

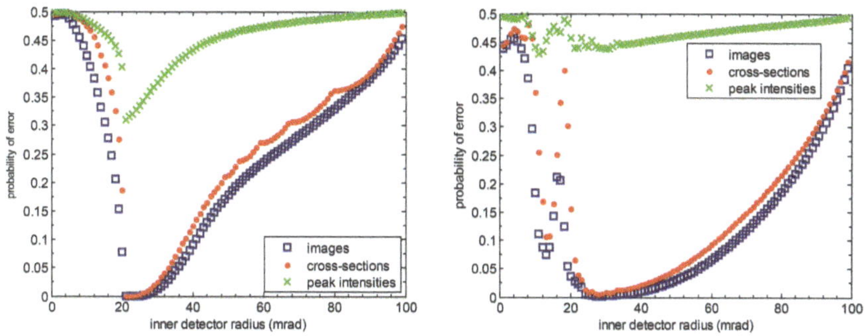

(a) $n_{\mathcal{H}_0} = 1$ and $D = 10^3$ e$^-$/Å^2 (b) $n_{\mathcal{H}_0} = 75$ and $D = 10^6$ e$^-$/Å^2

Figure 5.8 Probability of error for a Sr column as a function of the inner detector angle with (a) $n_{\mathcal{H}_0} = 1$ and an incident electron dose $D = 10^4$ e$^-$/Å^2, and (b) $n_{\mathcal{H}_0} = 75$ and $D = 10^6$ e$^-$/Å^2, with a fixed outer detector radius (100 mrad) using a binary hypothesis test. Reprinted from De Backer, De wael, et al. (2015), Copyright (2015), with permission from Elsevier.

concerns the different results for the three measures which are considered. For the linear imaging model discussed in Subsection 5.4.1, the scattering cross-sections have the same probability of error as the STEM images. Also for the detailed image simulation analyzed here, the probability of error for the scattering cross-sections (red dots) using Eq. (5.18) almost equals the probability of error for the STEM image (blue squares) for which Eq. (5.27) is used, whereas the probability of error for the peak intensities (green crosses), computed by using Eq. (5.9) is significantly larger. This means that the scattering cross-sections contain almost the same amount of information as the images themselves in terms of atom counting, provided that the scattering cross-sections monotonically increase with number of atoms. Thus, the detailed profiles of the atomic columns do not add extra information for atom counting. These differences can only be detected when using the images, since the scattering cross-sections sum all the pixel values in a Voronoi cell in the image into one number defined by Eq. (5.11). This result is very beneficial since the scattering cross-sections are now often used as a measure to analyze the number of atoms in an atomic column and, in addition, the scattering cross-sections are far more robust to compare with simulations. Scattering cross-sections are independent of the full width at half maximum (FWHM) of the source size, the defocus and other parameters (E et al., 2013; Martinez, De Backer, et al., 2014).

In Fig. 5.8, the probability of error is shown as a function of the inner detector radius with a fixed outer detector radius of 100 mrad. The incident dose is chosen such that the values for the probability of error cover the whole range of values between 0 and 0.5. The optimal inner detector angle is not affected by the selected electron dose. In Fig. 5.8(a), results for P_e are shown for $n_{\mathcal{H}_0} = 1$ for the three different measures for a Sr column. From this, it can be concluded that the optimal inner detector radius equals 21 mrad since the probability of error reaches a minimum here, suggesting that imaging in the low-angle ADF (LAADF) STEM regime is optimal for atom counting, since the objective aperture angle equals 20 mrad. However, when choosing $n_{\mathcal{H}_0} = 75$, the optimal inner detector radius increases to 28 mrad, as can be seen in Fig. 5.8(b). For this value of $n_{\mathcal{H}_0} = 75$, an inner detector radius of 21 mrad can definitely not be considered as an optimum, whereas for $n_{\mathcal{H}_0} = 1$ an inner detector angle of 28 mrad is near-optimal for atom counting. Because the optimal detector design depends on the choice of the hypotheses, we should move on toward multiple hypothesis testing as already suggested in Section 5.2.

5.4.5 Results from multiple hypothesis test

When using a multiple hypothesis test, all possible numbers of atoms up to a certain thickness of the atomic column are considered. Since it is demonstrated in Subsection 5.4.4 that for atom counting the scattering cross-sections contain almost the same amount of information as the images, we proceed here with the scattering cross-sections, which are used in practice, as shown in the previous chapters. It can also be shown that, in the case of a multiple hypothesis test, the general optimal detector design does not drastically change when using STEM images. Using Eq. (5.38), it is possible to determine the optimal detector design for the STEM detector in terms of inner and outer detector radii. Therefore, the probability of error is calculated for the different atomic columns in the $SrTiO_3$ crystal, i.e. the Sr, TiO, and O columns, for a multiple hypothesis test with $M = 15$ hypotheses, i.e. for a thickness up to 15 atoms, and for a test with 75 hypotheses, i.e. for a thickness up to $M = 75$ atoms and assuming equal prior probabilities for the different hypotheses. Since the TiO column exists of an equal number of Ti and O atoms, only counting of one type of atoms is considered. The results are shown in Fig. 5.9 for the scattering cross-sections. The optimal detector angles do not critically depend on the atom type. For a thickness of 15 atoms, the optimal detector design is LAADF STEM with an inner detector radius of 21 mrad and the maximal outer detector radius of 100 mrad considered in the simulations. The optimal inner detector angle changes for a thickness of 75 atoms. Here, a minimum

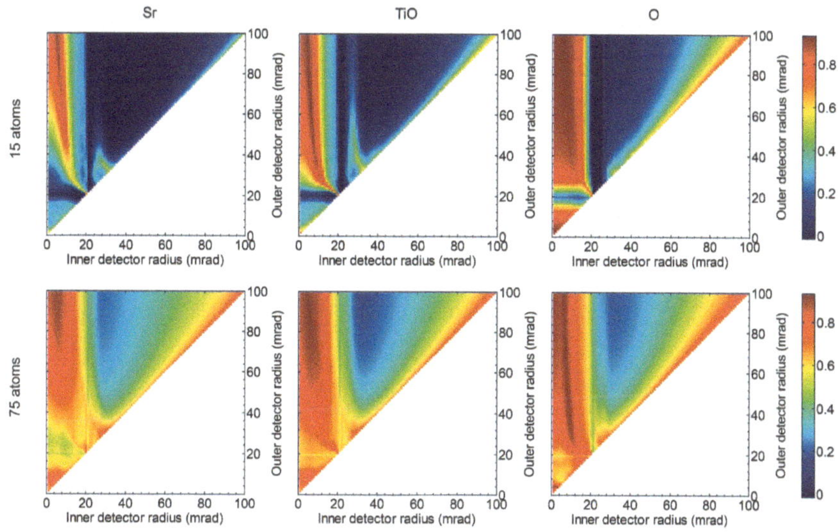

Figure 5.9 Probability of error for scattering cross-sections as a function of inner and outer detector angles for the three different atom columns in $SrTiO_3$ for an incoming electron dose $D = 10^5$ $e^-/Å^2$ using a multiple hypothesis test. The upper and lower row show the results when the goal is to find the optimal settings when counting up to $M = 15$ and $M = 75$ atoms, respectively. Reprinted from De Backer, De wael, et al. (2015), Copyright (2015), with permission from Elsevier.

for the probability of error for a Sr column is found at an inner detector angle of 28 mrad, for a TiO column at 33 mrad and for an O column at 30 mrad. Ideally, the design should be optimized for counting every atom type. This can be done by minimizing e.g. an averaged probability of error for the three atomic column types.

The actual number of the probability of error not only depends on the choice of the detector but also on the incident electron dose. Once the optimal detector design has been derived, one can then be interested to investigate the lowest possible incident electron dose for which the probability of error reaches an acceptable low value. Therefore, the probability of error is computed as a function of the incident electron dose for scattering cross-sections, for the three different atomic columns present in $SrTiO_3$. The results are shown in Fig. 5.10. It is clear from this figure, as one could expect, that the probability of error decreases for an increasing electron dose. The same trend was observed for the linear imaging model as shown in Fig. 5.5(b). If a maximum probability of error of 20% would be taken as acceptable, then an electron dose of 10^5 $e^-/Å^2$ would be necessary. This

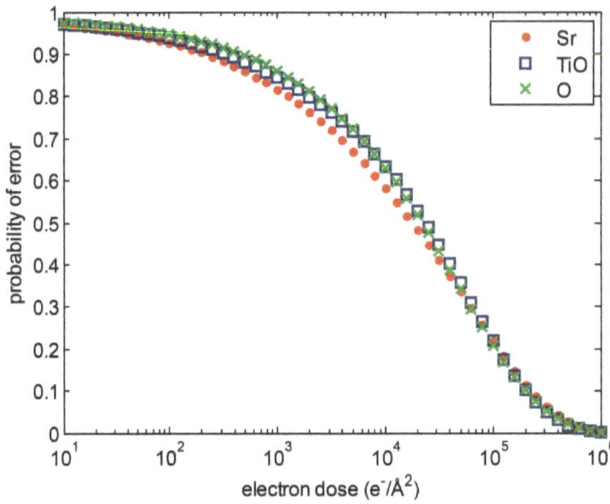

Figure 5.10 Probability of error as a function of incident electron dose for detector angles 30–100 mrad and thickness of $M = 75$ atoms using a multiple hypothesis test. Reprinted from De Backer, De wael, et al. (2015), Copyright (2015), with permission from Elsevier.

can be of great importance if one wants to reduce beam damage, but at the same time still obtain an acceptable precision to count atoms.

5.5. Discussion

In order to understand the different obtained optimal designs for the different thicknesses shown in Figs. 5.8 and 5.9, the values of the integrated scattering cross-sections are investigated for three different detector designs for a Sr column. In Fig. 5.11(a), the normalized integrated scattering cross-sections are shown as a function of thickness, i.e. the number of atoms in a Sr column, for three different detector collection ranges. The red dots show the behavior of the scattering cross-sections for a collection range of 21–100 mrad, the blue squares give the result for a detector collection range of 30–100 mrad, and the green crosses show the result for a detector ranging from 60–100 mrad. There is a clear difference in the results for the three detector settings, from which the optimal design shown in Fig. 5.9 can be understood. Up to a thickness of 20 atoms in a Sr column, there is a monotone increase in the values of the integrated scattering cross-sections. However, it is clear from the image that for a low inner detector angle of 21 mrad, i.e. LAADF STEM, this increase is much steeper, which means that for this detector setting $\Delta CS_{n,n+1}$ from Eq. (5.40) is much larger than

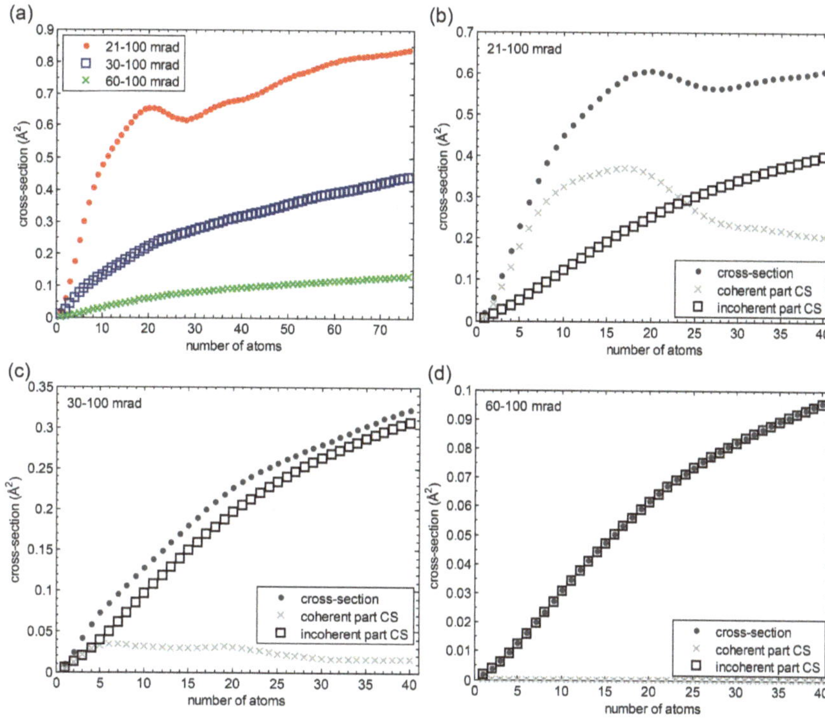

Figure 5.11 (a) Integrated scattering cross-sections (fraction of incident electron dose) as a function of the number of atoms in a Sr column. The coherent and incoherent contribution to the scattering cross-sections are shown for (b) 21–100 mrad, (c) 30–100 mrad, and (d) 60–100 mrad, respectively. Reprinted from De Backer, De wael, et al. (2015), Copyright (2015), with permission from Elsevier.

for the other detector setting shown. For this reason, the lower inner collection angle of 21 mrad turns out to be the optimal detector design for a Sr column of 15 atoms thick as can be seen from the upper left figure of Fig. 5.9. For Sr columns thicker than 20 atoms, however, there appears to be a different behavior in the scattering cross-sections for a detector collection range with a low inner detector angle of 21 mrad. The monotonic increase is interrupted by a sudden minimum, whereas for a higher inner detector angle of 30 mrad and 60 mrad the scattering cross-sections still increase as a function of thickness. This result shows us why for a thickness of 75 atoms in a Sr column, the detector collection range with a higher inner collection angle of 30 mrad becomes beneficial. Because the increase in scattering cross-sections for an inner detector angle of 30 mrad is steeper compared to an inner detector angle of 60 mrad, it is clear that this latter will lead to a higher probability of error and therefore does not result into the optimal design, as can be seen from the bottom left image in Fig. 5.9. In a similar manner, the obtained optimal detector designs for a TiO and pure O column given in Fig. 5.9 can be explained.

In order to understand why scattering cross-sections in the LAADF STEM regime no longer increase monotonically with the number of atoms as compared to the medium-angle ADF (MAADF) and HAADF regime, the coherent and incoherent contributions to the scattering cross-sections under different combinations of inner and outer detector angles are considered separately. The simulation software STEMsim allows one to save the coherent and incoherent contributions separately. Following Ishizuka (2002), both parts are described by different potentials which are normally summed up in order to obtain the complete signal. In Figs. 5.11(b)-(d), the coherent and incoherent contributions to the scattering cross-sections are separated for the different detector collection ranges of 21–100 mrad, 30–100 mrad, and 60–100 mrad, respectively. As can be seen from Fig. 5.11(b), the coherent contribution to the scattering cross-section causes the nonmonotonic increase. The relative contribution of the coherent part to the scattering cross-section decreases when increasing the inner detector angle as can be seen from Figs. 5.11(c) and (d). This analysis shows that the optimal detector design can be understood from the values of the scattering cross-sections, i.e. a combination of the coherent and incoherent contributions, as a function of the number of atoms in a projected atomic column.

5.6. Conclusions

In this chapter, the principles of detection theory have been used for quantifying the probability of error for atom counting from high-resolution STEM images (De Backer, De wael, et al., 2015). Therefore, binary hypothesis testing as well as multiple hypothesis testing has been worked out for atom counting. In this manner, the limits to the precision with which the number of atoms in a projected atomic column can be estimated are studied. The here-described method is put forward as a powerful tool that can be used to optimize the design of an experiment by varying experimental parameters, such as the detector collection range, according to the experimental possibilities. Furthermore, the exact optimal experiment design will depend on the material under study. Nevertheless, the conclusions from the study conducted in this chapter give some general guidelines on optimal experiment design for atom counting.

In this chapter, it is pointed out that the use of scattering cross-sections is afforded in the quantitative analysis of high-resolution STEM images when the goal is to count the number of atoms as precisely as possible. Scattering cross-sections perform almost equally well as compared to detailed STEM images of a projected atomic column and they outperform peak intensities. Furthermore, it has been shown that the probability of error increases for increasing number of atoms in the column. In other words, it becomes more difficult to count the number of atoms in a projected atomic column if the thickness increases. One could also use the proposed method of hypothesis testing to determine the minimally required electron dose in order to attain a pre-specified precision. In this manner, beam damage can be kept to a minimum. Finally, it is shown that the optimal inner detector angle increases for increasing thickness when deriving the optimal STEM detector design for atom counting.

Acknowledgments

This project has received funding from the European Research Council (ERC) under the European Union's Horizon 2020 research and innovation programme (Grant Agreement No. 770887 and No. 823717 ESTEEM3). The authors acknowledge financial support from the Research Foundation Flanders (FWO, Belgium) through Projects No. G.0502.18N, G.0267.18N, G.0368.15N, and a postdoctoral grant to A. D. B. The authors acknowledge funding from the University of Antwerp through a TOP project.

Appendix 5.A. Analytical expressions for the linear imaging model

Peak intensities

For the linear imaging model defined by Eq. (5.39), the expected value for the peak intensities $\lambda_{\mathcal{H}_i}^{\mathrm{PI}}$ can be reformulated as follows:

$$\mathbb{E}_{\mathcal{H}_i}\left[w^{\mathrm{PI}}\right] = \lambda_{\mathcal{H}_i}^{\mathrm{PI}} = N_p n_{\mathcal{H}_i} \Delta I^{\mathrm{PI}} \tag{5.A.1}$$

where N_p denotes the incident electron dose per pixel and ΔI^{PI} the pixel value at the position of the atomic column of the simulated Gauss corresponding to a single atom, i.e. the increase in peak intensity for a single atom. The sum of Kullback-Leibler divergences given by Eq. (5.31) then equals:

$$\begin{aligned} &D\left(p_{\mathcal{H}_1}, p_{\mathcal{H}_0}\right) + D\left(p_{\mathcal{H}_0}, p_{\mathcal{H}_1}\right) \\ &= \left(N_p n_{\mathcal{H}_1} \Delta I^{\mathrm{PI}} - N_p n_{\mathcal{H}_0} \Delta I^{\mathrm{PI}}\right)\left(\log\left(N_p n_{\mathcal{H}_1} \Delta I^{\mathrm{PI}}\right) - \log\left(N_p n_{\mathcal{H}_0} \Delta I^{\mathrm{PI}}\right)\right) \\ &= N_p \Delta I^{\mathrm{PI}}\left(n_{\mathcal{H}_1} - n_{\mathcal{H}_0}\right)\log\left(\frac{n_{\mathcal{H}_1}}{n_{\mathcal{H}_0}}\right). \end{aligned} \tag{5.A.2}$$

For the detection of the difference of a single atom, the expression for the sum of Kullback-Leibler divergences reduces to:

$$D\left(p_{\mathcal{H}_1}, p_{\mathcal{H}_0}\right) + D\left(p_{\mathcal{H}_0}, p_{\mathcal{H}_1}\right) = N_p \Delta I^{\mathrm{PI}} \log\left(\frac{n+1}{n}\right). \tag{5.A.3}$$

This expression linearly depends on the incident electron dose per pixel N_p and the intensity increase for a single atom ΔI^{PI} at the atomic column position. The intensity increase for a single atom ΔI^{PI} at the atomic column position for the linear imaging model as described in Eq. (5.A.1) depends on the atomic type Z and the STEM imaging conditions including the probe settings and detector angles. The dependence of the number of atoms is determined by $\log(n+1/n)$. This function decreases for an increasing number of atoms n.

The probability of error for the peak intensities is defined by Eq. (5.9) and takes also into account the width of the probability distribution. The expected values, and variances, of the peak intensities for the linear model, having a conditional probability function given by Eq. (5.7), equals Eq. (5.A.1). The intersection x^{PI} between both probability distributions given by Eq. (5.10) for the linear model equals:

$$x^{\mathrm{PI}} = N_p \Delta I^{\mathrm{PI}} \frac{n_{\mathcal{H}_0} - n_{\mathcal{H}_1}}{\log \frac{n_{\mathcal{H}_0}}{n_{\mathcal{H}_1}}} = \frac{N_p \Delta I^{\mathrm{PI}}}{\log \frac{n+1}{n}}. \tag{5.A.4}$$

Scattering cross-sections

Similar to the derivation of the sum of Kullback-Leibler divergences and the probability of error for the peak intensities, these measures can be defined for the scattering cross-sections. The expected scattering cross-section for the linear imaging model following Eqs. (5.12) and (5.13) is given by:

$$\mathbb{E}_{\mathcal{H}_0}\left[w^{\mathrm{CS}}\right] = \mathrm{CS}_n = \sum_{k=1}^{L}\sum_{l=1}^{L} N_p n_{\mathcal{H}_0} \Delta I_{kl} \Delta x^2 = N_p n_{\mathcal{H}_0} \Delta \mathrm{CS} \tag{5.A.5}$$

$$\mathbb{E}_{\mathcal{H}_1}\left[w^{\mathrm{CS}}\right] = \mathrm{CS}_{n+1} = \sum_{k=1}^{L}\sum_{l=1}^{L} N_p n_{\mathcal{H}_1} \Delta I_{kl} \Delta x^2 = N_p n_{\mathcal{H}_1} \Delta \mathrm{CS} \tag{5.A.6}$$

where ΔI_{kl} is the increase in intensity relative to the incident electron beam of the STEM image at pixel (k, l) for a single atom and $\Delta\mathrm{CS}$ corresponds to the normalized scattering cross-section for a single atom, i.e. the increase in scattering cross-section for a single atom. The sum of Kullback-Leibler divergences for the scattering cross-sections using the linear imaging model is given by:

$$\begin{aligned} & D\left(p_{\mathcal{H}_1}, p_{\mathcal{H}_0}\right) + D\left(p_{\mathcal{H}_0}, p_{\mathcal{H}_1}\right) \\ & \quad = \left(N_p n_{\mathcal{H}_1} \Delta\mathrm{CS} - N_p n_{\mathcal{H}_0} \Delta\mathrm{CS}\right)\left(\log\left(N_p n_{\mathcal{H}_1} \Delta\mathrm{CS}\right) - \log\left(N_p n_{\mathcal{H}_0} \Delta\mathrm{CS}\right)\right) \\ & \quad = N_p \Delta\mathrm{CS}\left(n_{\mathcal{H}_1} - n_{\mathcal{H}_0}\right)\log\left(\frac{n_{\mathcal{H}_1}}{n_{\mathcal{H}_1}}\right). \end{aligned} \tag{5.A.7}$$

Also here, this expression can be rewritten for the detection of the difference of a single atom:

$$D\left(p_{\mathcal{H}_1}, p_{\mathcal{H}_0}\right) + D\left(p_{\mathcal{H}_0}, p_{\mathcal{H}_1}\right) = N_p \Delta\mathrm{CS} \log\left(\frac{n+1}{n}\right). \tag{5.A.8}$$

This expression is similar to Eq. (5.A.3), but is now linearly dependent of the increase in scattering cross-section for a single atom $\Delta\mathrm{CS}$ instead of the image intensity increase for a single atom at the atomic column positions. The increase in scattering cross-section for a single atom $\Delta\mathrm{CS}$ for the linear imaging model, as described in Eq. (5.A.6), also depends on the atomic type Z and the STEM imaging conditions such as the probe settings and detector angles.

The probability of error can be derived for the linear imaging model using Eq. (5.18). The expected value for the scattering cross-sections equals Eq. (5.A.6), where the scattering cross-sections have a probability function given by Eq. (5.16). The intersection between the two probability functions under $\mathcal{H}_0$ and $\mathcal{H}_1$ given by Eq. (5.19) is then given by:

$$x^{CS} = N_p \Delta\mathrm{CS} \frac{n_{\mathcal{H}_0} - n_{\mathcal{H}_1}}{\log \frac{n_{\mathcal{H}_0}}{n_{\mathcal{H}_1}}} = \frac{N_p \Delta\mathrm{CS}}{\log \frac{n+1}{n}}. \tag{5.A.9}$$

STEM image

Finally, the sum of Kullback-Leibler divergences and probability of error for the linear imaging model can be defined for STEM images. Here, the expected values at the pixel intensities (k, l) of the STEM image for the linear imaging model correspond to:

$$\mathbb{E}_{\mathcal{H}_i}[w_{kl}] = \lambda_{\mathcal{H}_i,kl} = N_p n_{\mathcal{H}_i} \Delta I_{kl}. \tag{5.A.10}$$

The sum of Kullback-Leibler divergences for the linear imaging model for the STEM images equals:

$$\begin{aligned} &D\left(p_{\mathcal{H}_1}, p_{\mathcal{H}_0}\right) + D\left(p_{\mathcal{H}_0}, p_{\mathcal{H}_1}\right) \\ &= \sum_{k=1}^{K}\sum_{l=1}^{L}\left(N_p n_{\mathcal{H}_1}\Delta I_{kl} - N_p n_{\mathcal{H}_0}\Delta I_{kl}\right)\left(\log\left(N_p n_{\mathcal{H}_1}\Delta I_{kl}\right) - \log\left(N_p n_{\mathcal{H}_0}\Delta I_{kl}\right)\right) \\ &= \sum_{k=1}^{K}\sum_{l=1}^{L} N_p \Delta I_{kl}\left(n_{\mathcal{H}_1} - n_{\mathcal{H}_0}\right)\log\left(\frac{n_{\mathcal{H}_1}}{n_{\mathcal{H}_0}}\right) \\ &= N_p \frac{\Delta CS}{\Delta x^2}\left(n_{\mathcal{H}_1} - n_{\mathcal{H}_0}\right)\log\left(\frac{n_{\mathcal{H}_1}}{n_{\mathcal{H}_0}}\right). \end{aligned} \tag{5.A.11}$$

In the last line, the increase in scattering cross-section for a single atom as defined in Eq. (5.A.6) is introduced. Like for the peak intensities and the scattering cross-sections, also here the expression for the sum of Kullback-Leibler divergences is rewritten for detecting the difference of a single atom:

$$D\left(p_{\mathcal{H}_1}, p_{\mathcal{H}_0}\right) + D\left(p_{\mathcal{H}_0}, p_{\mathcal{H}_1}\right) = N_p \frac{\Delta CS}{\Delta x^2}\log\left(\frac{n+1}{n}\right). \tag{5.A.12}$$

The sum of Kullback-Leibler divergences for the STEM images depends on the scattering cross-sections and the magnification via the pixel size Δx.

The probability of error for the STEM images is given by Eq. (5.27). Whereas the sum of Kullback-Leibler divergences quantifies the distance between the two expected log-likelihood ratios, the probability of error quantifies the overlapping area between the distributions of the two log-likelihood ratios. The mean and variance of the normally distributed log-likelihood ratio are given by Eqs. (5.25) and (5.26):

$$\begin{aligned} \mu_{\mathcal{H}_i} &= \sum_{k=1}^{K}\sum_{l=1}^{L}\left(\lambda_{\mathcal{H}_i,kl}\log\frac{\lambda_{\mathcal{H}_1,kl}}{\lambda_{\mathcal{H}_0,kl}} - \lambda_{\mathcal{H}_1,kl} + \lambda_{\mathcal{H}_0,kl}\right) \\ &= N_p n_{\mathcal{H}_i}\frac{\Delta CS}{\Delta x^2}\log\frac{n+1}{n} - \frac{N_p \Delta CS}{\Delta x^2} \end{aligned} \tag{5.A.13}$$

$$\sigma^2_{\mathcal{H}_i} = \sum_{k=1}^{K} \sum_{l=1}^{L} \lambda_{\mathcal{H}_i,kl} \left(\log \frac{\lambda_{\mathcal{H}_1,kl}}{\lambda_{\mathcal{H}_0,kl}} \right)^2$$
$$= N_p n_{\mathcal{H}_i} \frac{\Delta \mathrm{CS}}{\Delta x^2} \left(\log \frac{n+1}{n} \right)^2 . \tag{5.A.14}$$

The difference of the expected values of the log-likelihood ratio of Eq. (5.A.13) again corresponds to the sum of Kullback-Leibler divergences as derived in Eq. (5.A.12).

CHAPTER SIX

Atom column detection

Jarmo Fatermans[a,b,c], Annick De Backer[a,b], Arnold J. den Dekker[c], and Sandra Van Aert[a,b,*]

[a]Electron Microscopy for Materials Science (EMAT), University of Antwerp, Antwerp, Belgium
[b]NANOlab Center of Excellence, University of Antwerp, Antwerp, Belgium
[c]imec-Vision Lab, University of Antwerp, Wilrijk, Belgium
*Corresponding author. e-mail address: sandra.vanaert@uantwerpen.be

Contents

6.1. Introduction

Reliable detection of atoms is of key importance to solve a wide range of scientific and technological problems. To reach this goal, electron microscopy is one of the most promising techniques because of the strong interaction of electrons with matter. In particular, because of several important developments in aberration-correction technology (Hawkes, 2015; Krivanek et al., 2009; Rose, 2009), scanning transmission electron microscopy (STEM) has become an excellent technique to visualize nanomaterials down to sub-angstrom resolution (Erni et al., 2009; Krivanek et al., 2010; Oltalan et al., 2010; Sawada et al., 2009; Urban, 2008), enabling

Advances in Imaging and Electron Physics, Volume 217
ISSN 1076-5670
https://doi.org/10.1016/bs.aiep.2021.01.006

the detection of even single atoms, and hereby solving challenging problems in materials science, such as detecting Au catalysts in a Si nanowire, reconstructing oxygen octahedrons in heterointerfaces, or directly observing dopant atoms (Allen et al., 2008; Hwang et al., 2013; Ishikawa, Mishra, et al., 2014; Jia, Mi, Faley, et al., 2009).

Detection of an atomic column or single atom from STEM images is often performed by means of a visual interpretation of the image. Materials that are stable under the electron beam can be imaged with a sufficiently high electron dose, typically of the order of more than 10^6–10^8 $e^-/Å^2$, resulting into high-quality STEM images exhibiting high signal-to-noise ratio (SNR) and high contrast. This allows one to resolve the individual atomic columns and therefore, for such beam-stable materials, the atomic columns can be detected in a merely visual manner. However, some materials, especially light-element nanomaterials, are sensitive to the electron beam and, as a result, they cannot withstand a high incoming electron dose. In order to avoid beam damage to occur, such beam-sensitive materials can only be imaged in STEM using a limited electron dose of the order of 10^3–10^5 $e^-/Å^2$. This leads to images exhibiting low SNR and weak contrast, and hence low contrast-to-noise ratio (CNR) (Findlay et al., 2014). Resolving the individual atomic columns from these images by visual inspection, can be unreliable and may lead to biased results, especially when no prior knowledge about the structure of the material is available. To overcome this problem, an objective and quantitative approach is needed.

A possible solution is offered by Bayes' theorem, which enables the calculation of the probability of a certain event in terms of conditions that can be related to that event (Jaynes, 2003; Sivia & Skilling, 2006; Stuart & Ord, 1994). In principle, this allows one to compare the probability of a certain number of atomic columns present in the image data to the probability of any other number of columns. In this way, the most probable result from the data under investigation can be selected given some prior knowledge. The idea of using Bayesian probability theory, aiming to obtain relevant information from image data, has been applied in the technique of maximum entropy reconstruction, where use is made of an entropy prior (Gull & Skilling, 1984; Jaynes, 1957; Skilling & Bryan, 1984). The maximum entropy method has been useful for image processing in several physical domains including radio astronomy and X-ray diffraction (Gull & Daniell, 1978). In electron microscopy, maximum entropy reconstruction has been used as a deconvolution or noise reduction technique where the parameters that need to be estimated correspond to single pixels (McGibbon et al., 1999; Nakanishi et al., 2006, 2002; Nellist & Pennycook, 1998; Sang et al., 2009; Watanabe et al., 2002). The number of unknowns in such a reconstruction, though, can be drastically reduced if knowledge about the projected structure is taken into account. In such a case, a model describing the image pixel values of an electron microscopy image should be constructed as a function of a number of relevant parameters, reducing the number of unknowns. In Chapter 2, it was explained that such parametric models, describing the projected atomic columns of atomic-resolution STEM images as Gaussian peaks superimposed on a constant background (den Dekker et al., 2005; Van Aert et al., 2005), are available due to the fact that image intensities are sharply peaked at the atomic column locations (Nellist, 2007; Van Dyck, 2002). It has been explained in the previous chapters that, by using model-based parameter estimation, accurate and precise estimates can be obtained of the structure parameters of the material under study. A danger in the approach of model-based parameter estimation, though, is the use of an incorrect model or parameterization, because it lacks a mechanism to provide information about the most optimal model describing the experimental data. This may lead to biased results as the correct parameterization may not have been considered in the analysis. Therefore, model-selection techniques, like for example the Integrated Classification Likelihood (ICL) criterion introduced in Chapter 4 for atom counting from STEM images, can be very useful to decide what model should be selected.

Bayesian probability theory can be applied, not only on the level of parameter estimation, like in the technique of maximum entropy reconstruction, but also to the concept of model selection. Bayesian model selection allows for selecting the optimal model from a set of competing models by comparing their probabilities of describing the experimental data. Interestingly, Bayesian selection techniques have been used in a wide variety of applications (Knuth et al., 2015) covering fields such as chemistry (Armstrong & Hibbert, 2009; Bermejo et al., 2004; Holmes, 2014; Shashilov et al., 2010; Vispa et al., 2016, 2017), nuclear and particle physics (Bergström, 2012; De Cruz et al., 2011), and astronomy and astrophysics (Corsaro & De Ridder, 2014; Debono, 2014; Feroz et al., 2011; Pounds & Vaughan, 2006; Trotta, 2008). In electron microscopy, the maximum a posteriori (MAP) probability rule has been developed as an objective and quantitative method to detect atom columns and even single atoms from high-resolution annular dark-field (ADF) STEM images (Fatermans et al., 2018, 2019). The principles of the MAP probability rule are based on earlier work of Sivia and Carlile (1992); Sivia et al. (1992, 1993) in the field of molecular spectroscopy. The method combines statistical parameter estimation theory and model-order selection in a Bayesian framework.

The MAP probability rule for atom column detection is not only applicable to ADF STEM images. Nowadays, there is an increasing interest to perform a simultaneous acquisition of both annular bright-field (ABF) and ADF STEM images by using ABF and ADF detectors operating at the same time in order to visualize atoms of a large range of atomic numbers for studying and interpreting materials at the atomic scale consisting of both light and heavy atoms. Although the ABF imaging mode allows for better visualization of light elements as compared to ADF imaging, it is, in general, not straightforward to visually detect light-element atomic columns. Light-element nanomaterials are typically sensitive to the electron beam, limiting the incoming electron dose and leading to images exhibiting low CNR. As such, the MAP rule is also important to determine the number of atomic columns for which there is most evidence in simultaneously acquired ABF and ADF image data (Fatermans et al., 2020).

This chapter is organized as follows. In Section 6.2, Bayes' theorem will be introduced. In Section 6.3, the methodology of the MAP probability rule to determine the most probable number of atomic columns from ADF STEM images is described in detail. Hereby, approximate analytical expressions are derived in Subsection 6.3.1. In addition, the algorithm of the MAP probability rule is explained in Subsection 6.3.2. In Section 6.4, the MAP probability rule is applied to experimental ADF STEM images of $SrTiO_3$ in Subsection 6.4.1, graphene in Subsection 6.4.2, a gold nanorod in Subsection 6.4.3, and ultrasmall nanoclusters in Subsection 6.4.4. In Section 6.5, analytical expressions for the MAP probability rule for ABF-ADF STEM are derived. This is followed by Section 6.6 where the MAP rule is applied to simultaneously acquired experimental ABF and ADF STEM images of $SrTiO_3$ in Subsection 6.6.1 and $LiMn_2O_4$ in Subsection 6.6.2. Finally, in Section 6.7, conclusions will be drawn.

6.2. Bayes' theorem

In principle, probability theory consists of a number of rules quantifying logical and consistent reasoning to express a belief of truth about a certain proposition (Cox, 1946). Hereby, a probability is assigned to a proposition expressing a certain degree of belief or plausibility that the given proposition is true (D'Agostini, 2003; Gregory, 2005; Jaynes, 2003;

Sivia & Skilling, 2006). Consistency is preserved if the probabilities describing our beliefs obey the rules of probability theory.

An important property of probabilities is that they are bounded below by 0 and above by 1, such that the probability of a proposition X to be true given some background information I can be written as

$$0 \leq p(X|I) \leq 1. \tag{6.1}$$

It is noted that Eq. (6.1) is conditional on the background information I. This is important because assigning a probability to a certain proposition will always depend on the available background information as there is no such thing as an absolute probability. Often, though, the term depicting the background information is omitted from equations in Bayesian theory. More properties of probability theory can be derived by the requirement of logical consistency. When the probability of a certain proposition X to be true is known, this automatically implies that also the probability of X to be false, $\bar{X}$, is known. In combination with Eq. (6.1), this gives rise to the following property:

$$p(X|I) + p(\bar{X}|I) = 1, \tag{6.2}$$

which is referred to as the sum rule. More generally, for a set of K propositions $\boldsymbol{X} = (X_1, X_2, \ldots, X_K)^T$, we expect the probabilities to be normalized, such that

$$\sum_{k=1}^{K} p(X_k|I) = 1. \tag{6.3}$$

Another quantitative statement following from logical consistency is the product rule, defining the joint probability of two or more propositions. Here, when the probability of a proposition X to be true is known and, in addition, the probability of a proposition Y to be true, given the truth of X, is also known, then the probability that both X and Y are true can be calculated. This statement can be formulated by the following expression:

$$p(X, Y|I) = p(Y|X, I)p(X|I). \tag{6.4}$$

Since X and Y are interchangeable, Eq. (6.4) can also be written as

$$p(X, Y|I) = p(X|Y, I)p(Y|I). \tag{6.5}$$

Furthermore, from Eq. (6.4), it follows that

$$p(X, Y|I) + p(X, \bar{Y}|I) = [p(Y|X, I) + p(\bar{Y}|X, I)]p(X|I). \tag{6.6}$$

Since the expression between square brackets at the right-hand side of Eq. (6.6) is equal to unity, following Eq. (6.2), we can write that

$$p(X|I) = p(X, Y|I) + p(X, \bar{Y}|I). \tag{6.7}$$

For a set of L propositions $\boldsymbol{Y} = (Y_1, Y_2, \ldots, Y_L)^T$, Eq. (6.7) can be generalized by making use of the property of Eq. (6.3):

$$p(X|I) = \sum_{l=1}^{L} p(X, Y_l|I). \tag{6.8}$$

By symmetry, we also have that for a set of K propositions $\boldsymbol{X} = (X_1, X_2, \ldots, X_K)^T$

$$p(Y|I) = \sum_{k=1}^{K} p(Y, X_k|I). \tag{6.9}$$

In Eqs. (6.8) and (6.9), a general form of so-called marginalization of joint probabilities for a discrete set of variables has been defined. In case of continuous variables, the summation is replaced by an integration.

The rules presented above form the basic algebra of probability theory as many other results can be derived from them. An important result following from these fundamental rules is Bayes' theorem, offering a way of calculating the probability of a certain proposition in terms of conditions that can be related to that proposition (Jaynes, 2003; Sivia & Skilling, 2006; Stuart & Ord, 1994). By equating the expressions given in Eqs. (6.4) and (6.5), Bayes' theorem can be obtained:

$$p(X|Y, I) = \frac{p(Y|X, I)p(X|I)}{p(Y|I)}. \tag{6.10}$$

Typically, the background information I is explicitly omitted from Eq. (6.10) resulting into a simplified expression:

$$p(X|Y) = \frac{p(Y|X)p(X)}{p(Y)}. \tag{6.11}$$

The relevance of Bayes' theorem for data analysis becomes more apparent when the proposition X is replaced by a quantity of interest Q and the proposition Y by the available data D:

$$p(Q|D) = \frac{p(D|Q)p(Q)}{p(D)}. \tag{6.12}$$

The power of Eq. (6.12) lies in the fact that the probability of the quantity Q given the available data D, $p(Q|D)$, can be calculated from a more accessible term $p(D|Q)$, which depicts the probability of observing the data D given the quantity Q. The various terms in Bayes' theorem have some formal names. The term on the left-hand side of Eq. (6.12) depicts the posterior probability of Q conditional on D and represents the state of knowledge after performing the experiment and obtaining the data. The term $p(D|Q)$ reflects the evidence that the available data D was observed given the quantity Q. Furthermore, the prior information is defined by $p(Q)$ which asserts the plausibility of Q before conducting the experiment. Finally, the denominator of Eq. (6.12) is a normalization constant since it is independent of the quantity of interest Q.

6.3. Maximum a posteriori probability for ADF STEM

In this section, the methodology of the maximum a posteriori (MAP) probability rule to detect single atoms from atomic-resolution STEM images is explained in detail (Fatermans et al., 2019). The method is built upon a combination of parameter estimation and model-order selection using the principles of Bayesian probability theory, as discussed in the previous section. The MAP probability rule allows for determining the structure of unknown nanomaterials in an automatic and objective manner and proves to be especially useful for the analysis of the structure of beam-sensitive and light-element nanomaterials (Fatermans et al., 2018). In Subsection 6.3.1, it is shown that approximate analytical expressions can be derived for the probability of the presence of a certain number of atomic columns in ADF STEM image data. The way of how these probabilities can be obtained and compared to each other is discussed in Subsection 6.3.2.

6.3.1 Bayesian approach

In Chapter 2, commonly used parametric expectation models were discussed to describe STEM images. In Eq. (2.51), the expectation model $f_{kl}(\boldsymbol{\theta})$, describing the expectation of the observed pixel value w_{kl} at position (x_k, y_l), with $\boldsymbol{\theta} = (\theta_1, \ldots, \theta_M)^T$ the vector of unknown structure parameters, was given for the most general case of assuming a different width for each estimated Gaussian peak. For this model, $\boldsymbol{\theta}$ consists of $M = 4N + 1$ parameters. In order to extract reliable structure information of nanomaterials from ADF STEM images using parametric models, such as the one presented by Eq. (2.51), the number of atomic columns N present in the image should be known, as each Gaussian peak in the model should correspond to a column in the image. Usually, for beam-stable materials, this number can be determined visually due to the high incoming electron dose that can be used to image these materials, leading to images exhibiting high SNR. For beam-sensitive nanostructures, though, the incoming electron dose is limited in order to avoid beam damage and, as a result, the images exhibit low SNR and weak contrast, and hence low CNR. Visual inspection of such images may lead to biased results. To overcome this problem, the number of atomic columns N can be reliably quantified by the MAP probability rule. It selects the number of columns N of which the probability given the observed image pixel values $\boldsymbol{w}$, $p(N|\boldsymbol{w})$, is maximal. The probability $p(N|\boldsymbol{w})$ depicts the posterior probability term in Bayes' theorem, given by Eq. (6.12). As such, for our purpose, $p(N|\boldsymbol{w})$ can be written as:

$$p(N|\boldsymbol{w}) = \frac{p(\boldsymbol{w}|N)p(N)}{p(\boldsymbol{w})}. \tag{6.13}$$

The term $p(\boldsymbol{w}|N)$ reflects the evidence that the image data $\boldsymbol{w}$ is generated by N atomic columns. The probability $p(N)$ expresses prior knowledge of the number of atomic columns N in the image, which, in what follows, has been chosen to be a uniform distribution, reflecting no a priori preference for any number of columns. The denominator in Eq. (6.13) can be explicitly written as

$$p(\boldsymbol{w}) = \sum_{N} p(\boldsymbol{w}|N)p(N) \tag{6.14}$$

and is a normalization constant, which is independent of the number of columns N as a summation over N is performed. Therefore, this term cancels out when comparing posterior probabilities as a function of N. As a result, Eq. (6.13) reduces to

$$p(N|\boldsymbol{w}) \propto p(\boldsymbol{w}|N). \tag{6.15}$$

By making use of the marginalization property of probability theory, given by Eq. (6.8), the right-hand side of Eq. (6.15) can be written as

$$p(\boldsymbol{w}|N) = \int p(\boldsymbol{w}, \boldsymbol{\theta}|N) d^M \boldsymbol{\theta}, \tag{6.16}$$

where the marginalized variables are the parameters $\boldsymbol{\theta}$ of the expectation model. Then, by following the product rule of Eq. (6.5), this expression becomes

$$p(\boldsymbol{w}|N) = \int p(\boldsymbol{w}|\boldsymbol{\theta}, N) p(\boldsymbol{\theta}|N) d^M \boldsymbol{\theta}. \tag{6.17}$$

The first term in the integral, $p(\boldsymbol{w}|\boldsymbol{\theta}, N)$, is the likelihood function which describes the probability of the observed image pixel values $\boldsymbol{w}$ for particular values of the parameters $\boldsymbol{\theta}$ of a model with N atomic columns. Therefore, it is an explicit function of the parameters $\boldsymbol{\theta}$. In essence, the likelihood function is a measure of the goodness of fit of the model with the experimental measurements or image pixel values. The other term in the integral, $p(\boldsymbol{\theta}|N)$, is the prior density of the parameters $\boldsymbol{\theta}$ for a model with N columns. In practice, calculation of the posterior probability $p(N|\boldsymbol{w})$, given by Eq. (6.15), can be reduced to calculating the marginal likelihood $p(\boldsymbol{w}|N)$ described by Eq. (6.17). In order to do so, explicit expressions for the likelihood function $p(\boldsymbol{w}|\boldsymbol{\theta}, N)$ and the prior density $p(\boldsymbol{\theta}|N)$ are required.

An expression for the likelihood function, $p(\boldsymbol{w}|\boldsymbol{\theta}, N)$, can be derived by taking into account knowledge about the statistical properties of the errors in the experimental measurements. In essence, this is covered by the joint probability (density) function, introduced in Chapter 2, describing the probability of a set of given observations. Since a STEM image of $K \times L$ pixels is formed by counting electrons scattered to the detector, the pixel values are inevitably subject to Poisson noise causing each observed image pixel value w_{kl} at position (x_k, y_l) to be Poisson distributed (Haight, 1967; Mood et al., 1974; van den Bos & den Dekker, 2001). For an increasing expectation value $f_{kl}(\boldsymbol{\theta})$ of w_{kl}, the Poisson distribution tends to be a normal distribution with mean $\mu_{kl} = f_{kl}(\boldsymbol{\theta})$ and standard deviation $\sigma_{kl} = \sqrt{f_{kl}(\boldsymbol{\theta})}$ (Papoulis & Pillai, 2002). Under the assumption that the pixel values are statistically independent, the likelihood function can be expressed as follows:

$$p(\boldsymbol{w}|\boldsymbol{\theta}, N) = \frac{e^{-\chi^2(\boldsymbol{\theta})/2}}{\prod_{k=1}^{K} \prod_{l=1}^{L} \sqrt{2\pi\sigma_{kl}^2}}, \tag{6.18}$$

where

$$\chi^2(\boldsymbol{\theta}) = \sum_{k=1}^{K}\sum_{l=1}^{L}\left(\frac{w_{kl} - \mu_{kl}}{\sigma_{kl}}\right)^2. \tag{6.19}$$

For simplicity, it can be assumed that $\sigma_{kl} \approx \sqrt{w_{kl}}$, so that σ_{kl} is independent of the parameters $\boldsymbol{\theta}$. The likelihood function then becomes

$$p(\boldsymbol{w}|\boldsymbol{\theta}, N) = \frac{e^{-\chi^2(\boldsymbol{\theta})/2}}{\prod_{k=1}^{K}\prod_{l=1}^{L}\sqrt{2\pi w_{kl}}}, \tag{6.20}$$

where

$$\chi^2(\boldsymbol{\theta}) = \sum_{k=1}^{K}\sum_{l=1}^{L}\frac{\left(w_{kl} - f_{kl}(\boldsymbol{\theta})\right)^2}{w_{kl}} \tag{6.21}$$

is a weighted sum-of-squared-residuals misfit between the data and the parametric model.

For the term $p(\boldsymbol{\theta}|N)$ in Eq. (6.17), different prior density functions can be constructed reflecting different types of prior knowledge. Here, $p(\boldsymbol{\theta}|N)$ is expressed as a product of uniform distributions over a predefined range for each parameter θ_m:

$$p(\boldsymbol{\theta}|N) = \begin{cases} \prod_{m=1}^{M} \frac{1}{\theta_{m_{max}} - \theta_{m_{min}}} & \text{for } m = 1, \ldots, M\text{: } \theta_{m_{min}} \leqslant \theta_m \leqslant \theta_{m_{max}} \\ 0 & \text{otherwise} \end{cases} \tag{6.22}$$

where the subscripts *max* and *min* refer to a predefined maximum and minimum value, respectively. The expression in Eq. (6.22) has been depicted in Fig. 6.1 for a single parameter θ_m. In case $\theta_{m_{min}} \leqslant \theta_m \leqslant \theta_{m_{max}}$ for $m = 1, \ldots, M$, $p(\boldsymbol{\theta}|N)$ for the parameter vector in Eq. (2.52) of the model given by Eq. (2.51) can be explicitly written as:

$$\frac{1}{[(\beta_{x_{max}} - \beta_{x_{min}})(\beta_{y_{max}} - \beta_{y_{min}})(\rho_{max} - \rho_{min})(\eta_{max} - \eta_{min})]^N(\zeta_{max} - \zeta_{min})}, \tag{6.23}$$

assuming that the predefined maximum and minimum values are the same for β_{x_n}, β_{y_n}, ρ_n, and η_n for each n. The choice for this type of prior simplifies the subsequent algebra significantly. More importantly, the amount of prior knowledge that is introduced can be kept minimal by using this uniform prior combined with a conservative choice of the predefined parameter ranges, thus avoiding biased results due to the incorporation of possibly invalid prior knowledge. Moreover, this form of prior can also be used in

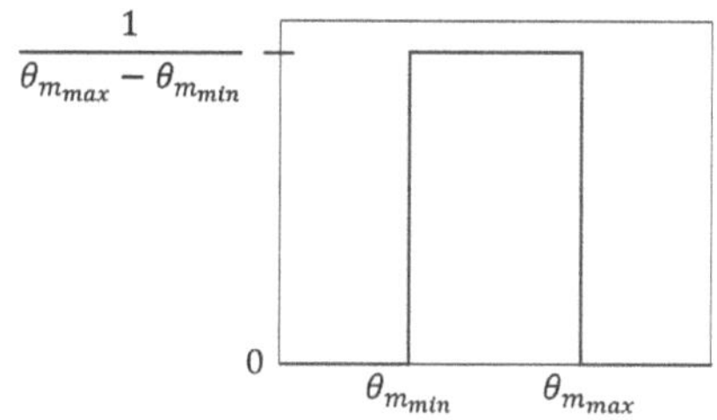

Figure 6.1 Plot of uniform prior distribution $p(\boldsymbol{\theta}|N)$, given by Eq. (6.22), for a single parameter θ_m.

a flexible way since the predefined parameter ranges can easily be adapted depending on the available prior knowledge.

Given the assumption of an a priori equiprobable number of atomic columns N, the evaluation of the MAP probability rule through Eq. (6.15) comes down to calculating the marginal likelihood $p(\boldsymbol{w}|N)$ in Eq. (6.17) and determining the number of columns N with the highest posterior probability $p(N|\boldsymbol{w})$. Approximating the likelihood function $p(\boldsymbol{w}|\boldsymbol{\theta}, N)$ by a normal distribution, given by Eq. (6.20), in combination with the choice of a uniformly distributed prior density function $p(\boldsymbol{\theta}|N)$, given by Eq. (6.22), allows one to derive an approximate analytical expression. Substituting Eq. (6.20) and Eq. (6.22) in Eq. (6.17) results into Eq. (6.15) being written as:

$$p(N|\boldsymbol{w}) \propto N! \cdot \left(\prod_{m=1}^{M} \frac{1}{\theta_{m_{max}} - \theta_{m_{min}}} \right) \int_{\mathbb{D}} \frac{e^{-\chi^2(\boldsymbol{\theta})/2}}{\prod_{k=1}^{K} \prod_{l=1}^{L} \sqrt{2\pi w_{kl}}} d^M\boldsymbol{\theta}, \tag{6.24}$$

where

$$\mathbb{D} = \{(\theta_1, ..., \theta_M) \in \mathbb{R}^M \text{ for } m = 1, ..., M: \theta_{m_{min}} \leqslant \theta_m \leqslant \theta_{m_{max}}\}. \tag{6.25}$$

The factor $N!$ is associated with the arbitrary labeling of the N Gaussian peaks in the model. It reflects the number of ways the indices of the various peaks can be permuted (Sivia & Skilling, 2006). The expression in Eq. (6.24) can be calculated by expanding the likelihood function by a second order Taylor series around the parameter vector $\hat{\boldsymbol{\theta}}$ that maximizes the likelihood function, known as the maximum likelihood estimate:

$$e^{-\chi^2(\boldsymbol{\theta})/2} \approx e^{-\chi^2(\hat{\boldsymbol{\theta}})/2} \times e^{-(\boldsymbol{\theta}-\hat{\boldsymbol{\theta}})^T \left[\frac{\partial \chi^2(\boldsymbol{\theta})}{\partial \boldsymbol{\theta}} \Big|_{\boldsymbol{\theta}=\hat{\boldsymbol{\theta}}} \right] /2} \times e^{-(\boldsymbol{\theta}-\hat{\boldsymbol{\theta}})^T \left[\frac{\partial^2 \chi^2(\boldsymbol{\theta})}{\partial \boldsymbol{\theta} \partial \boldsymbol{\theta}^T} \Big|_{\boldsymbol{\theta}=\hat{\boldsymbol{\theta}}} \right] (\boldsymbol{\theta}-\hat{\boldsymbol{\theta}})/4}. \tag{6.26}$$

Since $\frac{\partial \chi^2(\boldsymbol{\theta})}{\partial \boldsymbol{\theta}}\big|_{\boldsymbol{\theta}=\hat{\boldsymbol{\theta}}} = 0$, as $\hat{\boldsymbol{\theta}}$ minimizes $\chi^2(\boldsymbol{\theta})$, the second term in Eq. (6.26) is equal to one. Then, by writing $\chi^2(\hat{\boldsymbol{\theta}})$ as χ^2_{min} and $\frac{\partial^2 \chi^2(\boldsymbol{\theta})}{\partial \boldsymbol{\theta} \partial \boldsymbol{\theta}^T}\big|_{\boldsymbol{\theta}=\hat{\boldsymbol{\theta}}}$ as $\nabla\nabla\chi^2$, which depicts the Hessian matrix of $\chi(\boldsymbol{\theta})$ at $\hat{\boldsymbol{\theta}}$, Eq. (6.26) becomes:

$$e^{-\chi^2(\boldsymbol{\theta})/2} \approx e^{-\chi^2_{min}/2} \times e^{-(\boldsymbol{\theta}-\hat{\boldsymbol{\theta}})^T \nabla\nabla\chi^2 (\boldsymbol{\theta}-\hat{\boldsymbol{\theta}})/4}, \tag{6.27}$$

which means that Eq. (6.24) is given by

$$p(N|\boldsymbol{w}) \propto N! \cdot e^{-\chi^2_{min}/2} \cdot \left(\prod_{m=1}^{M} \frac{1}{\theta_{m_{max}} - \theta_{m_{min}}} \right) \int_{\mathbb{D}} \frac{e^{-(\boldsymbol{\theta}-\hat{\boldsymbol{\theta}})^T \nabla\nabla\chi^2 (\boldsymbol{\theta}-\hat{\boldsymbol{\theta}})/4}}{\prod_{k=1}^{K} \prod_{l=1}^{L} \sqrt{2\pi w_{kl}}} d^M \boldsymbol{\theta}. \tag{6.28}$$

This expression contains a Gaussian multiple integral which can be solved analytically under the assumptions that (i) the maximum likelihood estimate $\hat{\boldsymbol{\theta}}$ lies well within the support of the prior density function, described by Eq. (6.22), and (ii) the likelihood function has only one significant maximum (Sivia & Carlile, 1992). Then, the integral in Eq. (6.28) is well approximated by an integral over $\mathbb{R}^M$, resulting in the following expression for the posterior probability of the presence of N atomic columns in the image, given the observed image pixel values $\boldsymbol{w}$, for the model described by Eq. (2.51) with $M = 4N + 1$ parameters, including Eq. (6.23):

$$\begin{aligned} p(N|\boldsymbol{w}) \propto & \frac{N!}{[(\beta_{x_{max}} - \beta_{x_{min}})(\beta_{y_{max}} - \beta_{y_{min}})(\rho_{max} - \rho_{min})]^N} \\ & \times \frac{e^{-\chi^2_{min}/2}}{(\eta_{max} - \eta_{min})^N (\zeta_{max} - \zeta_{min})} \times \frac{(4\pi)^{M/2} [det(\nabla\nabla\chi^2)]^{-1/2}}{\prod_{k=1}^{K} \prod_{l=1}^{L} \sqrt{2\pi w_{kl}}}. \end{aligned} \tag{6.29}$$

When the terms which are independent of N are dropped, the expression for $p(N|\boldsymbol{w})$ reduces to

$$p(N|\boldsymbol{w}) \propto \frac{N! (4\pi)^{2N} e^{-\chi^2_{min}/2} [det(\nabla\nabla\chi^2)]^{-1/2}}{[(\beta_{x_{max}} - \beta_{x_{min}})(\beta_{y_{max}} - \beta_{y_{min}})(\rho_{max} - \rho_{min})(\eta_{max} - \eta_{min})]^N}. \tag{6.30}$$

The importance of Eq. (6.30) is that it allows one to compute the posterior probability of a certain number of atomic columns present in an ADF STEM image, for a model where the columns are assumed to be Gaussian shaped and to have different widths. It relies, in particular, on the assumption that $\sigma_{kl} \approx \sqrt{w_{kl}}$, as stated before. By an iterative procedure, better approximations of σ_{kl} can be obtained, but no significant effect on $p(N|\boldsymbol{w})$ could be observed, validating the approximation of $\sigma_{kl} \approx \sqrt{w_{kl}}$. The MAP probability rule compares the calculated posterior probabilities and selects the number of columns N with the highest probability. Similar expressions can be derived for other types of models. For example, for a model where the atomic columns are assumed to be Gaussian shaped and to have equal widths, the expectation model $f_{kl}(\boldsymbol{\theta})$ of pixel (k, l) at position (x_k, y_l) is given by:

$$f_{kl}(\boldsymbol{\theta}) = \zeta + \sum_{n=1}^{N} \eta_n exp\left(- \frac{(x_k - \beta_{x_n})^2 + (y_l - \beta_{y_n})^2}{2\rho^2} \right) \tag{6.31}$$

where the unknown parameter vector can be written as:

$$\boldsymbol{\theta} = (\beta_{x_1}, \ldots, \beta_{x_N}, \beta_{y_1}, \ldots, \beta_{y_N}, \eta_1, \ldots, \eta_N, \rho, \zeta)^T \tag{6.32}$$

containing $M = 3N + 2$ parameters. For such a model, the posterior probability becomes

$$p(N|\boldsymbol{w}) \propto \frac{N!(4\pi)^{1.5N} e^{-\chi^2_{min}/2} [det(\nabla\nabla\chi^2)]^{-1/2}}{[(\beta_{x_{max}} - \beta_{x_{min}})(\beta_{y_{max}} - \beta_{y_{min}})(\eta_{max} - \eta_{min})]^N}. \tag{6.33}$$

As another example, for an expectation model given by

$$f_{kl}(\boldsymbol{\theta}) = \zeta + \sum_{n=1}^{N} \eta exp\left(- \frac{(x_k - \beta_{x_n})^2 + (y_l - \beta_{y_n})^2}{2\rho^2} \right), \tag{6.34}$$

where the columns are assumed to be Gaussian shaped with equal widths and equal heights, with unknown parameter vector

$$\boldsymbol{\theta} = (\beta_{x_1}, \ldots, \beta_{x_N}, \beta_{y_1}, \ldots, \beta_{y_N}, \eta, \rho, \zeta)^T \tag{6.35}$$

containing $M = 2N + 3$ parameters, the posterior probability becomes

$$p(N|\boldsymbol{w}) \propto \frac{N!(4\pi)^{N} e^{-\chi^2_{min}/2} [det(\nabla\nabla\chi^2)]^{-1/2}}{[(\beta_{x_{max}} - \beta_{x_{min}})(\beta_{y_{max}} - \beta_{y_{min}})]^N}. \tag{6.36}$$

The methodology of the MAP probability rule can be extended to models where other shapes of atomic columns are assumed beside a Gaussian function. In these cases, an expression for the posterior probability as a function of the number of atomic columns can be derived in a similar way. Alternative shapes may be Lorentzian functions or a mixture of Lorentzian and Gaussian distributions, which has been revealed to be a better description of the source size distribution (Verbeeck et al., 2012). A Gaussian-Lorentzian mixture possesses a similar full width at half maximum (FWHM) as a Gaussian distribution, but considerably longer tails. It should be noted that the analytical expressions for $p(N|\boldsymbol{w})$ were derived under the assumption that the Poisson distribution that governs the image pixel values can be approximated by a normal distribution. The accuracy of this approximation, and therefore the accuracy of the expressions given by Eqs. (6.30), (6.33) and (6.36), increases with an increasing amount of detected electrons. The expressions will be most accurate if all image pixel values fully satisfy the normality assumption, but have shown to be robust to small violations of this assumption. Therefore, the MAP probability rule is an adequate method allowing to decide the number of atomic columns present in a STEM image.

6.3.2 Algorithm

As stated in the previous subsection, the MAP probability rule selects the most probable number of atomic columns by comparing posterior probabilities as a function of N, i.e. for different numbers of Gaussian peaks in the model describing the atomic columns in the image. The analysis is performed starting from an initial model containing N_0 peaks up to and including a model containing a value of N_{max} peaks. The parameters $\boldsymbol{\theta}$ of the initial model are optimized by minimizing the weighted sum-of-squared-residuals misfit $\chi^2(\boldsymbol{\theta})$, given by Eq. (6.21), subject to the constraint that $\boldsymbol{\theta}$ should belong to the support of the prior density function described by Eq. (6.22). Next, an extra peak is added to the initial configuration, resulting in a model containing $N_0 + 1$ peaks. Again, the parameters of this model are optimized by minimizing $\chi^2(\boldsymbol{\theta})$ subject to the constraint that $\boldsymbol{\theta}$ belongs to the support of the prior density function. To avoid ending up in a local minimum of $\chi^2(\boldsymbol{\theta})$, many different starting positions for the extra added peak need to be tested. To optimize the parameters associated with the other peaks, the estimated parameter values of the previous optimization, in this case of a model with N_0 peaks, are used as starting values. Next, another peak is added, in the same way as described above, in order to obtain the optimal parameter values of a model containing $N_0 + 2$ peaks. This procedure continues until the parameters of a model with N_{max} peaks are optimized. In order to determine the most probable number of atomic columns present in a STEM image, the posterior probability of $N = N_0, \ldots, N_{max}$ columns is computed relatively to the posterior probability of N_{max} columns as follows

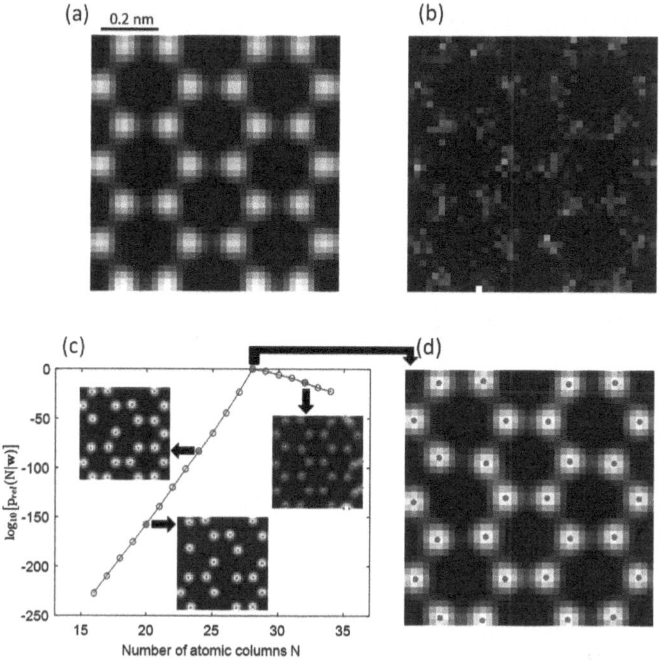

Figure 6.2 (a) Simulated ADF STEM image of graphene. (b) Image in (a) disturbed by Poisson noise with an incoming electron dose of $3 \cdot 10^5$ e$^-$/Å^2. The number of detected electrons equals an average of 276 e$^-$/Å^2. (c) MAP probability rule evaluated for the data shown in (b). Refined models with optimized parameters are shown in the inset taking into account 20, 24, and 32 atomic columns. (d) Most probable parametric model of the data in (b) containing 28 atomic columns as indicated by the MAP probability rule in (c). (Reprinted from Fatermans et al. (2019) with permission from Elsevier.)

$$p_{rel}(N|\boldsymbol{w}) = \frac{p(N|\boldsymbol{w})/p(N_{max}|\boldsymbol{w})}{\max_N \left[p(N|\boldsymbol{w})/p(N_{max}|\boldsymbol{w}) \right]}, \tag{6.37}$$

where the denominator is a normalization constant so that the maximum value of $p_{rel}(N|\boldsymbol{w})$ corresponds to one. The most probable number of atomic columns is then given by the value N that maximizes Eq. (6.37). Direct visualization of the probability of the number of columns in the image is possible by plotting $p_{rel}(N|\boldsymbol{w})$ on a logarithmic scale as a function of N resulting in a relative probability curve. The approach described above has been implemented in the freely available StatSTEM software (De Backer et al., 2016).

The procedure has been illustrated based on a simulated image of graphene, shown in Fig. 6.2(a). The image has been simulated using the

Table 6.1 Microscope parameter values for simulation of a STEM image of graphene using MULTEM. (Reprinted from Fatermans et al. (2019) with permission from Elsevier.)

Parameter	Symbol	Value
Acceleration voltage	V_0 (kV)	80
Defocus	ϵ (Å)	−20.0
Spherical aberration	C_s (mm)	0.0037
Spherical aberration of 5th order	C_5 (mm)	0
Semiconvergence angle	α (mrad)	24.8
Detector inner radius	β_1 (mrad)	26
Detector outer radius	β_2 (mrad)	50
Pixel size	$\Delta x = \Delta y$ (Å)	0.20
Number of scanned pixels	$K \times L$	40×40
FWHM of the source image	FWHM (Å)	0.7

MULTEM software (Lobato et al., 2016; Lobato & Van Dyck, 2015) and the simulation parameters are listed in Table 6.1. The MAP probability rule is applied to detect the carbon atoms of graphene from a Poisson disturbed simulated STEM image shown in Fig. 6.2(b), using a parametric model described by a superposition of Gaussian peaks with equal widths and equal heights given by Eq. (6.34). Such a model is acceptable since the simulated graphene consists of a monolayer of only carbon atoms. Fig. 6.2(c) shows the relative probability curve as a function of the number of columns N calculated by the decimal logarithm of Eq. (6.37). For certain values of N, the corresponding optimized models are shown. The logarithmic scale is necessary in order to plot the relative probabilities over a wide range of values for the number of columns N. It is noted that the relative probability curve, as shown in Fig. 6.2(c), is merely a way of visualizing the relative probabilities. In order to decide the most probable number of atomic columns in an image, one should be aware of the logarithmic scale and should consider the underlying probabilities of the curve. In this manner, the most probable structure found by the MAP probability rule from Fig. 6.2(b) is shown in Fig. 6.2(d), corresponding to the expected hexagonal lattice of graphene, as visually perceived by the noise-free simulated image in Fig. 6.2(a). Interestingly, besides determining the most probable atomic structure from the image data, the MAP probability rule also provides a quantitative statement of how much more probable a certain number of columns is as compared to another number. This information allows to make an optimal decision of the number of atomic columns that are present in the image and with what confidence. For example, from Fig. 6.2(c) it can be derived that the probability of the presence of 28 atomic columns is around 23 orders of magnitude larger than 27 atomic columns and around 72 times larger than 29 atomic columns. This clearly shows that, for the image data of Fig. 6.2(b), the distinction between 27 and 28 carbon atoms is much clearer than the distinction between 28 and 29 atoms. It is noted that it is not possible to provide such a quantitative statement of the probabilities of the number of atomic columns from merely visually interpreting the image.

6.4. Atom column detection from experimental ADF STEM images

The technique of STEM has become a widely used technique to visualize nanomaterials with sub-angstrom resolution due to improvements in aberration-correction technology (Hawkes, 2015; Krivanek et al., 2009; Rose, 2009). Yet, this does not guarantee visually easily interpretable images as for some materials the incoming electron dose should be limited in order to avoid beam damage. In this section, the validity and usefulness of the MAP probability rule is demonstrated to experimental ADF STEM images of samples of different shape, size, and atom type (Fatermans et al., 2018). The examples include $SrTiO_3$ in Subsection 6.4.1, graphene in Subsection 6.4.2, a gold nanorod in Subsection 6.4.3, and ultrasmall nanoclusters in Subsection 6.4.4.

6.4.1 $SrTiO_3$

At room temperature, $SrTiO_3$ adapts the perovskite structure whose general chemical formula is ABX_3, where "A" and "B" are cations of different sizes. In the perovskite structure, $SrTiO_3$ is an attractive material for applications in microelectronics because of its high storage capacity, good insulating properties and excellent optical transparency in the visible region (Higuchi et al., 1998; Kolodiazhnyi & Petric, 2005; Luo et al., 2004). In addition, $SrTiO_3$ has been used as a substrate for epitaxial growth of high temperature superconducting films (Rehn, 1992).

Experimental ADF images of $SrTiO_3$, imaged along the [100] direction, have been recorded by using a probe-corrected FEI Titan, operated at an acceleration voltage of 300 kV. The ADF regime has been selected by using a semiconvergence angle of 21.3 mrad and a detector collection

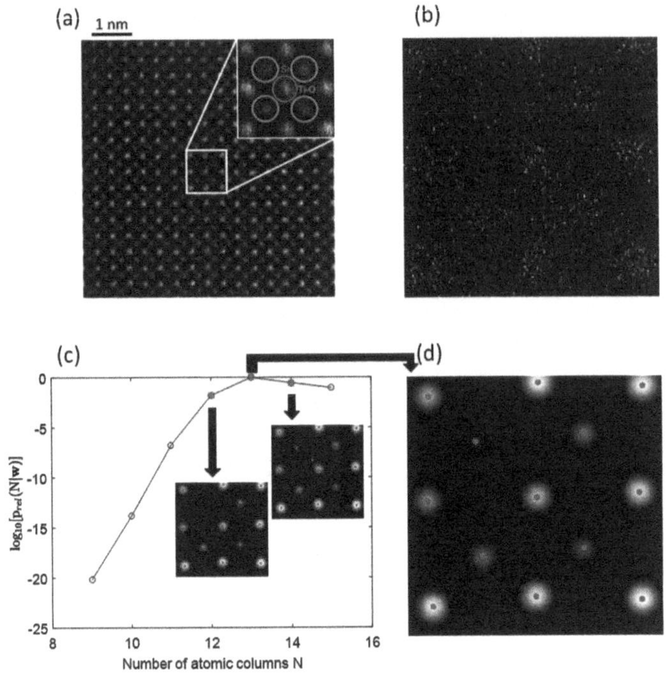

Figure 6.3 (a) Experimental ADF STEM image of $SrTiO_3$ [100] with an incoming electron dose of $(9.1 \pm 0.4) \cdot 10^5$ $e^-/Å^2$. The region indicated by the square has been magnified in the inset where the Sr columns and Ti-O columns are indicated in red and green, respectively. (b) Noisier counterpart of the inset in (a) with an incoming electron dose of $(1.08 \pm 0.05) \cdot 10^4$ $e^-/Å^2$ and an average number of detected electrons of 410 $e^-/Å^2$. (c) MAP probability rule evaluated for the experimental data shown in (b). Refined models with optimized parameters are shown in the inset taking into account 12 and 14 atomic columns. (d) Most probable parametric model of the data in (b) containing 13 atomic columns as indicated by the MAP probability rule in (c). (Reprinted with permission from Fatermans et al. (2018) by the American Physical Society.)

range of 58–197 mrad. Since $SrTiO_3$ is a beam-stable material, it can be imaged using a high incoming electron dose of around 10^6 $e^-/Å^2$. As a result, acquiring an image exhibiting high quality is feasible and may be used as a reference to verify the result obtained from applying the MAP probability rule to an image with low quality. A high CNR image is shown in Fig. 6.3(a) with an incoming electron dose of $(9.1 \pm 0.4) \cdot 10^5$ $e^-/Å^2$. From this image, the brighter Sr columns and darker Ti-O columns are easily recognizable. The inset indicates the locations of both types of columns. Fig. 6.3(b) shows a noisier counterpart of the inset of Fig. 6.3(a) with an approximately 100 times lower electron dose. Hereby, the dose was varied by defocussing the monochromator. The MAP probability rule has been applied to the image data shown in Fig. 6.3(b) in order to verify whether the expected crystal structure of $SrTiO_3$ in [100] direction can be retrieved from such low-quality data. For this, the algorithm described in Subsection 6.3.2 has been used. The evaluation of the MAP probability rule is shown in Fig. 6.3(c) for an increasing number of atomic columns. Since the width of an atomic column is mainly determined by the finite source size and to a lesser extent by the atom type (LeBeau & Stemmer, 2008), a parametric model described by a superposition of Gaussian peaks with equal widths, given by Eq. (6.31), has been used in this analysis. The most time-consuming step is not the optimization of the parameters of a single model, but depends strongly on the number of starting positions that are required for the optimization process to avoid ending up in a local minimum when adding a new peak during the execution of the algorithm. It is noted that computation time can be reduced by taking into account knowledge about the expected locations of atomic columns in the material under investigation, which has not been done in this case. Alternatively, in order to speed up calculations one might consider not optimizing the entire set of parameters when searching for a new column, since to some extent the parameters of the peaks that have been detected already do not vary a lot when an extra peak is added to the parametric model. Hereby, one needs to consider a balance between good fit and speed. From Fig. 6.3(c), it follows that the presence of 13 atomic columns in Fig. 6.3(b) is most probable. The refined model with optimized parameters, taking into account 13 atomic columns, is shown in Fig. 6.3(d), corresponding to the expected crystal structure. It has been shown in Section 2.4 of Chapter 2 that imprecisions of the estimated atomic column positions can be expected due to the limited dose conditions of Fig. 6.3(b). As stated in Subsection 6.3.2, the MAP probability provides a quantitative statement of how much more probable a certain number of columns is as compared to another number. Here, from Fig. 6.3(c), it can be derived that the probability of the presence of 13 atomic columns is around 63 times larger than 12 columns and around 3 times larger than 14 columns. The refined models containing 12 and 13 atomic columns are shown in the inset of Fig. 6.3(c). It is noted that in this analysis the MAP probability rule has been applied without introducing any prior knowledge concerning the atomic structure or chemical composition of $SrTiO_3$. In principle, also an ordinary peak-finding routine, which searches for local maxima by choosing an arbitrary threshold (De Backer et al., 2016), can be used to determine the number of atomic columns. This

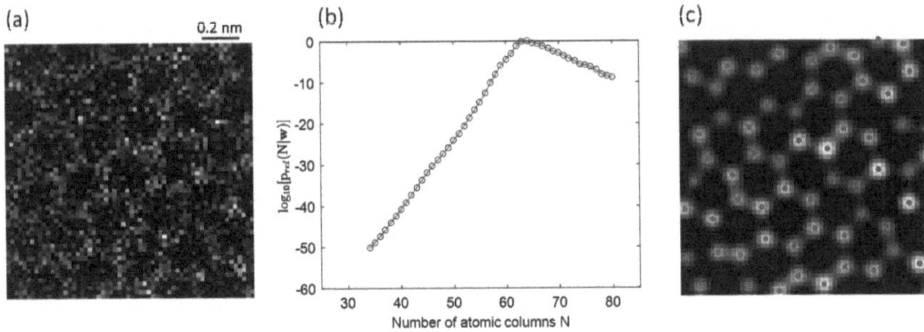

Figure 6.4 (a) Synthetic ADF image of graphene obtained from an experimental 4D STEM dataset with an incoming electron dose of around $3 \cdot 10^5$ $e^-/Å^2$ and an average detected electron dose of around 304 $e^-/Å^2$. (b) MAP probability rule evaluated for the experimental data shown in (a). (c) Most probable parametric model of the experimental data in (a) as indicated by the MAP probability rule in (b). (Reprinted with permission from Fatermans et al. (2018) by the American Physical Society.)

technique, though, failed to detect the expected crystal structure of $SrTiO_3$ from the image data shown in Fig. 6.3(b). By changing the threshold, one might obtain a better result, but this approach relies heavily on visual inspection which will lead to biased structure information.

6.4.2 Graphene

As shown earlier in the simulated image of Fig. 6.2(a), graphene consists of a single layer of carbon atoms arranged in a hexagonal lattice. In essence, it is a single layer of the most common allotrope of carbon, graphite, which can be regarded as a stack of graphene layers. With the isolation of graphene from graphite, the first two-dimensional (2D) material was discovered (Novoselov et al., 2004). This has led to the discovery of a whole family of 2D materials, including hexagonal boron nitride (h-BN) and molybdenum disulphide (MoS_2). As graphene possesses outstanding properties, it plays an important role in developing future applications in a wide variety of domains. Graphene is reported to be many times stronger than steel, yet incredibly lightweight and flexible (Lee et al., 2008). In addition, it is an efficient conductor of heat and electricity (Balandin et al., 2008) and also optical transparent (Sheehy & Schmalian, 2009; Zhu et al., 2014).

Fig. 6.4(a) shows a synthetic ADF image of graphene obtained from an experimental four-dimensional (4D) STEM dataset with an acceleration voltage of 80 kV, semiconvergence angle of 24.8 mrad and a detector collection range of 26–50 mrad. A synthetic image is obtained with a pixelated detector where the intensities of the pixels within a certain detector collection range are integrated. It is equivalent to a STEM image obtained with a conventional annular detector with the same collection angles. Graphene, which consists of light elements, carbon, cannot withstand a high incoming electron dose as it can easily be damaged by the electron beam. For this reason, acquiring images of graphene exhibiting high CNR is challenging as the electron dose needs to be limited. The property of beam sensitivity leads to typical rather low-quality ADF STEM images of graphene, such as the one shown in Fig. 6.4(a). Quantitative structure information though, can still be extracted from such images by using the MAP probability rule for determining the most probable atomic arrangement. Fig. 6.4(b) shows the evaluation of the MAP probability rule to detect the carbon atoms of graphene from the low-contrast data in Fig. 6.4(a), using a parametric model described by a superposition of Gaussian peaks with equal widths, given by Eq. (6.31). Hereby, the fact that graphene consists of a monolayer of carbon atoms has not been taken into account in the analysis, limiting the amount of prior knowledge concerning the atomic structure of graphene. This is opposed to the example of the simulated graphene in Fig. 6.2, where the model given by Eq. (6.34), with equal widths and equal heights, has been used in the analysis. The most probable parametric model that can be obtained by the MAP probability rule from the image data in Fig. 6.4(a) is shown in Fig. 6.4(c), detecting 64 carbon atoms and clearly resolving the hexagonal lattice of graphene. Hereby, imprecisions in the estimated atom positions may arise because of the finite dose conditions of Fig. 6.4(a). Moreover, for the atoms located at the edge of the image, additional imprecisions can be expected. This is related to the fact that these atoms are only partly present in the image data, causing a loss of information. The problem of imprecise estimations for atomic columns located at the edge of an image is often unavoidable since, for example, for large crystallographic structures there might always be atomic columns in the border region of the recorded images. Estimations of atomic columns located close to the border of an image are less precise and can, therefore, be considered to be less reliable. Similar as for the example of $SrTiO_3$ in the previous subsection, it is noted that the computation time of the algorithm described in Subsection 6.3.2 is mainly due to the number of starting positions to avoid the optimization process to end up in a local minimum when adding an extra peak to the model. The considerations to reduce computation time which were discussed in the previous subsection also apply for this example of graphene. In addition, it is noted that, as for the image data of $SrTiO_3$ in Fig. 6.3(a), an ordinary peak-finding routine failed to obtain the expected atomic arrangement of graphene from the image data shown in Fig. 6.4(a).

6.4.3 Gold nanorod

Metal nanorods hold some interesting optical properties which strongly depend on both the particle size and shape and which are not observed in the bulk material (Pérez-Juste et al., 2005). The possibility to tune their optical response by adjusting size and shape makes these nanorods excellent building blocks for electro-optical devices and contrast agents in biomedical applications (Becker et al., 2008; Chen et al., 2005; Gómez-Graña et al., 2013; Goris et al., 2013; Henglein, 2000; Liu & Guyot-Sionnest, 2004).

An experimental ADF STEM image of a gold nanorod is shown in Fig. 6.5(a) which was obtained by a double aberration-corrected FEI Titan operated at 300 kV with a semiconvergence angle of 21.8 mrad and a detector collection range of 62–190 mrad (Van Aert et al., 2013). In Section 4.2.3, this image was already the subject of an atom-counting analysis. It is noted that, due to the electron beam, especially surface atoms might not stay in the same position during the image acquisition (Williams & Carter, 2009). The MAP probability rule has been used to identify the presence of atoms near the edge of the nanorod, which are difficult to detect visually. In this region, the rod is only a few atoms thick which, in combination with the relatively low incoming electron dose of around $2 \cdot 10^5$ $e^-/Å^2$, leads to a low CNR. The following two-step procedure is suggested to determine the most probable structure of the gold nanorod. First, an ordinary peak-finding routine is applied on the image. This routine is able to quickly detect most of the thicker columns, but has difficulties in correctly detecting atoms near the edge of the particle. Therefore, the MAP probability rule is applied to investigate the presence of atomic columns near the edge. Hereby, the edge has been divided into different subregions. In each of these subregions, the MAP probability rule has been applied. Dividing the edge into subregions has been done in a manual way taking into account some overlap between neighboring subregions in order to avoid artefacts and imprecisions arising from atoms that are located right at the border of a subregion, as discussed in the previous subsection. Note that it can be seen from Fig. 6.5(a) that, at the left-hand side, there are columns located right on the border of the image. For these columns, it is unavoidable that the estimations will be less precise. Alternatively, one might consider dividing the region of the edge of the nanorod in a more automated way by using edge

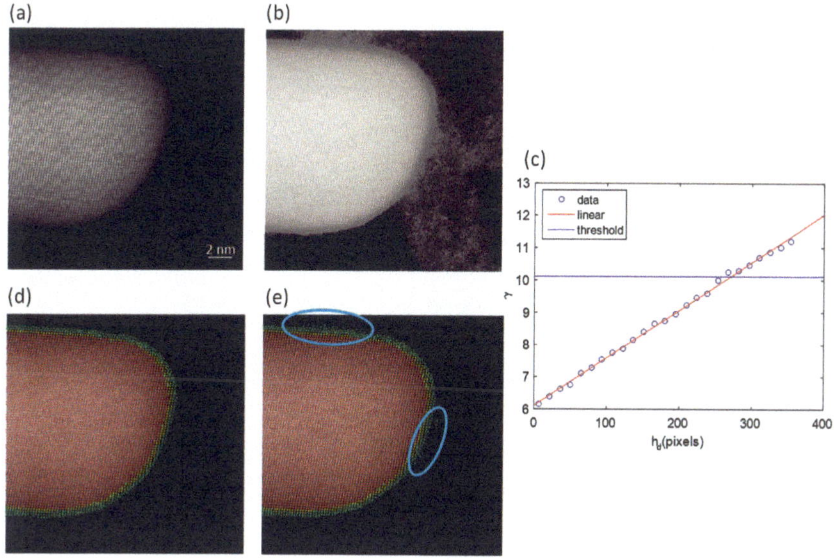

Figure 6.5 (a) Experimental ADF STEM image of a gold nanorod with an incoming electron dose of around $2 \cdot 10^5$ $e^-/Å^2$. (b) Same image as in (a) with adjusted contrast to highlight the variability of the carbon background. (c) Variogram of the background in (a). (d) Most probable atomic columns detected from (a) by a combination of an ordinary peak-finding routine and the MAP probability rule near the edge of the rod, where columns in red have been detected by the ordinary peak-finding routine, columns in yellow by both the ordinary peak-finding routine and the MAP probability rule, and columns in green by the MAP probability rule only. Hereby, the variability of the background has been taken into account. (e) Most probable atomic columns detected from (a) by the same procedure as in (d), but, hereby, neglecting the variability of the background. Regions where extra atoms have been detected as compared to (d) are indicated.

detection (Kook et al., 2016; Meng et al., 2018). During the analysis of the subregions, influence from atomic columns located in the inner part of the nanorod in the neighborhood of the edge has been taken into account. Hereby, the image data has been modeled by a superposition of Gaussian peaks with equal widths, given by Eq. (6.31), and the total intensity under a Gaussian peak has been chosen to be at least the total intensity scattered by a single gold atom, obtained from simulation with the MULTEM software (Lobato et al., 2016; Lobato & Van Dyck, 2015). In order to take thickness fluctuations in the carbon support into account, the background in each of the subregions is allowed to deviate from each other. As such, local fluctuations in the carbon support can be modeled more accurately. From

Fig. 6.5(a), it is not directly apparent that the carbon support is fluctuating, but by simply changing the contrast, though, as shown in Fig. 6.5(b), it is clear that the carbon support does not remain constant over the entire image. In order to acquire reliable structure information, the sizes of the defined subregions should be selected in such a way that the background in each subregion can adequately be modeled as a constant. This is the case when the size of a subregion is not larger than a certain maximum distance upon which the variability of the background does not affect atom detection. The so-called variogram, which was originally introduced in geology (Matheron, 1963), enables measuring the variability γ of the background as a function of distance h_d between pixels. This quantity is defined as

$$\gamma(h_d) = \frac{1}{2Z(h_d)} \sum_{z(h_d)=1}^{Z(h_d)} (\Delta s_{z(h_d)})^2, \tag{6.38}$$

where $Z(h_d)$ denotes the number of paired observations separated by a distance h_d. The difference between the $z(h_d)$th pair of observations is depicted as $\Delta s_{z(h_d)}$. The variogram of the carbon background in the image of the gold nanorod is shown in Fig. 6.5(c). It can be seen that the variability of the background γ increases as a function of distance. This indicates that, as the distance between pixels grows, the corresponding pixel values tend to become less similar. In Fig. 6.5(c), it is indicated that the measurements follow a linear trend. In order to determine the maximum size of the defined subregions, ensuring adequate modeling of these regions with a constant background, a threshold value on the variability of the carbon background needs to be chosen. This threshold is indicated by the horizontal line in Fig. 6.5(c) and defines the maximum variability which does not affect atom detection. A mathematical expression for this threshold can be defined by considering an area consisting of $K \times L$ pixels, where, besides a background ζ, no atomic columns are present. Here, the data is considered to be free of noise. If, due to, for example, surrounding areas with lower backgrounds, a background $\hat{\zeta}$ is fitted to the area under investigation, for which $\hat{\zeta} < \zeta$, there is a possibility of detecting a false atomic column. Detecting such a false column is avoided when adding a peak to the model does not improve the fit to the data. In this case, this means that the uniform sum-of squared-residuals misfit of zero columns, χ_0^2, is smaller than or equal to the misfit of one column, χ_1^2, including the fitted background $\hat{\zeta}$:

$$\chi_0^2 \leq \chi_1^2. \tag{6.39}$$

When an atomic column at position (x_k, y_l) is described as a Gaussian peak g with height η and width ρ:

$$g = \eta \exp\left(- \frac{(x_k - \beta_x)^2 + (y_l - \beta_y)^2}{2\rho^2} \right), \tag{6.40}$$

Eq. (6.39) can be explicitly written as

$$\sum_{k=1}^{K}\sum_{l=1}^{L}(\zeta_{kl} - \hat{\zeta}_{kl})^2 \leq \sum_{k=1}^{K}\sum_{l=1}^{L}(\zeta_{kl} - (\hat{\zeta}_{kl} + \hat{g}_{kl}))^2 = \sum_{k=1}^{K}\sum_{l=1}^{L}(\zeta_{kl} - \hat{\zeta}_{kl} - \hat{g}_{kl})^2. \tag{6.41}$$

When it is defined that $\zeta_{kl} - \hat{\zeta}_{kl} = \Delta\zeta_{kl} = \Delta\zeta$, as this difference is the same for all pixel values, since ζ and $\hat{\zeta}$ are both constant for all positions (x_k, y_l), Eq. (6.41) becomes

$$KL(\Delta\zeta)^2 \leq \sum_{k=1}^{K}\sum_{l=1}^{L}(\Delta\zeta - \hat{g}_{kl})^2. \tag{6.42}$$

From this, it follows that

$$\Delta\zeta \leq \frac{1}{2}\frac{\sum_{k=1}^{K}\sum_{l=1}^{L}\hat{g}_{kl}^2}{\sum_{k=1}^{K}\sum_{l=1}^{L}\hat{g}_{kl}}. \tag{6.43}$$

In a continuous form, Eq. (6.43) can be written as

$$\Delta\zeta \leq \frac{1}{2}\frac{\int\int \hat{g}^2 dxdy}{\int\int \hat{g} dxdy} = \frac{1}{2} \cdot \frac{\pi\hat{\eta}^2\hat{\rho}^2}{2\pi\hat{\eta}\hat{\rho}^2} = \frac{\hat{\eta}}{4}. \tag{6.44}$$

As the variability γ is calculated by half the average squared difference between pixel values, as given by Eq. (6.38), it can be written, in combination with Eq. (6.44), that

$$\Delta\gamma = \frac{1}{2}(\Delta\zeta)^2 \leq \frac{1}{2}\left(\frac{\hat{\eta}}{4}\right)^2 = \frac{\hat{\eta}^2}{32}. \tag{6.45}$$

Following from Eq. (6.45), as a rule of thumb, the size of each subregion should obey

$$\Delta\gamma \leq \frac{\eta_{min}^2}{32}, \tag{6.46}$$

where η_{min} denotes the minimum height of a Gaussian peak in the model. It is noted that Eq. (6.46) can also be obtained, in an analogous way as described above, by considering an area of $K \times L$ pixels where an atomic column is present, positioned on top of a background ζ. When a background $\hat{\zeta}$ is fitted, for which $\hat{\zeta} > \zeta$, there is a possibility that the atomic column cannot be detected. For such a situation, detecting the column happens when the misfit of one column is smaller than or equal to the misfit of zero columns, given by $\chi_1^2 \leq \chi_0^2$. It follows from the threshold in Fig. 6.5(c) given by Eq. (6.46) that the maximum distance $h_{d_{max}}$ upon which the variability γ of the background does not affect atom detection is given by 271 pixels, which corresponds to around 49 Å. As the size of each subregion has been chosen to be 10 Å by 10 Å, Eq. (6.46) holds and, as a result, the background in each subregion can adequately be modeled as a constant. The result of the analysis is shown in Fig. 6.5(d). Columns in red have been found by the ordinary peak-finding routine. Columns in yellow have been detected by both the ordinary peak-finding routine and the MAP probability rule. Columns in green have been detected by the MAP probability rule only. It is noted that at the edge of the gold nanorod there were no columns found by the ordinary peak-finding procedure which were not detected by the MAP probability rule. To illustrate the importance of adequately modeling the background, the analysis by the MAP probability rule has also been performed by assuming the constant background of all subregions to be equal. As a result, the background is modeled in a less accurate way. Fig. 6.5(e) shows the result of this analysis. As compared to Fig. 6.5(d), extra atoms, which seem to be detached from the rod, have been detected in the indicated regions. These extra detected atoms appear in regions where there is an increased background intensity, as apparent in Fig. 6.5(b). As such, they do not correspond to actual atoms, but are the result of estimating a too low background in these regions.

For radiation-sensitive materials, the incoming electron dose should be kept as low as possible. To illustrate the possibilities of the MAP probability rule when applied to low-dose images, a 10 nm thick gold nanorod has been simulated, which is shown in Fig. 6.6(a), using MULTEM with an incoming electron dose of merely 5000 $e^-/\text{Å}^2$. The simulation parameters are listed in Table 6.2. In order to make the simulation more realistic, the structure has been fully relaxed by using molecular dynamics simulations employing the embedded atom method (EAM) potential (Foiles et al., 1986) with the GPU Lammps package (Brown et al., 2012, 2011; Brown & Yamada, 2013). For the analysis, the background has been put to zero since no carbon support is present in the simulation and the edge of the rod has been divided into 49 equal subregions to speed up the calculations.

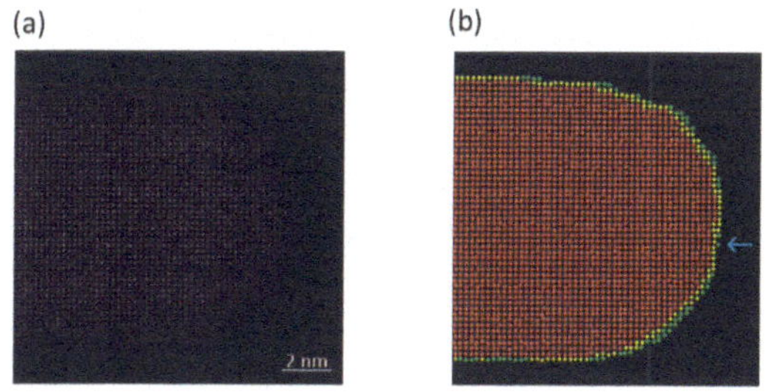

Figure 6.6 (a) Simulated ADF STEM image of a gold nanorod with the presence of Poisson noise with an incoming electron dose of 5000 $e^-/\text{Å}^2$. (b) Most probable atomic columns detected from (a), where the MAP probability rule has been applied to the border region of the nanorod. Columns in red have been detected by the ordinary peak-finding routine, columns in yellow by both the ordinary peak-finding routine and the MAP probability rule, and columns in green by the MAP probability rule only. The arrow indicates an atom detected at a position where no atom was present. (Reprinted with permission from Fatermans et al. (2018) by the American Physical Society.)

Table 6.2 Microscope parameter values for simulation of a STEM image of a gold nanorod using MULTEM.

Parameter	Symbol	Value
Acceleration voltage	V_0 (kV)	200
Defocus	ϵ (Å)	15.8
Spherical aberration	C_s (mm)	0.001
Spherical aberration of 5th order	C_5 (mm)	0
Semiconvergence angle	α (mrad)	24.0
Detector inner radius	β_1 (mrad)	60
Detector outer radius	β_2 (mrad)	165
Pixel size	$\Delta x = \Delta y$ (Å)	0.125
FWHM of the source image	FWHM (Å)	0.7

Fig. 6.6(b) shows the detected atomic columns from Fig. 6.6(a). From this analysis, it follows that all present atomic columns have been detected and that only one extra atom has been found at a position where the simulation did not include an atom. This atom is indicated by the blue arrow shown in Fig. 6.6(b). This probability for over- or underfitting is an inherent limitation of model-order selection methods and will, in general, increase with decreasing CNR. However, the present example demonstrates the good performance of the MAP probability rule when analyzing a low CNR image. Moreover, it can be seen from Fig. 6.6(b) that a substantial amount of atomic columns would not have been detected in the absence of the MAP

probability rule. This is an important result since it is well known that the exact surface morphology of nanoparticles can influence their physical properties.

6.4.4 Ultrasmall nanoclusters

The MAP probability rule is also of great importance to detect single atoms in ultrasmall nanoclusters. Such clusters can be thought of as fundamental building blocks leading to metamaterials with physical and chemical properties that are not available in nature (Binns, 2001; Claridge et al., 2009). The growth mechanisms for small nanoclusters are much more exotic than for bulk materials and therefore have a more complex structure.

As a first example, a series of experimental ADF STEM images of a small germanium cluster is shown in Figs. 6.7(a)-(c). The clusters have been imaged using a double aberration-corrected FEI Titan operated at 120 kV and the beam current was set to 40 pA (Bals et al., 2012). The MAP probability rule, in combination with atom counting (De Backer et al., 2013), can be used to verify whether or not there is a loss of atoms during the acquisition of the series of the images. In this analysis, one should, however, take into consideration that displacements of individual atoms are possible due to the interaction with the electron beam (Williams & Carter, 2009). This may cause atoms to move in or out of the field of view. The MAP probability rule is used to determine the most probable structure of the cluster, using the intensity of a single germanium atom as prior knowledge, obtained from an image simulation with MULTEM. The evaluation of the MAP probability rule is shown in Figs. 6.7(d)-(f) for an increasing number of atomic columns. The most probable parametric models are shown in Figs. 6.7(g)-(i), where the dots refer to estimated column positions. The scattering cross-sections of the columns, which are equal to the volumes under the estimated Gaussian peaks, have been calculated to count the number of atoms in a column, using the statistics-based atom-counting methodology described in Chapter 4 (De Backer et al., 2013). Hereby, the size of the cluster in Fig. 6.7 has been estimated to be 26 ± 2 atoms. As compared to earlier results obtained by detecting atomic columns by visual inspection of the images (Bals et al., 2012), the MAP probability rule detects more peaks, whereas the total number of atoms in each of the individual frames remains almost constant. The extra peaks correspond to atoms which are not perfectly aligned along the beam direction. This suggests that the MAP probability rule is able to disentangle strongly overlapping peaks resulting into more accurate structure information as compared to visual

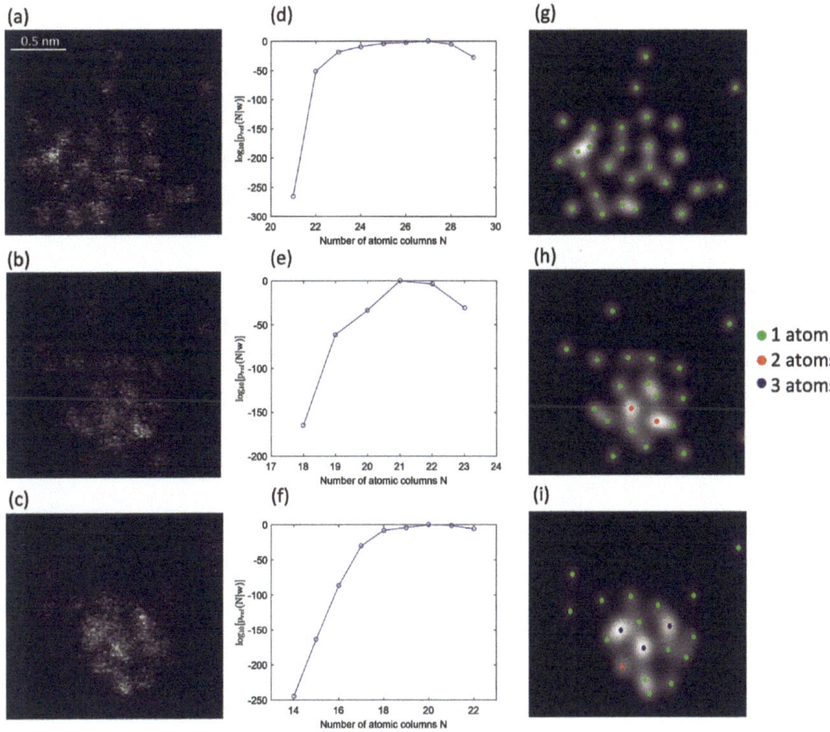

Figure 6.7 (a)-(c) Sequence of ADF STEM images of a germanium cluster with an incoming electron dose of around $6.8 \cdot 10^4$ $e^-/Å^2$. (d)-(f) MAP probability rule evaluated for the experimental data shown in (a)-(c). (g)-(i) Most probable parametric model of the experimental data in (a)-(c) as indicated by the MAP probability rule in (d)-(f) with atom counting results added in which green, red and blue dots correspond to one, two or three atoms, respectively.

inspection. This is of great importance to fully understand the dynamics of such a small nanocluster.

As a second example, a time series of experimental ADF STEM images of a small gold nanocluster is shown in Figs. 6.8(a)-(d), which were obtained from a double aberration-corrected FEI Titan operated at 120 kV using a detector range of 28–172 mrad. Gold particles are often investigated because of their tunable properties, such as optical behavior and catalytic activity (Alshammari et al., 2012; Daniel & Astruc, 2004). Interestingly, the structure of small gold nanoclusters deviates from the common gold cubic structure. The clusters tend to congregate to stable structures with a specific number of atoms determined by so-called magic numbers (Mackay, 1962).

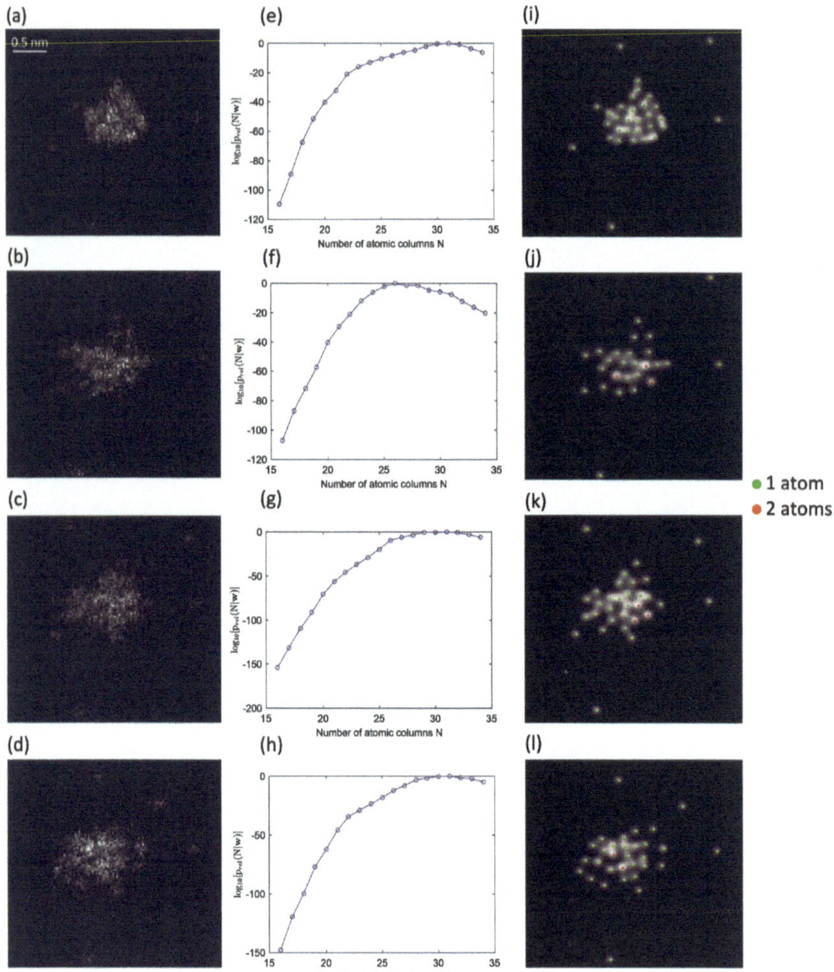

Figure 6.8 (a)-(d) Sequence of ADF STEM images of a gold cluster with an incoming electron dose of around 4500 e$^-$/Å^2. (e)-(h) MAP probability rule evaluated for the experimental data shown in (a)-(d). (i)-(l) Most probable parametric model of the experimental data in (a)-(d) as indicated by the MAP probability rule in (e)-(h) with atom counting results added in which green and red dots correspond to one or two atoms, respectively.

In order to confirm that the size of the nanocluster in Fig. 6.8 is determined by a magic number, the number of atoms needs to be counted. Similar as for the small germanium cluster in Fig. 6.7, the beam sensitivity hinders a visual interpretation since the individual atoms are not clearly distinguishable. In addition, the nanocluster changes shape during the acquisition. In order to obtain a reliable estimation of the structure of the nanocluster, the MAP probability rule has been used, in an analogous way as for the germanium cluster in Fig. 6.7, including the intensity of a single gold atom as prior knowledge obtained from simulation with MULTEM. The evaluation of the MAP probability rule is shown in Figs. 6.8(e)-(h) as a function of the number of columns N. The most probable parametric models are shown in Figs. 6.8(i)-(l). It follows from this analysis that the size of the cluster in Fig. 6.8 has been estimated to be 32 ± 2 atoms, corresponding to a magic number of 34 atoms.

6.5. Maximum a posteriori probability for ABF-ADF STEM

The proposed MAP probability rule for atom column detection, which has been introduced in Section 6.3, is not limited to be applied to ADF STEM images only. Also for simultaneously acquired ABF and ADF STEM images it can play an important role to reliably detect the number of atomic columns N present in the images. Hereby, in order to extract reliable structure information from simultaneously acquired ABF and ADF STEM images, one can use a model neglecting specimen tilt, such as in Eq. (2.53), or including tilt, such as in Eq. (2.55) (Fatermans et al., 2020). A reliable quantification of the number of atomic columns N by the MAP probability rule is important since a simultaneous acquisition of ABF and ADF STEM images is, in particular, interesting for characterizing materials containing light elements, which are typically beam sensitive. Therefore, a merely visual inspection of such images may lead to biased results. Analytical expressions for the MAP probability rule for simultaneously acquired ABF and ADF STEM images can be derived in a similar manner as described in Subsection 6.3.1 for ADF STEM starting from Bayes' theorem given by Eq. (6.13). As mentioned before, the MAP probability rule aims to select the number of columns that maximizes the posterior probability $p(N|\boldsymbol{w})$, which reflects the presence of N atomic columns given the observed image pixel values $\boldsymbol{w}$. For a simultaneous acquisition of ABF and ADF images, the observed image pixel values, $\boldsymbol{w}$, can be considered as the ensemble of the observed image pixel values of both images, $\{\boldsymbol{w}^{ABF}, \boldsymbol{w}^{ADF}\}$. By combining Eqs. (6.15) and (6.17), it has been shown in Subsection 6.3.1

that Bayes' theorem can be reduced to

$$p(N|\boldsymbol{w}) \propto \int p(\boldsymbol{w}|\boldsymbol{\theta}, N)p(\boldsymbol{\theta}|N)d^M\boldsymbol{\theta}, \tag{6.47}$$

where the first term in the integral, $p(\boldsymbol{w}|\boldsymbol{\theta}, N)$, is the likelihood function which describes the probability of the observed image pixel values $\boldsymbol{w}$ for particular values of the parameters $\boldsymbol{\theta}$ of a model with N atomic columns. For simultaneously acquired ABF and ADF STEM image data, the likelihood function can be expressed in a similar way as described in Subsection 6.3.1:

$$p(\boldsymbol{w}|\boldsymbol{\theta}, N) = \frac{e^{-\chi^2(\boldsymbol{\theta})/2}}{\prod_{k=1}^{K}\prod_{l=1}^{L} 2\pi\sqrt{w_{kl}^{ABF} w_{kl}^{ADF}}}, \tag{6.48}$$

where

$$\chi^2(\boldsymbol{\theta}) = \sum_{k=1}^{K}\sum_{l=1}^{L}\left[\frac{[w_{kl}^{ABF} - f_{kl}^{ABF}]^2}{w_{kl}^{ABF}} + \frac{[w_{kl}^{ADF} - f_{kl}^{ADF}]^2}{w_{kl}^{ADF}}\right] \tag{6.49}$$

is a weighted sum-of-squared-residuals misfit between the data and the parametric model. This expression follows from the fact that the noise can be considered to be uncorrelated in simultaneously acquired STEM images with different detector geometries (Brown et al., 2016; Hammel & Rose, 1995; Rose, 1974). The other term in the integral in Eq. (6.47), $p(\boldsymbol{\theta}|N)$, is the prior density of the parameters $\boldsymbol{\theta}$ for a model with N columns. Different expressions for the prior density function can be constructed reflecting different types of prior knowledge. Here, similarly as in Subsection 6.3.1, $p(\boldsymbol{\theta}|N)$ is expressed as a product of uniform distributions over a predefined range for each parameter θ_m given by Eq. (6.22). As mentioned in Subsection 6.3.1, approximate analytical expressions for $p(N|\boldsymbol{w})$ can be derived for several parametric models to determine the number of atomic columns for which there is most evidence in ADF image data (Fatermans et al., 2019). From Eqs. (6.47), (6.48) and (6.22), following the same methodology, $p(N|\boldsymbol{w})$ for simultaneously acquired ABF and ADF STEM images using the model given by Eq. (2.53), neglecting the effect of specimen tilt, results into

$$p(N|\boldsymbol{w}) \propto \frac{N!}{[(\beta_{x_{max}} - \beta_{x_{min}})(\beta_{y_{max}} - \beta_{y_{min}})(\rho_{max}^{ABF} - \rho_{min}^{ABF})]^N} \times \frac{e^{-\chi_{min}^2/2}(4\pi)^{3N}[det(\nabla\nabla\chi^2)]^{-1/2}}{[(\rho_{max}^{ADF} - \rho_{min}^{ADF})(\eta_{max}^{ABF} - \eta_{min}^{ABF})(\eta_{max}^{ADF} - \eta_{min}^{ADF})]^N}, \tag{6.50}$$

where $\chi_{min}^2 = \chi^2(\hat{\boldsymbol{\theta}})$, with $\hat{\boldsymbol{\theta}}$ the parameter vector that minimizes $\chi^2(\boldsymbol{\theta})$ given by Eq. (6.49) and where the term $\det(\nabla\nabla\chi^2) = \det\left(\frac{\partial^2\chi^2(\boldsymbol{\theta})}{\partial\boldsymbol{\theta}\partial\boldsymbol{\theta}^T}\Big|_{\boldsymbol{\theta}=\hat{\boldsymbol{\theta}}}\right)$ represents the determinant of the Hessian matrix of $\chi^2(\boldsymbol{\theta})$ evaluated at $\hat{\boldsymbol{\theta}}$. For the model given by Eq. (2.55), including the effect of tilt, the posterior probability becomes

$$p(N|\boldsymbol{w}) \propto \frac{N!(\alpha_{max} - \alpha_{min})^{-N}}{[(\beta_{x_{max}} - \beta_{x_{min}})(\beta_{y_{max}} - \beta_{y_{min}})(\rho_{max}^{ABF} - \rho_{min}^{ABF})]^N} \times \frac{e^{-\chi_{min}^2/2}(4\pi)^{3.5N}[det(\nabla\nabla\chi^2)]^{-1/2}}{[(\rho_{max}^{ADF} - \rho_{min}^{ADF})(\eta_{max}^{ABF} - \eta_{min}^{ABF})(\eta_{max}^{ADF} - \eta_{min}^{ADF})]^N}. \tag{6.51}$$

By using Eqs. (6.50) or (6.51), depending on the underlying parametric model, the MAP probability rule allows to select the most probable number of atomic columns present in simultaneously acquired ABF and ADF STEM images, and hence the most probable atomic structure, by evaluating $p(N|\boldsymbol{w})$ as a function of N. It should be noted that the MAP probability rule is not limited to the models given by Eqs. (2.53) and (2.55). Similar expressions as Eqs. (6.50) and (6.51) can be derived for other types of models as well.

6.6. Atom column detection from experimental ABF-ADF STEM images

In this section, the proposed methodology to detect atomic columns by the MAP probability rule from simultaneously acquired ABF and ADF STEM images is applied to two experimental examples exhibiting low CNR. In Section 6.6.1, $SrTiO_3$ is analyzed, whereas, in Section 6.6.2, $LiMn_2O_4$ is investigated. Hereby, the effect of specimen tilt will be taken into account as small tilts might be present, causing a possible shift of the measured atomic column locations in the ABF image as compared to the locations in the ADF image, as mentioned in Section 2.5 of Chapter 2.

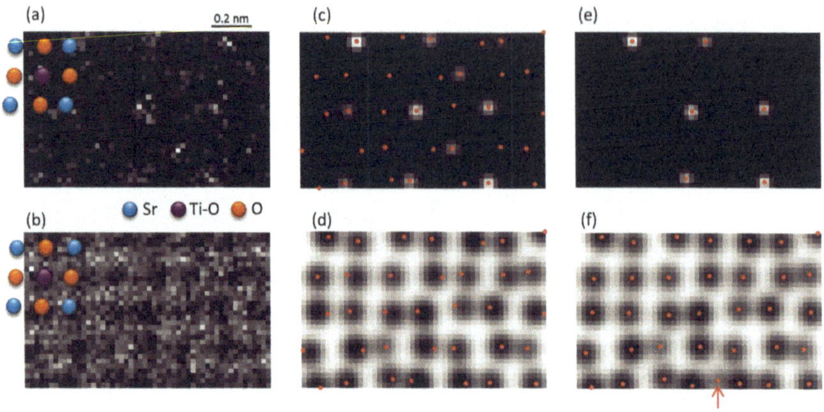

Figure 6.9 (a) Synthetic ADF and (b) ABF STEM image of $SrTiO_3$ obtained from an experimental 4D STEM dataset. (c) and (d) Most probable refined models of the experimental ADF data in (a) and the ABF data in (b), respectively, obtained from the MAP probability rule by analyzing the data in (a) and (b) simultaneously taking into account specimen tilt. The red dots indicate the estimated column locations corresponding to the fitted ADF peak positions. (e) and (f) Most probable refined models by separately analyzing (a) and (b), where the red dots correspond to the fitted peak positions in the image data. The arrow in (f) indicates the detection of an extra column at a position where no column is expected. (Reprinted from Fatermans et al. (2020) with permission from Elsevier.)

6.6.1 $SrTiO_3$

Figs. 6.9(a) and (b) show a synthetic ADF and ABF STEM image of $SrTiO_3$, respectively, obtained from an experimental 4D STEM dataset using a Medipix3 Merlin camera attached to a probe-corrected FEI Titan, where the Sr, Ti-O and O columns have been indicated. The detector collection range for the ADF image has been chosen to be equal to 88–98 mrad, whereas the collection range for the ABF image is 12–22 mrad. The MAP probability rule for simultaneously acquired ABF and ADF STEM data has been applied to these images in order to obtain the number of atomic columns for which there is most evidence in the image data, and, as a result, the most probable atomic structure. For this, the parametric model including the effect of specimen tilt given by Eq. (2.55) has been used, since tilt might be present, where equal widths for the Gaussian peaks in the ABF image, ρ^{ABF}, and ADF image, ρ^{ADF}, have been used. Hereby, it has been assumed that the shift of the ABF peaks from the ADF peak locations cannot exceed 50 pm, which is already a high value for typical shifts due to specimen tilt (Gao et al., 2018; Zhou et al.,

2016). Figs. 6.9(c) and (d) show the most probable parametric models indicated by the MAP probability rule by simultaneously analyzing the ADF and ABF STEM image data of Figs. 6.9(a) and (b), respectively, consisting of 35 atomic columns, corresponding to the expected number of columns of $SrTiO_3$ in [100] direction. As compared to the detection of the atomic columns of $SrTiO_3$ from only an ADF STEM image in Subsection 6.4.1, the O columns can, in this case, also be detected due to the simultaneous acquisition of an ABF STEM image. It is noted that in the analysis no prior knowledge about the expected locations of the atomic columns in $SrTiO_3$ have been taken into account. The estimated column positions which are shown in Figs. 6.9(c) and (d) by red dots correspond to the fitted ADF peak positions since measured atomic column locations in ADF STEM are considered to be reliable indicators of the true column positions because of the insensitivity to specimen tilt (Maccagnano-Zacher et al., 2008; So & Kimoto, 2012). In Figs. 6.9(e) and (f), the most probable refined models of the image data in Figs. 6.9(a) and (b), respectively, are shown, obtained by the MAP probability rule by separately analyzing the ADF and ABF image data. For this, a parametric model such as given by Eq. (6.31) has been used, where equal widths for the estimated Gaussian peaks have been assumed. It can be seen from Fig. 6.9(f) that the expected atomic columns in $SrTiO_3$ are found. In this process, though, one atomic column, which has been indicated by the red arrow in Fig. 6.9(f), has been detected at a position where no column is expected. Although from analyzing the ABF image data independently of the ADF image data the atomic columns can be detected quite reliably, the ABF peak locations are sensitive to specimen tilt, which causes them to shift from their true column positions. When only analyzing an ABF STEM image, the estimated column positions are prone to this effect, which may lead to biased structure information.

6.6.2 $LiMn_2O_4$

As another example, simultaneous ADF and ABF STEM images of $LiMn_2O_4$ have been acquired which are shown in Figs. 6.10(a) and (b), respectively, together with the spinel structure of $LiMn_2O_4$. The images have been obtained from a probe-corrected FEI Titan operated at 300 kV with a semiconvergence angle of 19.8 mrad. The detector collection ranges for the ABF and ADF images are 8–17 mrad and 44–190 mrad, respectively. Analyzing Li-containing materials is of relevance since they have great technological interest because of their importance in battery devices (Li, Wang, et al., 2009; Nitta et al., 2015). Therefore, the direct visualization of the Li

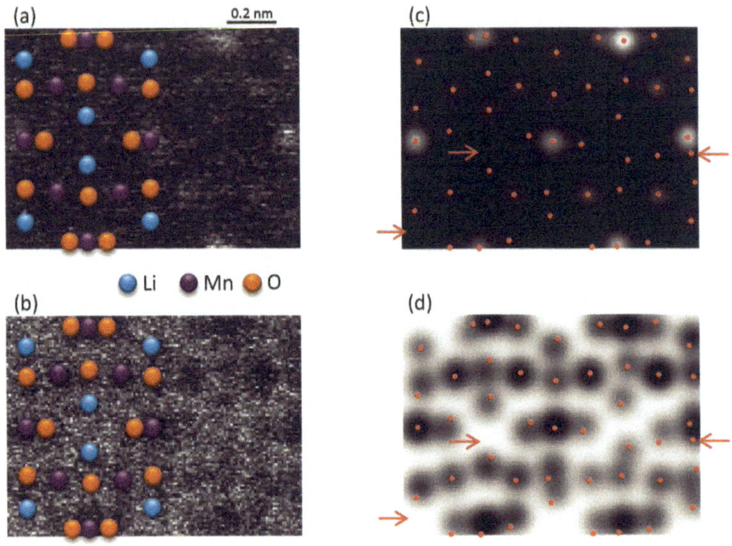

Figure 6.10 (a) Experimental ADF and (b) ABF STEM image of $LiMn_2O_4$ overlain with the spinel atomic arrangement. (c) Most probable refined model of the experimental ADF data in (a) and (d) most probable model of the ABF data in (b) obtained from the MAP probability rule by analyzing the data in (a) and (b) simultaneously taking into account specimen tilt. The red dots indicate the estimated column locations corresponding to the fitted ADF peak positions. The arrows in (c) and (d) indicate the expected locations of two columns that were not detected and the detection of an extra column at a position where no column is expected. (Reprinted from Fatermans et al. (2020) with permission from Elsevier.)

atoms has been an important research topic in STEM and has been achieved by imaging in the ABF mode (Gu et al., 2011; He et al., 2011; Huang et al., 2011; Lee et al., 2011; Oshima et al., 2010). Nevertheless, materials containing light elements, such as Li, are sensitive to the electron beam. For such materials, beam damage is limited by imaging the specimen with a low incoming electron dose, typically varying between around 10^3 to around 10^5 $e^-/Å^2$ in atomic-resolution STEM, leading to images exhibiting low CNR. This is also the case for the images shown in Figs. 6.10(a) and (b) where the determination of the number of atomic columns and their positions is not straightforward by merely visually interpreting the image data. In order to overcome this problem, the MAP probability rule for simultaneously acquired ABF and ADF STEM images is applied to Figs. 6.10(a) and (b) by using the parametric model given by Eq. (2.55) including the effect of specimen tilt, since some tilt might have been present during the acquisition of the images. Similarly as for the example of $SrTiO_3$ in Fig. 6.9, equal widths for the Gaussian peaks in the ABF and ADF image have been used and the shift of the ABF peaks from the ADF peak locations has been chosen not to be larger than 50 pm. Figs. 6.10(c) and (d) show the most probable parametric models indicated by the MAP probability rule for the ADF and ABF STEM image data of Figs. 6.10(a) and (b), respectively, consisting of 46 atomic columns. The estimated column locations, shown by red dots, correspond to the fitted ADF peak positions. The detected number of atomic columns is slightly less than the expected 47 columns considering the spinel structure of $LiMn_2O_4$ since two columns were not detected and one extra column position was found in the experimental data. These columns have been indicated in Figs. 6.10(c) and (d) by red arrows. The probability for over- or underfitting is an inherent limitation of model-order selection methods, like the MAP probability rule, and will generally increase with decreasing CNR. Despite this, the MAP probability rule has been able to retrieve a reliable representation of the atomic column locations in $LiMn_2O_4$ without including any prior information about the expected spinel atomic arrangement.

6.7. Conclusions

In this chapter, the methodology of the MAP probability rule to detect single atoms from atomic-resolution STEM images has been explained in detail. The method is built upon model-based parameter estimation, which was the topic of Chapter 2, and Bayesian probability theory, whose basic algebraic calculation rules have been thoroughly elaborated. The combination of statistical parameter estimation and model-order selection allows the MAP probability rule to perform automatic and objective structure quantification of unknown nanomaterials and to detect atomic columns and even single atoms with high reliability. For this purpose, it has been shown that, for a variety of parametric models, approximate analytical expressions can be derived for the probability of the presence of a certain number of atomic columns in ADF STEM image data. The MAP probability rule selects the number of columns that maximizes this probability. Furthermore, the MAP probability rule also enables one to quantify how more likely a certain atomic structure is as compared to other structures. Obtaining automatic and objective structure information is especially useful for the analysis of radiation-sensitive and light-element nanostructures. Typically, images of such materials exhibit low CNR due to the use of a limited incoming electron dose in STEM of merely 10^3–10^5 $e^-/\text{Å}^2$ in order to avoid beam damage. Visual inspection of such images is unreliable, as it is challenging to distinguish individual atomic columns, and might lead to biased results. The validity and usefulness of applying the MAP probability rule to images exhibiting low CNR have been demonstrated by analyzing experimental and simulated ADF STEM images of samples of different shape, size, and atom type. Hereby, it has been shown that important structural information can be obtained by using the MAP probability rule, which would otherwise not have been available.

Furthermore, the application of the MAP probability rule is not limited to ADF STEM images only, but can also be applied to simultaneously acquired ABF and ADF STEM images. Hereby, the effect of specimen tilt, which shifts the ABF peak locations from the true atomic column locations, has been taken into account since small tilts of the electron beam with respect to the crystal zone axis can easily be present in STEM experiments. The validity of the MAP probability rule within the framework of simultaneously fitting ABF and ADF image data has been illustrated by investigating experimental STEM data. It has been shown that the proposed methodology allows to obtain a reliable estimation of atomic column locations without including prior information about the expected positions of these columns.

Acknowledgments

This project has received funding from the European Research Council (ERC) under the European Union's Horizon 2020 research and innovation programme (Grant Agreement No. 770887 and No. 823717 ESTEEM3). The authors acknowledge financial support from the Research Foundation Flanders (FWO, Belgium) through Projects No. G.0502.18N, G.0267.18N, G.0368.15N, and a postdoctoral grant to A. D. B. The authors acknowledge funding from the University of Antwerp through a TOP project.

CHAPTER SEVEN

Image-quality evaluation and model selection with maximum a posteriori probability

Jarmo Fatermans[a,b,c], **Annick De Backer**[a,b], **Arnold J. den Dekker**[c], **and Sandra Van Aert**[a,b,*]

[a]Electron Microscopy for Materials Science (EMAT), University of Antwerp, Antwerp, Belgium
[b]NANOlab Center of Excellence, University of Antwerp, Antwerp, Belgium
[c]imec-Vision Lab, University of Antwerp, Wilrijk, Belgium
*Corresponding author. e-mail address: sandra.vanaert@uantwerpen.be

Contents

7.1. Introduction

In Chapter 6, the maximum a posteriori (MAP) probability rule has been introduced as a quantitative method to determine the number of atomic columns for which there is most evidence in annular dark-field (ADF) scanning transmission electron microscopy (STEM) image data and in simultaneously acquired annular bright-field (ABF) and ADF STEM image data. The method is based on a combination of statistical parameter estimation (De Backer et al., 2016; den Dekker et al., 2013, 2005; Van Aert et al., 2005, 2012; van den Bos, 2007) and model-order selection (Claeskens & Hjort, 2008; Myung, 2000; Stoica & Selen, 2004) from which analytical expressions could be derived using a Bayesian approach (Fatermans et al.,

Advances in Imaging and Electron Physics, Volume 217
ISSN 1076-5670
https://doi.org/10.1016/bs.aiep.2021.01.007

2019). It was shown that the MAP probability rule is able to automatically and objectively determine the most probable structure of unknown nanomaterials and to detect single atoms with high reliability (Fatermans et al., 2020, 2018). Moreover, the proposed method quantifies how more likely an obtained atomic structure is as compared to other structures.

Since the MAP probability rule can be used to quantify how more likely a certain number of atomic columns is as compared to another number, it offers a way to determine atom detectability, which is defined as the probability to detect an atom from an image. The relation of atom detectability with STEM image-quality measures, such as signal-to-noise ratio (SNR) and contrast-to-noise ratio (CNR) (Welvaert & Rosseel, 2013), can be evaluated. In general, atom detectability is indeed related to image quality, as, for example, one expects the detectability of atomic columns in an image to increase with increasing image quality. In this manner, the probability to detect an atomic column from STEM image data, which can be provided by the MAP probability rule, is correlated with the quality of the image. As such, for STEM images, a new image-quality measure, the integrated CNR (ICNR), has been proposed that better correlates with atom detectability than conventional image-quality measures (Fatermans et al., 2020, 2019).

The MAP probability rule for atom column detection from STEM images manifests itself as a combination of parameter estimation and the concept of model selection. Hereby, besides estimating the unknown parameters of the expectation model, also the number of parameters that most closely represents the underlying process that generated the experimental observations needs to be determined. In other words, one aims to select the best model from a set of candidate models given experimental data. In typical model-selection methods, a tradeoff between high goodness of fit and low complexity of the models under investigation needs to be performed. This is important since model fit can be easily improved by increased model complexity without necessarily bearing any interpretable relationship with the underlying data-generating process (Claeskens & Hjort, 2008; Corsaro & De Ridder, 2014; Myung, 2000; Stoica & Selen, 2004). The MAP probability rule itself can be considered as a model-selection criterion, such as the Akaike Information Criterion (AIC) (Akaike, 1974) and Bayesian Information Criterion (BIC) (Schwarz, 1978). In particular, the MAP probability rule is closely related to the BIC, since both methods are derived from a Bayesian approach. In principle, atom detection from STEM images can be performed with any model-selection criterion, but, interestingly, the MAP probability rule takes into account more dimensions of model complexity as compared to other commonly used model-selection criteria. This allows for a superior performance to detect the correct number of atomic columns from ADF STEM images (Fatermans et al., 2019). In addition, the MAP probability rule offers a more flexible way to detect atoms from images than other criteria. This is due to the fact that the prior can be tuned, resulting into a different value for the complexity of the model under consideration. Moreover, by using the MAP probability rule, it is clear what prior knowledge has been taken into account during the analysis, which is not always straightforward for other model-selection criteria.

This chapter is organized as follows. In Section 7.2, it is explained how the MAP probability rule can be used to evaluate the correlation between measures of image quality and atom detectability in STEM images. In Subsection 7.2.1, the correlation with conventional measures, such as SNR and CNR, is investigated, whereas in Subsection 7.2.2 a new image-quality measure which correlates better with atom detectability is introduced. In Section 7.3, the relation of the MAP probability rule to model selection is investigated, starting with an overview of typical model-order selection criteria in Subsection 7.3.1. Then, Subsection 7.3.2 compares the performance of the MAP probability rule in correctly detecting atoms from STEM images to that of other model-selection criteria. This is followed in Subsection 7.3.3 by explicitly showing the connection between the MAP probability rule and the BIC. In Subsection 7.3.4, the effect of incorporating a certain amount of prior knowledge in the analysis is explained. Finally, in Section 7.4, conclusions will be drawn.

7.2. Atom detectability

In general, one expects that the detectability of atomic columns in images will increase with increasing image quality. Fig. 7.1 depicts a set of Poisson noise disturbed simulated ADF STEM images of a single atom with varying background and incoming electron dose. Intuitively, it is clear that detecting the atom will be easiest from high-quality images with a high electron dose and low background, and most difficult from low-quality images with low dose and high background. In this section, the MAP probability rule will be used as a tool to evaluate the relation between image quality and atom detectability. Hereby, it is noted that when only the value of the incoming electron dose is available no information is provided

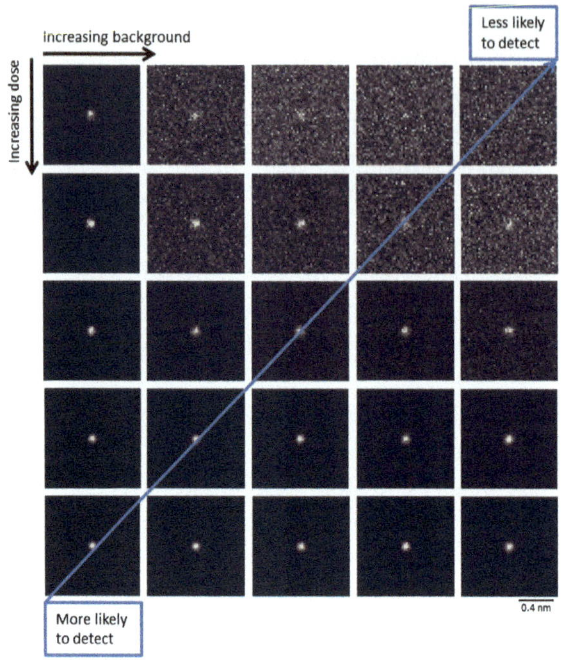

Figure 7.1 Set of simulated Poisson noise disturbed ADF STEM images of a single atom with varying incoming electron dose and background.

about the detectability of atomic columns in an image. Depending on the sample, the number of electrons that reaches the detector can be very different for the same incoming dose. This is shown by the horizontal lines of Fig. 7.1 where the same incoming electron dose is used, but due to the different backgrounds that are present in the images, the perceived atom detectability may vary significantly. Alternatively, one might consider using detected electron dose instead to evaluate the relation with atom detectability. Unfortunately, also the detected number of electrons does not provide information about the probability of detecting atomic columns from an image. In Fig. 7.1, the number of detected electrons increases while moving to the right-hand side, but, as visually perceived, this does not guarantee better atom detectability. In addition, a disadvantage of using detected electron dose is that it typically only provides an averaged value, losing information about variations in the detected number of electrons in the image. Therefore, other measures than incoming or detected electron dose are required to evaluate the relation between image quality and atom detectability. In Subsection 7.2.1, SNR and CNR are investigated. This is followed by

Subsection 7.2.2 where a new image-quality measure is proposed that correlates better with atom detectability from STEM images.

7.2.1 SNR and CNR

In this section, the relation of SNR and CNR as image-quality measures to atom detectability is evaluated by using the MAP probability rule. First, SNR is investigated. As local fluctuations of the background or differences in atomic column thicknesses or composition can occur, each column in a STEM image can possess a different SNR value. When the background in the image can be considered to be constant, the SNR of a column at position (β_x, β_y) can be written as

$$SNR = \frac{\eta + \zeta}{[\eta + \zeta]^{1/2}}, \tag{7.1}$$

where η and ζ denote the height of the atomic column and the background of the image, respectively, for a parametric model based on a superposition of Gaussian peaks, such as the ones discussed in Chapter 2. It is noted that the definition in Eq. (7.1) provides a rather simple description of SNR. There exist several alternative ways of defining SNR (Welvaert & Rosseel, 2013). To investigate the relation of SNR to atom detectability, a set of ADF STEM images of 12.5 Å by 12.5 Å with a pixel size of 0.25 Å has been simulated of an individual Au atom using MULTEM (Lobato et al., 2016; Lobato & Van Dyck, 2015) to which an arbitrary constant background has been added accounting for the contribution of electrons scattered by an amorphous substrate. The remaining simulation parameters are listed in Table 7.1. Each simulated image has been generated 1000 times containing random Poisson noise. The SNR of the atom in the image has been altered by changing the incoming electron dose ranging from 10^3 $e^-/Å^2$ to 10^5 $e^-/Å^2$ resulting in a higher SNR value for a higher incoming electron dose. For detecting the Au atom from the noise disturbed images by the MAP probability rule, a model assuming the image of the atom to be Gaussian shaped has been used, where a constant background ζ and width ρ, height η and x- and y-coordinate β_x and β_y of the atom need to be estimated. The prior density $p(\boldsymbol{\theta}|N)$ has been chosen as a product of uniform distributions for each parameter, in correspondence with Subsection 6.3.1 of Chapter 6, where the parameters ζ and η range from 0 up to the maximum pixel intensity in the simulated image, whereas the parameters ρ, β_x and β_y range according to the field of view of the image, i.e. from 0 Å up

Table 7.1 Microscope parameter values for simulation of a set of ADF STEM images of 12.5 Å by 12.5 Å using MULTEM. (Reprinted from Fatermans et al. (2019) with permission from Elsevier.)

Parameter	Symbol	Value
Acceleration voltage	V_0 (kV)	120
Defocus	ϵ (Å)	0
Spherical aberration	C_s (mm)	0.001
Spherical aberration of 5th order	C_5 (mm)	0
Semiconvergence angle	α (mrad)	21.3
Detector inner radius	β_1 (mrad)	28
Detector outer radius	β_2 (mrad)	172
FWHM of the source image	FWHM (Å)	0.7

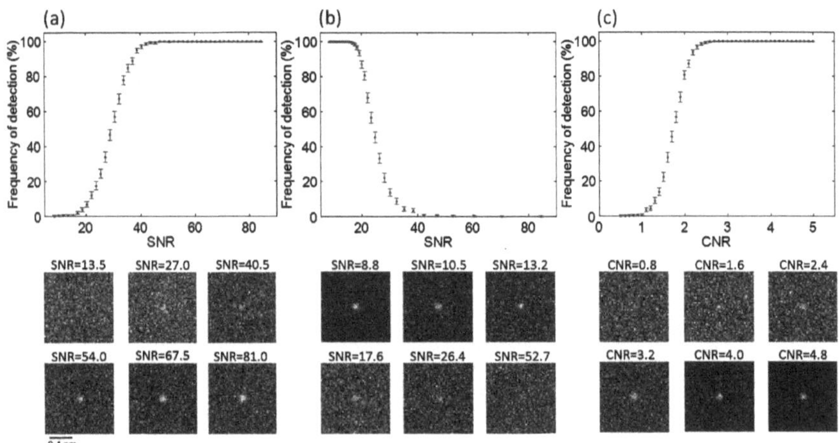

Figure 7.2 (a) Observed detection rate of an Au atom by the MAP probability rule from simulated ADF STEM images of 12.5 Å by 12.5 Å with a pixel size of 0.25 Å as a function of SNR using the simulation parameters listed in Table 7.1 where the SNR has been altered by changing the incoming electron dose ranging from 10^3 $e^-/Å^2$ to 10^5 $e^-/Å^2$ and (b) by changing the constant background for a fixed incoming electron dose of 10^4 $e^-/Å^2$. (c) Detection rate from the same set of images as in (b) as a function of CNR. Below the figures, the simulated images disturbed by Poisson noise for different values of SNR and CNR are shown. (Reprinted from Fatermans et al. (2019) with permission from Elsevier.)

to 12.5 Å. Fig. 7.2(a) shows the observed detection rates of the MAP probability rule as a function of SNR. The error bars show the 95% confidence Wilson score intervals of a binomial distribution (Wilson, 1927). At the bottom of Fig. 7.2, simulated STEM images containing Poisson noise for different SNR values are depicted. Alternatively, the SNR value of the Au atom in the simulated image can be altered by changing the added constant background independently of the incoming electron dose, simulating the effect of the atom to be positioned on a substrate with varying thickness. By altering the SNR by this procedure for a fixed incoming electron dose of 10^4 $e^-/\text{Å}^2$ and by using the same approach for the MAP probability rule as for Fig. 7.2(a), Fig. 7.2(b) shows that the detection rate decreases with increasing SNR value, as opposed to Fig. 7.2(a). Apparently, a high SNR value does not guarantee high atom detectability. This is also visually perceived by the images below Fig. 7.2(b) showing Poisson disturbed simulated STEM images for different SNR values. The reason for this behavior lies in the fact that the SNR measure given by Eq. (7.1) only considers the total sum of the height of the atomic column η and the background of the image ζ. Therefore, SNR does not take image contrast into account and it is possible that the SNR of an image is high while contrast is low. Contrast, however, also affects the visual perception and detectability of objects in an image, as is apparent from Fig. 7.1. An alternative measure to describe the quality of an image is given by the CNR (Welvaert & Rosseel, 2013). When a parametric model is used to describe the background and atomic columns in an ADF STEM image, the CNR of a column can be defined as

$$CNR = \frac{\eta}{[\eta + \zeta]^{1/2}}. \tag{7.2}$$

Note that the definition of CNR in Eq. (7.2) is closely related to the definition of SNR in Eq. (7.1), but in case of CNR the background ζ is subtracted before taking the ratio. As such, the CNR also takes image contrast into account. Therefore, CNR relates better to atom detectability than SNR. This is confirmed by Fig. 7.2(c) where the observed detection rate is shown as a function of the CNR for the same set of images as for Fig. 7.2(b). From this, it is seen that the detection rate increases with increasing CNR. It is noted that the same behavior is observed for the set of images of Fig. 7.2(a).

7.2.2 Integrated CNR

As followed from the previous section, CNR is a more intuitive image-quality measure than SNR when it comes to detecting atoms from ADF STEM images. In this section, the relation between the CNR measure and atom detectability is investigated for variations of atom type, incoming

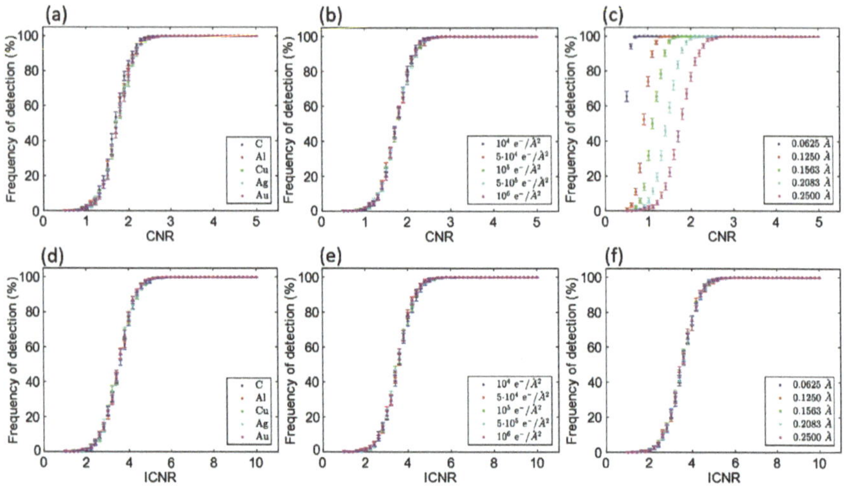

Figure 7.3 Observed detection rate of an individual atom by the MAP probability rule from simulated HAADF STEM images of 12.5 Å by 12.5 Å with a pixel size of 0.25 Å for varying atom types with an incoming electron dose of 10^6 $e^-/Å^2$ as a function of (a) CNR and (d) ICNR. Detection rate of an Au atom from simulated HAADF STEM images of 12.5 Å by 12.5 Å with a pixel size of 0.25 Å for varying incoming electron dose as a function of (b) CNR and (e) ICNR, and from images with an incoming electron dose of 10^5 $e^-/Å^2$ for varying pixel size as a function of (c) CNR and (f) ICNR. (Reprinted from Fatermans et al. (2019) with permission from Elsevier.)

electron dose and image pixel size. First, the detection rate of the MAP probability rule for different types of individual atoms as a function of CNR is shown in Fig. 7.3(a) for simulated ADF STEM images of 12.5 Å by 12.5 Å with a pixel size of 0.25 Å, using MULTEM with an incoming electron dose of 10^6 $e^-/Å^2$, where each simulated image has been generated 1000 times containing Poisson noise. The remaining simulation parameters can be found in Table 7.1 in Subsection 7.2.1. The CNR of the atom in the image has been altered by adding a constant background. For detecting the atom from the noise disturbed images by the MAP probability rule, the same approach as in Fig. 7.2 has been followed with an equivalent choice for the prior density $p(\boldsymbol{\theta}|N)$. It follows from Fig. 7.3(a) that the relation between CNR and detection rate depends hardly, yet slightly, on the atom type. Next, the relation between CNR and detection rate to varying incoming electron dose is investigated. For this, images of 12.5 Å by 12.5 Å with a pixel size of 0.25 Å of an individual Au atom have been simulated with different incoming electron doses. Fig. 7.3(b) shows that the relation between CNR and the atom detection rate is rather robust to incoming electron dose. On the other hand, a different pixel size influences the detection rate as a function of CNR significantly, as shown in Fig. 7.3(c) for images of an Au atom with an incoming electron dose of 10^5 $e^-/\text{Å}^2$ and varying pixel size. This is due to the fact that, for a fixed incoming electron dose/Å^2, an increased image pixel size leads to an increased electron dose/pixel. The increased value for electron dose/pixel results into a higher value for the background ζ and height η of the Au atom, since both ζ and η scale with electron dose/pixel, and hence into a higher CNR given by Eq. (7.2) and vice versa. For this reason, the integrated CNR (ICNR) is proposed whose relationship with atom detectability is independent of atom type, incoming electron dose and, in particular, the pixel size of the image. The ICNR of an atomic column in ADF STEM images is defined as the ratio of the total intensity of electrons scattered by the column, the so-called scattering cross-section (De Backer et al., 2016; E et al., 2013; Van Aert et al., 2013; Van Aert, Verbeeck, et al., 2009), to the square root of the sum of the scattering cross-section and the integrated background under the column. When a parametric model based on a superposition of Gaussian peaks is used to describe the background and atomic columns, the ICNR of a column can be calculated as

$$ICNR = \frac{2\pi \eta \rho^2}{[2\pi \eta \rho^2 + \pi (3\rho)^2 \zeta]^{1/2}}, \tag{7.3}$$

where ρ denotes the estimated width of the atomic column. In this work, ICNR values are calculated by expressing ρ in units of pixels. In order to estimate the integrated background under the column, the area under the column has been considered to be a circle with a radius of 3ρ, since 99.46% of the volume under the Gaussian peak describing the column is contained within this distance. It should be noted that Eq. (7.3) is only valid for individual atomic columns that are well separated in the image. For denser crystallographic structures, the contrast of a column depends, not only on the height of the column η and on the background in the image ζ, but also on the heights of and the distances from the surrounding columns. For the investigation of the relation between ICNR and atom detectability to varying atom type, incoming electron dose and pixel size, the same procedure for the simulated images has been followed as for the investigation of CNR. It is shown in Figs. 7.3(d) and (e) that the detection rate does not change with atom type or incoming electron dose, respectively, as long as the ICNR value remains unchanged. The functional relationship between detection rate and ICNR is also independent of pixel size, as shown in

Fig. 7.3(f), as opposed to the relationship between detection rate and CNR, as shown in Fig. 7.3(c). The results indicate that ICNR is a more robust measure for atom detectability from high-resolution high-angle (HAADF) STEM images than CNR.

The proposed image-quality measure for an individual atomic column given by Eq. (7.3) has been introduced specifically for ADF STEM images and can, in principle, explicitly be written as

$$ICNR = \frac{2\pi \eta^{ADF} (\rho^{ADF})^2}{[2\pi \eta^{ADF} (\rho^{ADF})^2 + \pi (3\rho^{ADF})^2 \zeta^{ADF}]^{1/2}}, \tag{7.4}$$

where η^{ADF}, ρ^{ADF}, and ζ^{ADF} denote the estimated height and width of the column, and the background in the ADF image, respectively. A similar expression for ICNR can also be defined for ABF STEM images:

$$ICNR = \frac{2\pi |\eta^{ABF}| (\rho^{ABF})^2}{[2\pi |\eta^{ABF}| (\rho^{ABF})^2 + \pi (3\rho^{ABF})^2 (\zeta^{ABF} + \eta^{ABF})]^{1/2}}, \tag{7.5}$$

where, similarly as for Eq. (7.4), η^{ABF}, ρ^{ABF}, and ζ^{ABF} denote the estimated height and width of the column, and the background in the ABF image, respectively. Note that η^{ABF} in Eq. (7.5) has a negative value, because columns in ABF images are displayed as dark spots on a bright background. Although, at first sight, Eqs. (7.4) and (7.5) may look different, they, in fact, represent the same formula. This can be better understood by considering the maximum pixel values, w_{max}^{ADF} and w_{max}^{ABF}, and minimum pixel values, w_{min}^{ADF} and w_{min}^{ABF}, of an ADF and ABF image, respectively, in the area around the column. As such, the following expressions hold:

$$\begin{aligned} w_{max}^{ADF} - w_{min}^{ADF} &\approx \eta^{ADF} \\ w_{max}^{ABF} - w_{min}^{ABF} &\approx |\eta^{ABF}| \\ w_{min}^{ADF} &\approx \zeta^{ADF} \\ w_{min}^{ABF} &\approx \zeta^{ABF} + \eta^{ABF}. \end{aligned} \tag{7.6}$$

From Eq. (7.6), both Eqs. (7.4) and (7.5) can be rewritten as

$$ICNR \approx \frac{2\pi (w_{max} - w_{min}) \rho^2}{[2\pi (w_{max} - w_{min}) \rho^2 + \pi (3\rho)^2 w_{min}]^{1/2}}, \tag{7.7}$$

where w_{max} represents either w_{max}^{ADF} or w_{max}^{ABF}, w_{min} either w_{min}^{ADF} or w_{min}^{ABF}, and ρ either ρ^{ADF} or ρ^{ABF} depending on what type of image is analyzed.

Table 7.2 Microscope parameter values for simulation of an ABF and ADF STEM image of an individual Si atom on a 5 nm thick C support using MULTEM. (Reprinted from Fatermans et al. (2020) with permission from Elsevier.)

Parameter	Symbol	Value
Acceleration voltage	V_0 (kV)	120
Defocus	ϵ (Å)	0
Spherical aberration	C_s (mm)	0.001
Spherical aberration of 5th order	C_5 (mm)	0
Semiconvergence angle	α (mrad)	21.0
ABF detector inner radius	β_1^{ABF} (mrad)	10
ABF detector outer radius	β_2^{ABF} (mrad)	20
ADF detector inner radius	β_1^{ADF} (mrad)	25
ADF detector outer radius	β_2^{ADF} (mrad)	100
Pixel size	$\Delta x = \Delta y$ (Å)	0.25
Number of scanned pixels	$K \times L$	50 × 50
FWHM of the source image	FWHM (Å)	0.7

For investigating atom detectability from ABF and ADF STEM images, an ABF and ADF image of an individual Si atom, deposited on a 5 nm thick amorphous C support, have been simulated using MULTEM. The amorphous C support has been obtained by placing C atoms on random positions within the volume of the layer, reaching a density of 2.2 g × cm^{-3} (Ricolleau et al., 2013). During this procedure, a new C atom is only included if its distance with the other atoms extends a chosen minimal distance of 0.14 nm, which is slightly lower than the C-C bond length in graphene. The parameters of this simulation are listed in Table 7.2. A varying incoming electron dose has been applied to the simulated images in order to obtain a set of images containing different ICNR values, where a higher electron dose results into a higher ICNR value and a lower dose into a lower ICNR value. It is noted that in this procedure the simulated ABF and ADF STEM data are not necessarily subject to the same incoming electron dose. In this way, a great variety of ICNR values for the ABF and ADF images can be obtained. The ICNR of the ADF STEM images has been calculated using Eq. (7.4), whereas Eq. (7.5) has been used for the ABF STEM images. Each image obtained in this manner has been generated 100 times containing random Poisson noise. Figs. 7.4(a) and (b) show the observed detection rate of the Si atom, i.e. the number of times the atom can be successfully detected from the noise disturbed simulated images by simultaneously analyzing the ABF and ADF STEM image data, using the

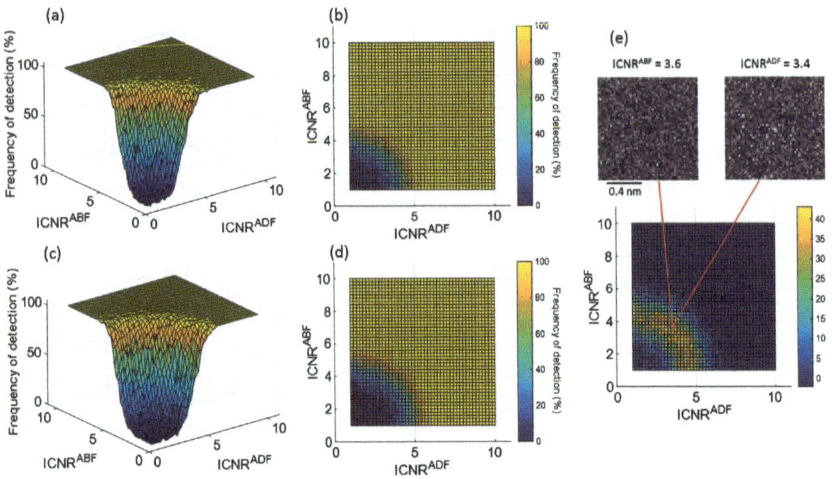

Figure 7.4 (a) Observed detection rate of a Si atom by the MAP probability rule from Poisson noise disturbed simulated ADF and ABF STEM images as a function of $ICNR^{ADF}$ and $ICNR^{ABF}$ by simultaneous model fitting. (b) Top view of (a). (c) Observed detection rate by separate model fitting. (d) Top view of (c). (e) Difference of detection between (b) and (d), where the inset shows the Poisson noise disturbed ABF and ADF images for which the largest difference between (b) and (d) is attained.

methodology described in Section 2.5 of Chapter 2. The detection rate is displayed as a function of the ICNR of the ADF and ABF images, depicted as $ICNR^{ADF}$ and $ICNR^{ABF}$, respectively, and is calculated by the MAP probability rule given by Eq. (6.50) evaluated for $N = 0$ and $N = 1$ atomic columns using a parametric model assuming the images of the atom to be Gaussian shaped, where the backgrounds ζ^{ADF} and ζ^{ABF}, and widths ρ^{ADF} and ρ^{ABF}, heights η^{ADF} and η^{ABF} and x- and y-coordinate β_x and β_y of the atom need to be estimated. It is clear from Figs. 7.4(a) and (b) that the detection rate decreases with decreasing ICNR values. Moreover, the circular symmetry indicates that atom detectability is independent of the exact $ICNR^{ADF}$ and $ICNR^{ABF}$ values as long as the square root of their quadratic sum remains unchanged. As a result, an alternative ICNR expression, directly correlating with atom detectability, can be proposed for determining the combined quality of atomic columns in ABF and ADF STEM images:

$$ICNR = \sqrt{(ICNR^{ADF})^2 + (ICNR^{ABF})^2}. \qquad (7.8)$$

Figs. 7.4(c) and (d) show the observed detection rate of the Si atom from the noise disturbed images by analyzing the ABF and ADF image data without taking into account that both images depict the same atom. Hereby, a model is used, such as the one given by Eq. (2.53), where ζ^{ADF}, ζ^{ABF}, ρ^{ADF}, ρ^{ABF}, η^{ADF}, η^{ABF}, β_x^{ADF}, β_y^{ADF}, β_x^{ABF} and β_y^{ABF} need to be estimated, where β_x^{ADF} and β_y^{ADF}, and β_x^{ABF} and β_y^{ABF} denote the x- and y-coordinate of the fitted Gaussian in the ADF and ABF images, respectively. Note that this model contains two extra parameters as compared to the model used to obtain Figs. 7.4(a) and (b). Qualitatively, Figs. 7.4(c) and (d) show the same behavior as Figs. 7.4(a) and (b), but the overall detection rate of the Si atom is lower. This has been explicitly shown in Fig. 7.4(e) depicting the difference between Figs. 7.4(b) and (d). The positive values for the difference of detection indicate that the detection rate can be up to 40% higher for simultaneous model fitting. The inset of Fig. 7.4(e) shows the Poisson noise disturbed ABF and ADF STEM images for which there is the largest difference between the simultaneous and separate analysis. From this investigation, it follows that simultaneously analyzing ABF and ADF STEM images is advantageous, as compared to a separate analysis, since it positively affects atom detectability.

Since the proposed image-quality measure of ICNR is closely related to atom detectability, it forms a direct explanation of why certain atomic columns in image data can or cannot be detected. The MAP probability rule is able to retrieve the atomic columns present in the image data when the ICNR values of the columns are sufficiently high. For example, for the STEM image data given in Fig. 6.9, the average ICNR values of each column type have been calculated, using Eq. (7.8), to be around 8.7, 6.0, and 7.4 for the Sr, Ti-O, and O columns, respectively. From the analysis given in Fig. 7.4, it follows that for such ICNR values the MAP probability rule can reliably detect atomic columns from simultaneously acquired ABF and ADF STEM data. When the ABF and ADF STEM images are separately analyzed though, it can be seen from Fig. 6.9(e) that from only analyzing the ADF STEM image only some of the Sr columns can be detected. This is due to the fact that the average ICNR value of the Sr columns in the ADF image, calculated from Eq. (7.4), is around 4.0, which typically restricts a 100% detection rate, as shown in Fig. 7.3. The average ICNR values in the ADF image of the other types of columns in $SrTiO_3$ are even lower, around 1.9 for the Ti-O and 0.6 for the O columns, restricting detection of these columns even further. The average ICNR values of the atomic columns in the ABF image of Fig. 6.9, though, calculated from Eq. (7.5), are higher, around 7.7, 5.7, and 7.4 for the Sr, Ti-O, and O columns, respectively, improving their detectability. As another example, in Fig. 6.10, the average

ICNR values from the combined ABF and ADF STEM signal of the Li, Mn, and O columns have been estimated by Eq. (7.8) to be around 7.7, 15.0, and 7.8, respectively, which, following the analysis of Fig. 7.4, allows for reliable atomic column detection by the MAP probability rule. Nevertheless, two atomic columns were not detected, although expected by the spinel structure of $LiMn_2O_4$. The reason for this is related to the ICNR values of these individual columns. It has been estimated by Eq. (7.8) that these atomic columns exhibit ICNR values of around 5.4. From the analysis performed in Fig. 7.4, it followed that for such ICNR values typically a 100% detection rate cannot be attained.

Besides comparing the atom detectability for a simultaneous and separate analysis of ABF and ADF STEM images, also the deviation from the true atom position and its standard deviation can be evaluated as a function of atom detectability or image quality expressed by ICNR. For this, the set of images of the Si atom on a 5 nm amorphous C support is used and analyzed in the same way as described before. Hereby, a model containing one atomic column has been considered. The deviation d_i of the estimated position of the Si atom $(\hat{\beta}_x, \hat{\beta}_y)$ from the true atom position (β_x, β_y) for the ith pair of noise disturbed simulated ABF and ADF STEM images for a certain $ICNR^{ADF}$ and $ICNR^{ABF}$ value from simultaneously fitting the ABF and ADF image data is investigated. The expression for d_i is given by

$$d_i = \sqrt{(\hat{\beta}_x - \beta_x)^2 + (\hat{\beta}_y - \beta_y)^2}. \tag{7.9}$$

Fig. 7.5(a) shows a top view image of the average deviation $\overline{d}$ as a function of $ICNR^{ADF}$ and $ICNR^{ABF}$. As such,

$$\overline{d} = \frac{1}{I}\sum_{i=1}^{I} d_i, \tag{7.10}$$

with $I = 100$ for each combination of $ICNR^{ADF}$ and $ICNR^{ABF}$. The standard deviation of $\overline{d}$, $\sigma_{\overline{d}}$, is given by

$$\sigma_{\overline{d}} = \frac{\sigma}{\sqrt{I}}, \tag{7.11}$$

with

$$\sigma = \sqrt{\frac{\sum_{i=1}^{I}(d_i - \overline{d})^2}{I-1}}, \tag{7.12}$$

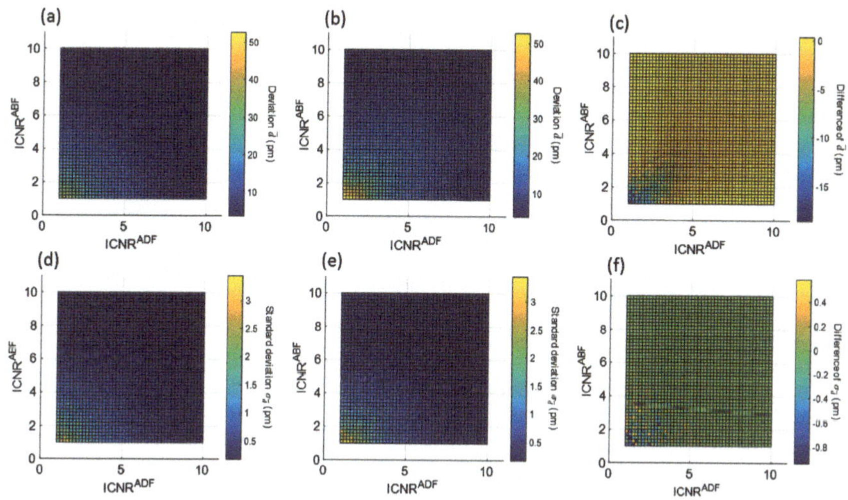

Figure 7.5 (a) Deviation of locating a simulated Si atom from its true position from Poisson noise disturbed simulated ADF and ABF STEM images as a function of $ICNR^{ADF}$ and $ICNR^{ABF}$ by simultaneous model fitting. (b) Deviation by separate model fitting. (c) Difference of deviation between (a) and (b). (d) Standard deviation of locating the Si atom by simultaneous model fitting. (e) Standard deviation by separate model fitting. (f) Difference of standard deviation between (d) and (e).

denoting the standard deviation of d_i. It can be concluded from Fig. 7.5(a) that for high ICNR values the deviation of the estimated location from the true position of the Si atom is small. For decreasing ICNR values, the deviation increases. Fig. 7.5(b) shows the results obtained from a separate analysis of the ABF and ADF image data. Here, $\overline{d}$ has been calculated using a weighted average where the weights have been chosen to be equal to the reciprocal of the variances,

$$\overline{d} = \frac{\frac{\overline{d^{ABF}}}{(\sigma_{\overline{d^{ABF}}})^2} + \frac{\overline{d^{ADF}}}{(\sigma_{\overline{d^{ADF}}})^2}}{\frac{1}{(\sigma_{\overline{d^{ABF}}})^2} + \frac{1}{(\sigma_{\overline{d^{ADF}}})^2}}. \tag{7.13}$$

The expression in Eq. (7.10) has been used to compute $\overline{d^{ABF}}$ and $\overline{d^{ADF}}$, where d_i, given by Eq. (7.9), represents either d_i^{ABF} or d_i^{ADF}, corresponding to the deviation from the true atom position (β_x, β_y) of the estimated position of the Si atom obtained from the ABF image, $(\hat{\beta}_x^{ABF}, \hat{\beta}_y^{ABF})$, or ADF image, $(\hat{\beta}_x^{ADF}, \hat{\beta}_y^{ADF})$, respectively. The standard deviations $\sigma_{\overline{d^{ABF}}}$ and $\sigma_{\overline{d^{ADF}}}$ are calculated from Eq. (7.11). Fig. 7.5(b) shows the same qualitative behavior as Fig. 7.5(a), but the overall deviation from the true atom position is smaller for Fig. 7.5(a). This is illustrated in Fig. 7.5(c) showing the difference of deviation between Figs. 7.5(a) and (b). The negative differences indicate that the deviation from the true atom position is smaller for simultaneously analyzing the ABF and ADF STEM images as compared to analyzing these images in a separate manner. The standard deviation of locating the Si atom from ABF and ADF image data has been displayed in Fig. 7.5(d) and (e) for simultaneously and separately modeling the ABF and ADF images, respectively. For Fig. 7.5(d), the standard deviation $\sigma_{\overline{d}}$ is given by Eq. (7.11). The standard deviation for Fig. 7.5(e) is calculated using

$$\sigma_{\overline{d}} = \sqrt{\frac{1}{\frac{1}{(\sigma_{\overline{d}ABF})^2} + \frac{1}{(\sigma_{\overline{d}ADF})^2}}}, \tag{7.14}$$

corresponding to the standard error of a weighted mean. It can be seen from Figs. 7.5(d) and (e) that $\sigma_{\overline{d}}$ increases for decreasing ICNR values and vice versa. The difference between Figs. 7.5(d) and (e) is shown in Fig. 7.5(f), where the negative values indicate that the standard deviation of locating the atom is, in general, smaller for a simultaneous analysis as compared to a separate analysis of ABF and ADF images. This shows that, beside a smaller spatial deviation, also the standard deviation improves by simultaneously analyzing ABF and ADF STEM images.

7.3. Relation to model selection

7.3.1 Model-order selection criteria

The heart of model selection consists of selecting one model from a set of competing models that represents most closely the underlying process that generated the experimental data. For this purpose, a criterion measuring how well the model fits the data is required. Such a criterion of goodness of fit quantifies the descriptive adequacy of a model, which is possible by, for example, a maximum likelihood evaluation. However, a model-selection criterion based solely on goodness of fit automatically selects the model which fits best to the data. This is undesired, since model fit can be easily improved by increased model complexity, referring to the flexibility of a model to fit the observed data. In this way, a model might be selected without necessarily bearing any interpretable relationship with the underlying data-generating process. For this reason, typical model-selection methods take into account both the goodness of fit and the complexity of the models under investigation (Myung, 2000).

There are at least three important factors that contribute to the complexity of a model (Myung & Pitt, 1997). The first one is the number of parameters. In general, a model with many parameters describes data better than a model with few parameters due to its higher flexibility, and hence complexity. Next, model complexity is also related to functional form, which is described as the way in which the parameters are combined in the model. A model with a more complex functional form is able to describe a wider range of data and can be considered to be more flexible than a model with a less complex functional form. The last dimension of model complexity is covered by the extension of the parameter space. A model of which the parameters can fluctuate over a wide range of values can describe a wider range of data. Therefore, such a model is considered to be more complex than a model of which the parameters can fluctuate over only a small range. All of these three aspects can significantly and independently influence model fit.

In Subsection 4.1.2 of Chapter 4, some typical model-selection criteria were already mentioned within the concept of atom counting. Here, for clarity, some of these criteria will be repeated within the framework of atom column detection. As mentioned by Eq. (4.14), model-selection criteria are typically written as twice the negative log-likelihood function, accounting for the goodness of fit, plus a penalty term that accounts for model complexity C (Claeskens & Hjort, 2008):

$$-2\log(\hat{L}) + 2C, \tag{7.15}$$

with

$$\hat{L} = L(\hat{\boldsymbol{\theta}}), \tag{7.16}$$

where $\hat{\boldsymbol{\theta}}$ denotes the maximum likelihood estimator. Different criteria have been proposed in the literature and describe the complexity of the model in a different way, often taking only one dimension of model complexity into account, i.e. the number of parameters (Stoica & Selen, 2004). Some common criteria have been listed below. The Akaike Information Criterion (AIC) (Akaike, 1974) can be written as

$$AIC = -2\log(\hat{L}) + 2M \tag{7.17}$$

with M the number of parameters in the model. A very similar criterion is the Generalized Information Criterion (GIC) (Broersen & Wensink, 1993), where the contribution of the penalty term can be modified by the parameter d:

$$GIC = -2\log(\hat{L}) + dM. \tag{7.18}$$

A clear guideline on how to choose the value of d is lacking. Different choices of $d = 3$ (Broersen & Wensink, 1996) and $d = 4$ (Stoica & Selen, 2004) are reported in the literature. These choices cause the GIC to penalize more heavily for the complexity of the model as compared to the AIC. Another common criterion is the Bayesian Information Criterion (BIC) (Schwarz, 1978):

$$BIC = -2\log(\hat{L}) + M\log(W) \tag{7.19}$$

with W the sample size, i.e. the number of datapoints. For a STEM image, W is equal to $K \times L$ pixels. As opposed to the AIC and GIC, the penalty term is dependent on the sample size. When $W > 8$, the BIC accounts more for the complexity of the model than the AIC. An alternative criterion is the Hannan-Quinn Information Criterion (HQC) (Hannan & Quinn, 1979) which replaces the $\log(W)$ factor in the BIC by the slower diverging quantity $\log[\log(W)]$:

$$HQC = -2\log(\hat{L}) + M\log[\log(W)]. \tag{7.20}$$

As mentioned before, model-selection methods consider a tradeoff between high goodness of fit and low model complexity in order to select the model which most closely describes the underlying process that generated the experimental data. In essence, the MAP probability rule, as described in detail in Chapter 6, is also a model-selection method where the posterior probability of the presence of N atomic columns in a STEM image of $W = K \times L$ pixels can be written in general terms as

$$p(N|\boldsymbol{w}) \propto \frac{\text{goodness of fit}}{\text{model complexity}}. \tag{7.21}$$

The numerator, goodness of fit, is the likelihood function evaluated at the maximum likelihood estimate $\hat{\boldsymbol{\theta}}$, as explained in Subsection 6.3.1 of Chapter 6:

$$p(\boldsymbol{w}|\hat{\boldsymbol{\theta}}, N) = \frac{e^{-\chi^2(\hat{\boldsymbol{\theta}})/2}}{\prod_{k=1}^{K}\prod_{l=1}^{L}\sqrt{2\pi w_{kl}}}, \tag{7.22}$$

with

$$\chi^2(\hat{\boldsymbol{\theta}}) = \sum_{k=1}^{K}\sum_{l=1}^{L} \frac{\left(w_{kl} - f_{kl}(\hat{\boldsymbol{\theta}})\right)^2}{w_{kl}}. \tag{7.23}$$

The full expression for the posterior probability $p(N|\boldsymbol{w})$ for a model with M parameters describing the atomic columns in a STEM image by Gaussian peaks with different widths is given in Subsection 6.3.1 of Chapter 6 as

$$\begin{aligned} p(N|\boldsymbol{w}) \propto & \frac{N!}{[(\beta_{x_{max}} - \beta_{x_{min}})(\beta_{y_{max}} - \beta_{y_{min}})(\rho_{max} - \rho_{min})]^N} \\ & \times \frac{e^{-\chi^2_{min}/2}}{(\eta_{max} - \eta_{min})^N(\zeta_{max} - \zeta_{min})} \times \frac{(4\pi)^{M/2}[det(\nabla\nabla\chi^2)]^{-1/2}}{\prod_{k=1}^{K}\prod_{l=1}^{L}\sqrt{2\pi w_{kl}}}. \end{aligned} \tag{7.24}$$

The prior density $p(\boldsymbol{\theta}|N)$ for such a model is given in Subsection 6.3.1 as

$$\begin{aligned} p(\boldsymbol{\theta}|N) = & \frac{1}{[(\beta_{x_{max}} - \beta_{x_{min}})(\beta_{y_{max}} - \beta_{y_{min}})(\rho_{max} - \rho_{min})]^N} \\ & \times \frac{1}{(\eta_{max} - \eta_{min})^N(\zeta_{max} - \zeta_{min})}. \end{aligned} \tag{7.25}$$

By considering Eq. (7.24), a mathematical expression for the model complexity in Eq. (7.21) can be provided. From Eq. (7.22), where $\chi^2(\hat{\boldsymbol{\theta}}) = \chi^2_{min}$, and from the expression for the prior density in Eq. (7.25), it follows that the posterior probability in Eq. (7.24) can be written as

$$p(N|\boldsymbol{w}) \propto \frac{p(\boldsymbol{w}|\hat{\boldsymbol{\theta}}, N)}{[det(\nabla\nabla\chi^2)]^{1/2}/[N!(4\pi)^{M/2}p(\boldsymbol{\theta}|N)]}. \tag{7.26}$$

Since the goodness of fit is given by $p(\boldsymbol{w}|\hat{\boldsymbol{\theta}}, N)$, the model complexity in Eq. (7.21) for the MAP probability rule is given by

$$\text{model complexity} = \frac{[det(\nabla\nabla\chi^2)]^{1/2}}{N!(4\pi)^{M/2}p(\boldsymbol{\theta}|N)}. \tag{7.27}$$

The term $\nabla\nabla\chi^2$ in Eq. (7.27) is related to the observed Fisher information matrix $\hat{J}$ (Dodge, 2003), as described in Subsection 2.2.2 of Chapter 2, which holds the information that is contained by the observed data about the unknown parameters $\boldsymbol{\theta}$. It is described by the Hessian matrix of minus the logarithm of the likelihood function $p(\boldsymbol{w}|\boldsymbol{\theta}, N)$, given by Eq. (7.22), evaluated at the maximum likelihood estimate $\hat{\boldsymbol{\theta}}$. As such,

$$\hat{J} = -\frac{\partial^2 log[p(\boldsymbol{w}|\boldsymbol{\theta}, N)]}{\partial\boldsymbol{\theta}\partial\boldsymbol{\theta}^T}\bigg|_{\theta=\hat{\theta}} = \frac{\partial^2[\chi^2(\boldsymbol{\theta})/2]}{\partial\boldsymbol{\theta}\partial\boldsymbol{\theta}^T}\bigg|_{\theta=\hat{\theta}} \tag{7.28}$$

where $\frac{\partial^2\chi^2(\boldsymbol{\theta})}{\partial\boldsymbol{\theta}\partial\boldsymbol{\theta}^T}\big|_{\theta=\hat{\theta}}$ can be written in short as $\nabla\nabla\chi^2$. The term $\nabla\nabla\chi^2$ is an $M \times M$ dimensional matrix and, therefore, it depends on the number of parameters. In addition, it explicitly contains the expectation model f_{kl}, describing the intensity of pixel (k, l) at position (x_k, y_l), as follows from Eq. (7.23). Therefore, the term $\nabla\nabla\chi^2$ in Eq. (7.27) takes into account two dimensions of model complexity, which are the number of parameters on the one hand and the functional form of the model on the other hand. The third dimension of model complexity, which is the extension of the parameter space, is described by the prior density $p(\boldsymbol{\theta}|N)$. By choosing $p(\boldsymbol{\theta}|N)$ as a product of uniform distributions for each parameter individually, large ranges for the possible values of the parameters correspond with a small value for $p(\boldsymbol{\theta}|N)$. Since the model complexity in Eq. (7.27) is inversely proportional to $p(\boldsymbol{\theta}|N)$, model complexity increases as the extension of the parameter space increases. As a result, the complexity term of the MAP probability rule depends on three dimensions of model complexity as opposed to other more common criteria, such as the AIC (Akaike, 1974), the GIC (Broersen & Wensink, 1993), which is an extension of AIC by introducing a general parameter d, the BIC (Schwarz, 1978) and the HQC (Hannan & Quinn, 1979), whose mathematical forms are given by Eqs. (7.17), (7.18), (7.19) and (7.20), respectively. Here, the expressions are given as an overview for comparison:

$$\begin{aligned} AIC &= -2log(\hat{L}) + 2M \\ GIC &= -2log(\hat{L}) + dM. \\ BIC &= -2log(\hat{L}) + Mlog(W) \\ HQC &= -2log(\hat{L}) + Mlog[log(W)]. \end{aligned} \tag{7.29}$$

It is clear from Eq. (7.29) that the complexity terms of these criteria only depend on the number of parameters M. By taking the logarithm of Eq. (7.26) multiplied by -2, the posterior probability $p(N|\boldsymbol{w})$ can be written in the same form as the model-selection criteria in Eq. (7.29):

$$\begin{aligned} -2log[p(N|\boldsymbol{w})] = &-2log(\hat{L}) + log[det(\nabla\nabla\chi^2)] - 2log(N!) \\ &- Mlog(4\pi) - 2log[p(\boldsymbol{\theta}|N)] + cst, \end{aligned} \tag{7.30}$$

where $p(\boldsymbol{w}|\hat{\boldsymbol{\theta}}, N)$ has been written as $\hat{L}$. The term *cst* in Eq. (7.30) refers to a constant coming from the proportionality of Eq. (7.26). From this expression, it can be directly seen that the MAP probability rule holds a more complex penalty term than the criteria given in Eq. (7.29).

The remainder of this section is organized as follows. In Subsection 7.3.2, the performance of the MAP probability rule in detecting atoms from STEM images is compared to that of other selection criteria. This is followed by Subsection 7.3.3 where the relation between the MAP probability rule and the BIC is explicitly shown. In Subsection 7.3.4, the effect of the prior density on atom detection is discussed.

7.3.2 Performance in detecting atoms

In this subsection, the performance of the MAP probability rule in detecting atoms from ADF STEM images is compared to the AIC, GIC, BIC and HQC, which were introduced in Eq. (7.29). For the GIC, $d = 3$ (Broersen & Wensink, 1996) and $d = 4$ (Stoica & Selen, 2004) are considered, which are referred to, in what follows, as GIC3 and GIC4, respectively. A set of 1000 images of Au atoms has been simulated by MULTEM with dimensions of 12.5 Å by 12.5 Å and a pixel size of 0.25 Å. The remaining simulation parameters are listed in Table 7.1 in Subsection 7.2.1. Hereby, the number of Au atoms in an individual image is uniformly distributed between 1 and 5 atoms. The atoms are randomly positioned within the field of view of the image according to a uniform distribution and the incoming electron dose of the image can uniformly fluctuate between $5 \cdot 10^3$ $e^-/Å^2$ and 10^5 $e^-/Å^2$, affecting the peak intensities of the atoms. Depending on the incoming electron dose, a constant background has been added to an individual image such that all 1000 simulated images have the same ICNR value. Hereby, each simulated image has been disturbed by Poisson noise. The performances of the model-selection criteria are evaluated by comparing the detected number of atoms in the image with the true number of atoms by calculating the average frequency of various selected orders (Stoica et al., 2004). For the analysis, a superposition of Gaussian peaks with equal widths and equal intensities has been used to model the Au atoms in the image. Since the width ρ has been considered to be a fixed value, the parameters to be estimated are the background ζ, height η and x- and y-coordinates β_{x_n} and β_{y_n}. For the MAP probability rule, the prior density $p(\boldsymbol{\theta}|N)$ was chosen as a product of uniform distributions where the background ζ ranges from 0 up to the maximum pixel intensity in the simulated image and where the distributions for the height η and x- and

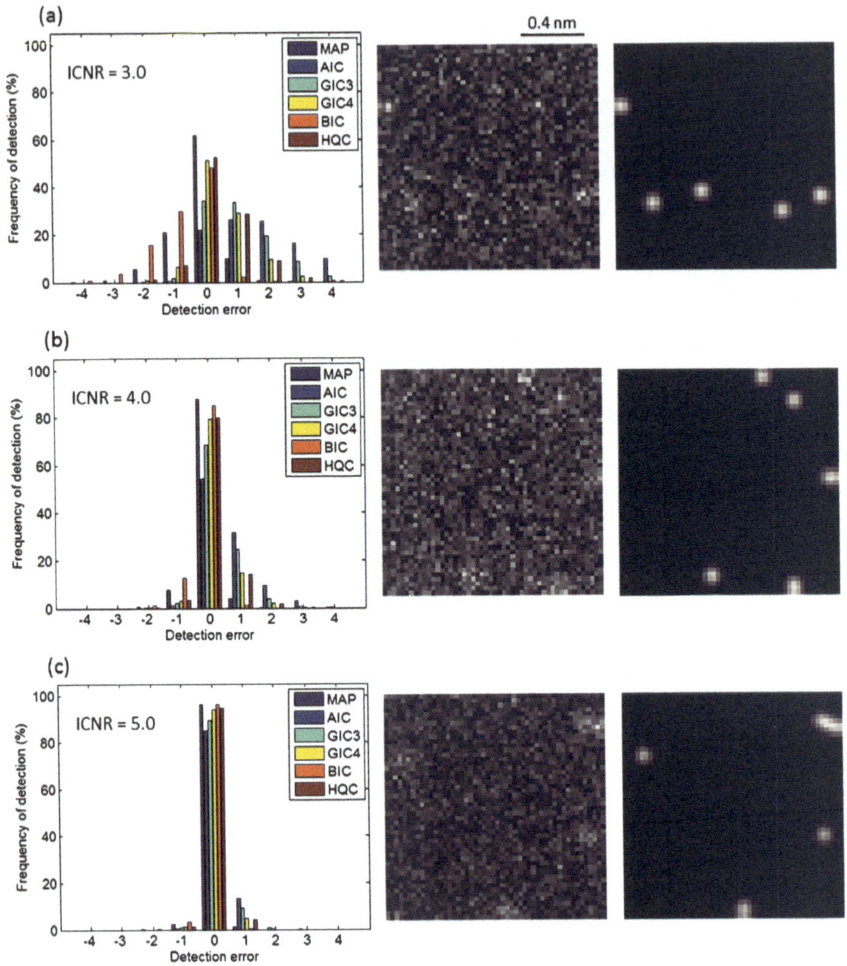

Figure 7.6 Average frequency of various selected orders for different model selection-criteria for detecting the number of Au atoms from a set of simulated HAADF STEM images of 12.5 Å by 12.5 Å with a pixel size of 0.25 Å and with (a) an ICNR value of 3.0, (b) an ICNR value of 4.0, and (c) an ICNR value of 5.0. At the right-hand side, randomly generated simulated images of 5 atoms disturbed by Poisson noise with the noise-free images as references are shown for each ICNR value. (Reprinted from Fatermans et al. (2019) with permission from Elsevier.)

y-coordinates β_{x_n} and β_{y_n} exactly correspond with the uniform distributions that were used to generate these parameters. The performances of the model-selection criteria are shown in Fig. 7.6 for three different ICNR values, defined by Eq. (7.3). A positive value of the detection error refers to the detection of too many atoms, whereas a negative value refers to the detection of too few atoms. Next to the graphs, randomly generated simulated images of 5 atoms disturbed by Poisson noise with the noise-free images as references are shown for each ICNR value. From Fig. 7.6(a), it can be seen that the MAP probability rule outperforms the other criteria for an ICNR value of 3.0. The AIC, GIC3, GIC4, and HQC have a tendency to detect too many atoms, whereas the BIC often detects too few atoms. This behavior is related to the different ways the model-selection criteria penalize the complexity of the model. From Subsection 7.2.2, showing that the detection rate increases when the ICNR of an image increases, it is expected that the frequency of detecting the correct number of atoms increases for increasing ICNR. This is confirmed in Fig. 7.6(b) and (c) for an ICNR value of 4.0 and 5.0, respectively. Moreover, in these cases, the performances of the different criteria tend to become more equal.

7.3.3 Relation between MAP and BIC

Previously, it has been explained that the MAP probability rule is related to the concept of model selection. In particular, it can be expected that the MAP probability rule has a close correspondence with the BIC as they both follow from Bayesian probability theory. The relation between the MAP probability rule and the BIC can be indicated starting from Eq. (7.30). The second term at the right-hand side of this expression can be written as

$$\begin{aligned} log[det(\nabla\nabla\chi^2)] &= log[det(W \cdot \frac{1}{W}\nabla\nabla\chi^2)] \\ &= Mlog(W) + log[det(\frac{1}{W}\nabla\nabla\chi^2)], \end{aligned} \tag{7.31}$$

where use is made of the fact that $det(sA_M) = s^M det(A_M)$ for a scalar s and $M \times M$ matrix A_M. As such, Eq. (7.30) can be rewritten as

$$\begin{aligned} -2log[p(N|\boldsymbol{w})] = &-2log(\hat{L}) + Mlog(W) + log[det(\frac{1}{W}\nabla\nabla\chi^2)] \\ &- 2log(N!) - Mlog(4\pi) - 2log[p(\boldsymbol{\theta}|N)] + cst. \end{aligned} \tag{7.32}$$

Interestingly, the first two terms of Eq. (7.32) correspond to the BIC given in Eq. (7.29). This indicates a relation between the MAP probability rule and the BIC, which is not surprising as both techniques are based on a Bayesian approach. In order to understand this relation, the behavior of the terms in Eq. (7.32) has been investigated as a function of the sample size by simulating a set of ADF STEM images of 12.5 Å by 12.5 Å of a single Au atom. The STEM simulations have been obtained from MULTEM and the simulation parameters are listed in Table 7.1 in Subsection 7.2.1. In order to acquire a set of simulated images containing an increasing number of pixels W, the pixel size has been decreased starting from 0.5 Å up to and including 0.05 Å. As such, 50 images have been simulated where for each image 20 random Poisson noise configurations have been applied. Hereby, the incoming electron dose has been set to 625 e^-/pixel to keep the amount of detected electrons per pixel constant, irrespective of the pixel size, so that the behavior of the terms of the MAP probability rule is only dependent on the number of pixels W. Note that since the number of e^-/pixel is kept constant throughout the analysis this means that the electron dose/Å^2 increases when the image pixel size decreases. For the analysis of the MAP probability rule, a model where the atom is assumed to be Gaussian shaped has been used consisting of $M = 5$ parameters: a constant background ζ, width ρ, height η and x- and y-coordinate β_x and β_y. The prior density $p(\boldsymbol{\theta}|N)$ has been chosen to be a product of uniform distributions where the parameters ζ and η range from 0 up to the maximum pixel intensity in the simulated image, whereas the parameters ρ, β_x and β_y range according to the field of view of the image, i.e. from 0 Å up to 12.5 Å. The different terms contributing to the MAP probability rule given by Eq. (7.32) are shown in Fig. 7.7. The term $-2log(\hat{L})$ depends on W and increases with an increasing number of pixels, but, for clarity, it has not been visualized since it would dominate the graphs as it has values up to three orders of magnitude larger than the other terms. It can be seen from Fig. 7.7(a) that the term $Mlog(W)$ increases as the number of pixels W increases, as expected. The terms $log[det(\frac{1}{W}\nabla\nabla\chi^2)]$, $-2log(N!)$, $-Mlog(4\pi)$ and $-2log[p(\boldsymbol{\theta}|N)]$ remain constant as W increases. As such, also the sum of these terms remains constant for more pixels. This sum has been depicted as $\mathcal{O}(1)$ in Fig. 7.7, denoting a term that tends to a constant as $W \to \infty$. In this way, Eq. (7.32) can be written for $W \to \infty$ as

$$\begin{aligned} -2log[p(N|\boldsymbol{w})] &= -2log(\hat{L}) + Mlog(W) + \mathcal{O}(1) \\ &\approx -2log(\hat{L}) + Mlog(W) \end{aligned} \tag{7.33}$$

implying that the MAP probability rule is asymptotically equivalent with the BIC since $-2log(\hat{L})$ and $Mlog(W)$ are the dominant terms when $W \to \infty$. Interestingly, the behavior of the Hessian matrix of $\chi^2(\boldsymbol{\theta})$ evaluated at $\hat{\boldsymbol{\theta}}$, $\nabla\nabla\chi^2$, and of the prior density, $p(\boldsymbol{\theta}|N)$, chosen as a product of

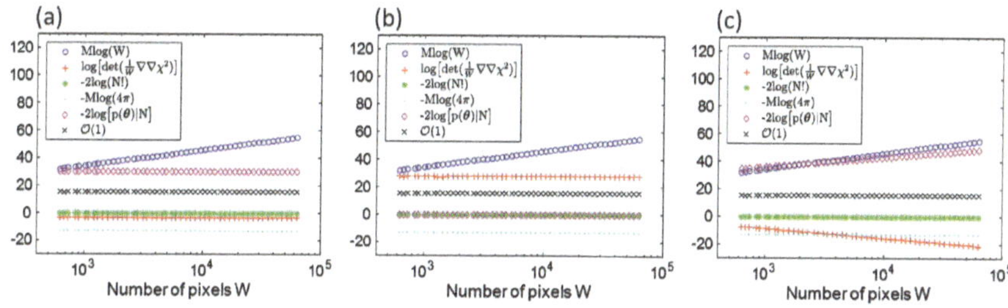

Figure 7.7 Behavior of the individual terms of the MAP probability rule in function of the number of pixels W for a set of simulated ADF STEM images of 12.5 Å by 12.5 Å of a single Au atom with a pixel size ranging from 0.5 Å until 0.05 Å and an incoming electron dose of 625 e^{-}/pixel. For clarity, the term $-2log(\hat{L})$ has not been visualized. Three different parameterizations of the model describing the STEM images have been used. In (a), the background ζ and height η have been fitted in electron counts and the width ρ and x- and y-coordinate β_x and β_y in Å, in (b), the parameters have been normalized between 0 and 1, and in (c), ζ and η have been fitted in electron counts and ρ, β_x and β_y in pixels. (Reprinted from Fatermans et al. (2019) with permission from Elsevier.)

uniform distributions for each parameter, is dependent on the parameterization of the model. Fig. 7.7(a) results from fitting the parameters ζ and η in pixel intensities, depicted in electron counts, and ρ, β_x and β_y in Å. A different parameterization of $\boldsymbol{\theta}$ might be to use a normalized model where the parameters can fluctuate between 0 and 1. Fig. 7.7(b) shows the behavior of the various terms of Eq. (7.32) of such a model. Here, the terms $log[det(\frac{1}{W}\nabla\nabla\chi^2)]$ and $-2log[p(\boldsymbol{\theta}|N)]$ have been shifted as compared to Fig. 7.7(a). Yet, they show the same constant behavior as a function of the number of pixels W. The behavior of these terms does not necessarily remain constant as a function of W under all parameterizations of $\boldsymbol{\theta}$ as shown in Fig. 7.7(c). Here, the parameters ζ and η have been fitted in electron counts and ρ, β_x and β_y in pixels. From Fig. 7.7(c), it is apparent that both $log[det(\frac{1}{W}\nabla\nabla\chi^2)]$ and $-2log[p(\boldsymbol{\theta}|N)]$ depend on the number of pixels W. Although $\nabla\nabla\chi^2$ and $p(\boldsymbol{\theta}|N)$ are not invariant under reparameterization of $\boldsymbol{\theta}$, yet the term $\mathcal{O}(1)$, which is the sum of $log[det(\frac{1}{W}\nabla\nabla\chi^2)]$, $-2log(N!)$, $-Mlog(4\pi)$ and $-2log[p(\boldsymbol{\theta}|N)]$, remains invariant under all presented parameterizations, as shown in Fig. 7.7. Therefore, the model complexity of Eq. (7.27) as described by the MAP probability rule remains unchanged under different parameterizations of the model. Moreover, as the goodness of fit is also independent of the model description, the MAP probability rule, given by the evaluation of the posterior probability $p(N|\boldsymbol{w})$, remains invariant under reparameterization of $\boldsymbol{\theta}$. It is noted that in general the MAP estimate following from $p(N|\boldsymbol{w})$ is not invariant.

7.3.4 Influence of prior density

Since the prior density $p(\boldsymbol{\theta}|N)$ is part of how the MAP probability rule determines model complexity, as shown in Eq. (7.27), atom detection in STEM images depends on the predefined parameter ranges when $p(\boldsymbol{\theta}|N)$ is defined as a product of uniform distributions for each parameter individually, as given by Eq. (7.25). Ideally, the result of a robust detection method should not depend strongly on the prior. In this section, the influence of different a priori chosen parameter ranges to atom detection is investigated as a function of ICNR. For this, ADF STEM images of 12.5 Å by 12.5 Å with a pixel size of 0.25 Å with different ICNR values have been simulated of an individual Au atom using MULTEM, where the atom is located in the middle of each simulated image, i.e. $\beta_x = \beta_y = 6.25$ Å. The remaining simulation parameters can be found in Table 7.1 in Subsection 7.2.1. Each image has been generated 1000 times containing random Poisson noise and the ICNR of the atom in the image has been altered by adding a constant background while keeping the incoming electron dose fixed to 10^4 e$^-$/Å^2. For detecting the Au atom from the noise disturbed images by the MAP probability rule, a model assuming the atom to be Gaussian shaped has been used, where a constant background ζ and width ρ, height η and x- and y-coordinate β_x and β_y of the atom need to be estimated. First, the parameters ζ and η have been chosen to range from 0 up to the maximum pixel intensity in the simulated images, whereas the parameters ρ, β_x and β_y range according to the field of view of the image, i.e. from 0 Å up to 12.5 Å. In order to investigate the effect of a different choice of the predefined parameter ranges, the ranges of β_x and β_y have been reduced, corresponding to taking into account more and more informative prior knowledge about the location of the Au atom. Fig. 7.8 shows that the detection rate as a function of ICNR is not influenced by different predefined ranges on β_x and β_y when the ICNR is high. For lower ICNR values, though, the detection rate of the Au atom increases for smaller predefined ranges on β_x and β_y. This shows that when more correct prior knowledge can be taken into account, it is beneficial to do so since it increases the chance of detecting atoms from low ICNR STEM images. As compared to the other model-selection criteria considered in this chapter, the MAP probability rule offers a more flexible way to detect atoms from ADF STEM images due to the fact that the prior can be tuned, resulting into a different value

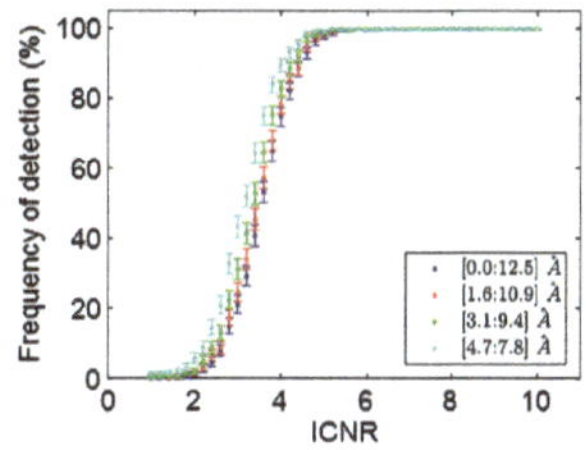

Figure 7.8 Detection rate of an Au atom from simulated ADF STEM images of 12.5 Å by 12.5 Å with a pixel size of 0.25 Å and an incoming electron dose of 10^4 $e^-/Å^2$ as a function of ICNR for different predefined ranges of β_x and β_y. (Reprinted from Fatermans et al. (2019) with permission from Elsevier.)

for the complexity of the model under consideration. Moreover, by using the MAP probability rule, it is clear what prior knowledge has been taken into account during the analysis, which is not always straightforward for other model-selection criteria.

7.4. Conclusions

In this chapter, it has been shown that the MAP probability rule can be effectively used as a tool to evaluate the relation between STEM image-quality measures and atom detectability. This has resulted into the introduction of the ICNR as a new image-quality measure that better correlates with atom detectability than conventional measures such as SNR and CNR. Atomic columns resulting from single images with ICNR values of less than around 5.0 become challenging to accurately detect since typically the detection rate of 100% drops rapidly starting from this value. The ICNR as a measure for image quality is not only valid for single images, but may also be used to estimate the atom detectability of atomic columns from simultaneously acquired ABF and ADF STEM images. The advantage of simultaneous model fitting, as compared to combining quantitative measurements from analyzing the ABF and ADF STEM images separately, has been discussed as a function of ICNR. It has been demonstrated that a simultaneous analysis improves atom detectability, as well as atom position accuracy and precision especially for ICNR values of less than around 5.0.

In addition, the relation of the MAP probability rule with model selection has been thoroughly investigated. Hereby, it has been explicitly shown that the MAP probability is asymptotically equivalent with the BIC. Interestingly, the complexity term of the MAP probability rule depends on three dimensions of model complexity, namely the number of parameters, the functional form of the model and the extension of the parameter space, as opposed to other model-selection criteria which typically depend on the number of parameters only. This extended complexity term allows the MAP probability rule to perform more accurately in detecting atoms from STEM images, especially from images exhibiting low ICNR values. Furthermore, the MAP probability rule allows for a clear and flexible incorporation of prior knowledge, which is often not the case for other model-selection methods.

Acknowledgments

This project has received funding from the European Research Council (ERC) under the European Union's Horizon 2020 research and innovation programme (Grant Agreement No. 770887 and No. 823717 ESTEEM3). The authors acknowledge financial support from the Research Foundation Flanders (FWO, Belgium) through Projects No. G.0502.18N, G.0267.18N, G.0368.15N, and a postdoctoral grant to A. D. B. The authors acknowledge funding from the University of Antwerp through a TOP project.

CHAPTER EIGHT

General conclusions and future perspectives

Annick De Backer[a,b,d], Jarmo Fatermans[a,b,c,d], Arnold J. den Dekker[c], and Sandra Van Aert[a,b,*]

[a]Electron Microscopy for Materials Science (EMAT), University of Antwerp, Antwerp, Belgium
[b]NANOlab Center of Excellence, University of Antwerp, Antwerp, Belgium
[c]imec-Vision Lab, University of Antwerp, Wilrijk, Belgium
*Corresponding author. e-mail address: sandra.vanaert@uantwerpen.be

Contents

In this work, atomic-scale aberration-corrected (scanning) transmission electron microscopy ((S)TEM) is pushed toward accurate and precise measurements of unknown structure parameters using advanced statistical techniques in a model-based framework. For this purpose, statistical parameter estimation theory has extensively been used for measuring unknown structure parameters, including atomic column positions and the number of atoms, and to detect atomic columns as accurately and precisely as possible from experimental images. The starting-point of statistical parameter estimation is the availability of a parametric model describing the expectations of the images. It has been assumed that the image intensities in high-resolution (S)TEM images can be described by a superposition of Gaussian functions peaked at the atomic column positions. The unknown parameters of this model, i.e. the position coordinates, the width, and the height of these Gaussian functions, are estimated by fitting this model to the experimental images using a criterion of goodness of fit, the least squares sum, which quantifies the similarity between the experimental images and the model.

In the first part, the accuracy and precision with which structure parameters can be estimated using statistical parameter estimation theory have been evaluated for a focal series of high-resolution TEM images recorded

[d] Both authors contributed equally to this work.

Advances in Imaging and Electron Physics, Volume 217
ISSN 1076-5670
https://doi.org/10.1016/bs.aiep.2021.01.008

for exit wave reconstruction, for annular dark-field (ADF) STEM images, and for simultaneously acquired annular bright-field (ABF) and ADF STEM images. Herein, the attainable precision with which the atomic column positions can be estimated using a focal series of high-resolution TEM images has been investigated (De Backer et al., 2011). In theory, it was expected that the precision would considerably improve using the maximum likelihood estimator based on the full series of focal images. As such, the theoretical lower bound on the variances of the unknown atomic column positions can be attained. However, this approach is numerically demanding. Therefore, maximum likelihood estimation has been compared with the results obtained by fitting a model to the reconstructed exit wave rather than to the full series of focal images. By means of simulations, it has been shown that the reconstructed complex exit wave contains the same amount of information concerning the atomic column positions as the full series of focal images. Only for thin samples, which act as weak phase objects, this information can be retrieved from the phase of the reconstructed complex exit wave, which is often used only in quantitative studies of exit waves reconstructed from a focal series of images.

Next, the accuracy and precision has been evaluated for the atomic column positions and the scattering cross-sections for ADF STEM images (De Backer et al., 2016). These scattering cross-sections correspond to the volume under the estimated Gaussian function and have been shown to be an excellent measure for interpreting the ADF STEM intensities in terms of number of atoms in a projected atomic column. Using detailed ADF STEM multislice simulations, different noise realizations have been obtained for different incident electron doses. It turned out that accurate and precise measurements can be obtained for the estimated parameters of interest. Moreover, in this study it has been shown that the theoretical lower bound on the variance could be attained for the estimates of the scattering cross-sections and position coordinates using the least squares estimator. This result is of great importance, since this means that the position coordinates and scattering cross-sections are estimated with the highest possible precision from the ADF STEM images.

Nowadays, there is an increasing interest of simultaneously acquiring ABF and ADF STEM images. Such a simultaneous acquisition combines the improved light-element visualization of ABF STEM with the easily interpretable image contrast of ADF STEM, allowing to visualize atoms of a large range of atomic numbers for studying and interpreting materials at the atomic scale consisting of both light and heavy atoms. In order to obtain accurate and precise measurements from such combined image data by using statistical parameter estimation theory, the commonly used parametric models for describing individual STEM images need to be extended (Fatermans et al., 2020). For ABF-ADF STEM, it has been shown that it is important to take into account the effect of specimen tilt. Small tilts of the electron beam with respect to the crystal zone axis can easily be present in STEM experiments and shift the ABF peak locations from the true atomic column locations. Not considering the effect of specimen tilt when modeling simultaneously acquired ABF-ADF STEM images may cause inaccuracies of the order of 0.1 Å on the atom column location estimates.

It has been shown that by using statistical parameter estimation theory high-precision measurements for structure parameters can be obtained using the least squares estimator. However, direct implementation of the least squares estimator in which all parameters are estimated at the same time, limits the field of view for which the parameters of the expectation model can be estimated. Therefore, an efficient algorithm has been proposed for the least squares estimation of the structure parameters (De Backer et al., 2016). This algorithm allows one to analyze larger nanostructures in a quantitative manner. Such analysis has been employed in order to reconstruct the 3D position and atom type of the atoms in an Au@Ag nanorod (Goris et al., 2013). In addition, displacement maps have been formed for the atomic column positions in a five-fold-twinned decahedral Au nanoparticle (Goris et al., 2015) and for the Al columns within $LaAlO_3$. These examples have illustrated the increased possibilities for a quantitative analysis at the atomic scale with high precision by the availability of the new algorithm. In order to provide end-users with this well-established quantification method, a user-friendly software package, StatSTEM, is developed which is freely available under a GNU public license (De Backer et al., 2016).

In the second part of this work, the focus lies on nanoparticle atom counting from ADF STEM images using advanced statistical methods (De Backer et al., 2013; Van Aert et al., 2013). First, the possibilities for counting the number of atoms of single element crystalline nanostructures in the dimension perpendicular to the recorded image using a statistics-based method have been investigated. In order to count the number of atoms, it is assumed that the total intensity of scattered electrons scales with the number of atoms per atomic column. These so-called scattering cross-sections are quantitatively determined using model-based statistical parameter estimation theory. The distribution describing the probability that the scattering cross-sections are generated by atomic columns containing a specific number of atoms is inferred on the basis of the experimental scattering cross-sections. This probability distribution, a Gaussian mixture model, consists of normal components which are smeared out due to a combination of experimental instabilities and detection noise. Finally, the number of atoms can be quantified using the estimated Gaussian mixture model. This method has been discussed in detail together with a thorough study on the inherent limitations. It has been demonstrated that the number of atomic columns available in the observed STEM image, the number of components in the estimated Gaussian mixture model, the width of the components of the Gaussian mixture model, and the typical shape of the criterion to assess the number of components in the Gaussian mixture model directly affect the accuracy and precision with which the number of atoms in a particular atomic column can be estimated. Nevertheless, it has been shown that single atom sensitivity is feasible taking the latter aspects into consideration.

Furthermore, it has been illustrated that the statistics-based atom-counting method can be combined with another complementary method. Here, the method has been combined with a simulation-based method. In the literature, both methods are independently proposed. The simulation-based method uses libraries of simulated scattering cross-sections which can be compared to normalized experimental scattering cross-sections. Although this approach is straightforward, it critically depends on accurate knowledge of the experimental detector settings and a priori structure information. In practice, these parameters require careful measurement or calibration and are not always available, hence limiting the applicability of this method. The statistics-based method can be used without prior knowledge concerning the structure, is independent of simulations, and provides the percentage of atomic columns for which the number of atoms is identified with single atom sensitivity. However, the reliability also depends on different parameters which can be linked with the quality of the recorded images. Therefore, it has been concluded that through a combination of the simulation- and statistics-based method, being completely independent of each other, one is able to count atoms with trustworthy single atom sensitivity.

One of the applications of the atom-counting method that has been investigated, is the determination of the number of atoms in an Au nanorod (Van Aert et al., 2013). In this analysis, the recorded image intensities have been normalized with respect to the incident beam by taking the detector sensitivity into account. This places STEM experiments on the same absolute intensity scale as used in theoretical image simulations. In order to validate the accuracy of the counting results, the experimental mean scattering cross-sections have been compared to the scattering cross-sections obtained from detailed high-angle ADF (HAADF) STEM simulations. An excellent match of the experimental and simulated scattering cross-sections within the expected 5–10% error range has been found. The statistical model-based method enables to quantify the precision of the atom-counting results. For this Au nanorod, it has been found that in 80% of all atomic columns the number of atoms has been measured without error. Furthermore, it has been illustrated that the atom-counting results from a single projection image can be used to retrieve a three-dimensional (3D) atomic model of the structure (De Backer et al., 2017).

The atom-counting method has also been applied to a more challenging nanostructure, a Pt/Ir mixed alloy nanoparticle (De Backer, Martinez, et al., 2015). Conducting the atom-counting method on such an industrial bimetallic catalyst provides a severe characterization challenge because of the beam sensitivity and limited size of these catalysts. It has been shown that only through a combination of thorough statistical methods and detailed image simulations the limits of atom counting can be diagnosed for this type of particles. This combined analysis has also brought forward the inherent limitations of the method which have been discussed earlier in a theoretical framework. From this study, it has been concluded that the interplay of several effects can greatly impact the reliability of atom counting in quantitative ADF STEM. This includes parameters such as electron dose and lateral sampling, but also aspects often beyond the control of the experimentalist, including particle tilt and contrast contributions from the carbon support.

In addition, it is noted that designing the optimal experiment for nanoparticle atom counting can greatly enhance the reliability of the atom-counting results, which is of great importance when studying beam-sensitive materials. Therefore, the principles of detection theory have been used to quantify the probability of error for atom counting for optimal experiment design (De Backer, De wael, et al., 2015). Binary, as well as multiple, hypothesis testing have been investigated in order to determine the limits to the precision with which the number of atoms in a projected atomic column can be estimated. The probability of error has been calculated for STEM images, scattering cross-sections, and peak intensities.

Based on this analysis, it has been concluded that the scattering cross-sections perform equally well as images and prove more robust than peak intensities. The optimal STEM detector design for atom counting has also been derived using the expression for the probability of error. It has been shown that for very thin objects a low-angle ADF (LAADF) detector is optimal and that for thicker objects the optimal inner detector angle of the annular detector increases. This technique of hypothesis testing has also been used in order to determine the minimally required electron dose in order to attain a pre-specified precision.

The third part of this work involved the use of advanced statistical techniques for atomic column detection from STEM images. In principle, in order to extract reliable structure information of nanomaterials from atomic-resolution STEM images using model-based parameter estimation, the number of atomic columns in the image should be known. The detection of an atomic column or single atom, in case the column consists of only one atom, from STEM images often depends on a visual interpretation of the image. Materials that are stable under the electron beam can be imaged with a sufficiently high electron dose, typically of the order of more than 10^6–10^8 $e^-/Å^2$, resulting into high-quality images exhibiting high signal-to-noise ratio (SNR) and high contrast. This allows one to resolve the individual atomic columns and, therefore, for such beam-stable materials, the atomic columns can be detected in a merely visual manner. However, some materials, especially light-element nanomaterials, are sensitive to the electron beam and, as a result, they cannot withstand a high incoming electron dose, since the high-energy electrons in the beam cause damage to the material. In order to avoid this beam damage to occur, such beam-sensitive materials can only be imaged in STEM using a limited electron dose of around 10^3–10^5 $e^-/Å^2$. This leads to recorded images exhibiting low SNR and weak contrast, and hence low contrast-to-noise ratio (CNR). Visually inspecting these images, trying to resolve the individual atomic columns, can be unreliable and may lead to biased results, especially when no prior knowledge about the structure of the material is available. To overcome this problem, an objective and quantitative approach is needed for detecting atomic columns from STEM images.

Interestingly, such a quantitative method for atomic column detection from STEM images is possible since parameter estimation theory can be generalized by combining its principles with model-order selection (Fatermans et al., 2018). Hereby, besides estimating the unknown parameters of the expectation model, also the number of parameters that most closely represents the underlying process that generated the experimental observations needs to be determined. Typical model-selection methods perform a trade-off between high goodness of fit and low model complexity, since model fit can be easily improved by increased model complexity without necessarily bearing any interpretable relationship with the underlying data-generating process. As a result, the maximum a posteriori (MAP) probability rule has been proposed as a quantitative method to detect atomic columns from atomic-resolution ADF STEM images (Fatermans et al., 2018, 2019). The method is built upon model-based parameter estimation and Bayesian probability theory. This combination of statistical parameter estimation and model-order selection allows the MAP probability rule to perform automatic and objective structure quantification of unknown nanomaterials and to detect atomic columns and even single atoms with high reliability. For this purpose, it has been shown that, for a variety of parametric models, approximate analytical expressions can be derived for the probability of the presence of a certain number of atomic columns in ADF STEM image data (Fatermans et al., 2019). The MAP probability rule selects the number of columns that maximizes this probability. Furthermore, the MAP probability rule also enables one to quantify how more likely a certain atomic structure is as compared to other structures. The validity and usefulness of applying the MAP probability rule to low-quality image data have been demonstrated by analyzing experimental and simulated ADF STEM images of samples of different shape, size, and atom type (Fatermans et al., 2018). Hereby, it has been shown that important structural information can be obtained by using the MAP probability rule, which would otherwise have been undiscovered. The methodology of the MAP probability rule is not only applicable to ADF STEM, but can also be used to detect atomic columns from simultaneously acquired ABF-ADF STEM images (Fatermans et al., 2020). This is of relevance because, although the ABF imaging mode allows for better visualization of light elements as compared to ADF imaging, it is, in general, not straightforward to visually detect light-element atomic columns since light-element nanomaterials are typically sensitive to the electron beam, limiting the incoming electron dose and leading to images exhibiting low CNR. As for ADF STEM, it has been illustrated that by using the MAP probability rule to analyze simultaneously acquired experimental ABF and ADF STEM data, a reliable estimation of atomic column locations can be obtained without including prior information about the expected positions of these columns.

Furthermore, besides detecting atomic columns from STEM images, it has also been shown in this work that the MAP probability rule can be effectively used as a way to evaluate the relation between STEM image-quality measures, such as SNR and CNR, and atom detectability, which is defined as the probability to detect an atom from an image (Fatermans et al., 2019). In general, atom detectability is indeed related to image quality, since, for example, one expects the detectability of atomic columns in an image to increase with increasing image quality. In this manner, the probability to detect an atomic column from STEM image data, which can be provided by the MAP probability rule, is correlated with the quality of the image. As such, for STEM images, a new image-quality measure, the integrated CNR (ICNR), has been proposed that better correlates with atom detectability than conventional image-quality measures (Fatermans et al., 2020, 2019). Atomic columns resulting from single images with ICNR values of less than around 5.0 become challenging to accurately detect since, typically, the detection rate of 100% drops rapidly starting from this value.

Moreover, the relation of the MAP probability rule with model selection has been thoroughly investigated (Fatermans et al., 2019). In particular, the MAP probability rule is closely related to the BIC, since both methods are derived using a Bayesian approach. In principle, atom detection from STEM images can be performed with any model-selection criterion, but, interestingly, the MAP probability rule takes into account more dimensions of model complexity as compared to other commonly used model-selection criteria. More specifically, the complexity term of the MAP probability rule depends on three dimensions of model complexity, namely the number of parameters, the functional form of the model, and the extension of the parameter space, as opposed to other model-selection criteria which typically depend on the number of parameters only. This extended complexity term allows the MAP probability rule to more accurately detect atoms from STEM images, especially from images exhibiting low ICNR values. Furthermore, the MAP probability rule allows for a clear and flexible incorporation of prior knowledge, which is often not the case for other model-selection methods.

As a general conclusion, it can be stated that advanced statistical techniques are ideal tools for quantitative electron microscopy at the atomic scale. Although this work provides a comprehensive overview of some well-established quantification techniques that are currently available in electron microscopy, it is noted that recently new methodologies have been developed which were not discussed in detail in this work. Furthermore, it is stressed that the results presented in this work merely form an intermediate endpoint in the development of quantitative methods. This paragraph provides an overview on future improvements that can be expected and on first results of ongoing work in the search for novel statistical tools in the field of quantitative electron microscopy.

For example, so far, the here-described methods for counting the number of atoms perform excellently for single-element crystalline structures which are still relatively stable under the incident electron beam. Furthermore, the simulation- and statistics-based approach for atom counting have their own advantages and disadvantages limiting the precision and accuracy that can be attained for the characterization of challenging nanostructures. In this work, both approaches are independently presented next to each other in order to validate counting results. However, it is shown that the development of a true hybrid method which combines the benefits of a statistical analysis and a comparison with simulations into the estimation of the probability distribution of the scattering cross-sections, enhances the performance of atom counting from ADF STEM images (De wael et al., 2017). An important aspect when aiming for atom-counting results with the highest attainable precision is to balance atom-counting fidelity and the risk of knock-on damage. For this purpose, a statistical framework combined with physics-based modeling of the dose-dependent process will be very helpful in order to theoretically predict the optimal electron dose for a reliable quantification of nanostructures in their native state (Van Aert et al., 2019).

Furthermore, a giant leap forward is expected if the number of atoms can be counted in any mixed element nanostructure, since technologically important materials often consist of more than one chemical element. In this case, different types of elements together constitute an atomic column. For mixed columns, all types of elements will contribute differently to the scattering cross-section thus significantly complicating atom counting as compared for single-element nanostructures. Already some very important steps have been taken for counting the number of atoms in heterogeneous nanostructures. In order to unscramble the mixed elements atomic columns, novel quantitative methods are being explored. The introduction of the so-called atomic lensing model enabled the determination of the 3D atomic structure of an Au@Ag core-shell nanorod (van den Bos et al., 2016, 2019). In addition, it is expected that significant advances can be made in the development of 3D reconstruction algorithms that will enable one to measure atom positions in 3D.

The quantitative analysis tools described in this work focus on stationary structures as a result of which the insight into the dynamics is lacking. Since the atomic structure of a nanomaterial can evolve over time via adatom dynamics, surface diffusion and reconstruction, beam effects, or during in situ experiments, a quantitative analysis at a single time point is often insufficient to understand the atomic structure-properties relationship. The introduction of a quantitative method to measures atomic scale dynamics of nanoparticles will be of great importance for revealing and quantifying the atomic structure when it evolves over time (Altantzis et al., 2019; De wael et al., 2020).

Regarding the proposed quantitative method for determining the structure of unknown nanomaterials in an automatic and objective manner and to detect atomic columns and even single atoms from atomic-resolution STEM images with high reliability, the MAP probability rule, also improvements can still be made. Although, this technique has proved to enable accurate and precise measurements of the atomic arrangement of the projected atomic columns or individual atoms from STEM images, it is currently only applicable for analyzing images with a small field of view containing a limited number of projected atomic columns. In addition, the methodology of the MAP probability rule, as described in this work, is currently derived in an analytical manner making use of certain approximations and assumptions. It is expected that also a numerical approach is possible by using Monte Carlo calculations. A numerical evaluation of the MAP probability rule would allow us to verify under what conditions the analytical approximations and assumptions still hold, and under what conditions they start to break down. This numerical approach of the MAP probability rule would be especially interesting for analyzing materials under extremely low-dose conditions. Furthermore, although in this work the proposed technique of MAP probability has only been applied to STEM, it is not limited to this imaging mode as the foundations of the MAP probability rule are very general and applicable to different domains. Therefore, it may be possible to apply the MAP probability rule also to other signals that can be collected in the electron microscope, such as electron energy loss spectroscopy (EELS) and energy-dispersive X-ray spectroscopy (EDX) signals, or to the entire STEM angular range for applications in four-dimensional (4D) STEM. In this manner, more optimal quantitative information can be obtained about the specimen under investigation.

The advanced statistical techniques presented in this work open up new possibilities for the characterization of nanomaterials at the atomic scale.

Future developments of novel quantification methods will provide once more a wealth of possibilities to study nanomaterials on a new level. This continuous process helps step by step unraveling the unique link between material structure and properties at the atomic scale.

Acknowledgments

This project has received funding from the European Research Council (ERC) under the European Union's Horizon 2020 research and innovation programme (Grant Agreement No. 770887 and No. 823717 ESTEEM3). The authors acknowledge financial support from the Research Foundation Flanders (FWO, Belgium) through Projects No. G.0502.18N, G.0267.18N, G.0368.15N, and a postdoctoral grant to A. D. B. The authors acknowledge funding from the University of Antwerp through a TOP project.

References

Abakumov, A. M., Tsirlin, A. A., & Antipov, E. V. (2013). 2.01 - Transition-metal perovskites. In K. Poeppelmeier, & J. Reedijk (Eds.), *Comprehensive inorganic chemistry II* (second edition) (pp. 1–40). Amsterdam: Elsevier.

Abramov, Y. A., Tsirelsoon, V. G., Zavodnik, V. E., Ivanov, S. A., & Brown, I. D. (1995). The chemical bond and atomic displacements in $SrTiO_3$ from X-ray diffraction analysis. *Acta Crystallographica. Section B, 51*, 942–951.

Adrian, M., Dubochet, J., Lepault, J., & McDowell, A. W. (1984). Cryo-electron microscopy of viruses. *Nature, 308*, 32–36.

Akaike, H. (1974). A new look at the statistical model identification. *IEEE Transactions on Automatic Control, 19*(6), 716–723.

Akashi, T., Takahashi, Y., Tanigaki, T., Shimakura, T., Kawasaki, T., Furutsu, T., Shinada, H., Müller, H., Haider, M., Osakabe, N., & Tonomura, A. (2015). Aberration corrected 1.2-MV cold field-emission transmission electron microscope with a sub-50-pm resolution. *Applied Physics Letters, 106*, Article 074101.

Alania, M., Altantzis, T., De Backer, A., Lobato, I., Bals, S., & Van Aert, S. (2017). Depth sectioning combined with atom-counting in HAADF STEM to retrieve the 3D atomic structure. *Ultramicroscopy, 177*, 36–42.

Alem, N., Yazyev, O. V., Kisielowski, C., Denes, P., Dahmen, U., Hartel, P., Haider, M., Bischoff, M., Jiang, B., Louie, S. G., & Zettl, A. (2011). Probing the out-of-plane distortion of single point defects in atomic thin hexagonal boron nitride at the picometer scale. *Physical Review Letters, 106*, Article 126102.

Allen, J. E., Hemesath, D. E., Perea, D. E., Lensch-Falk, J. L., Li, Z. Y., Yin, F., Gass, M. H., Wang, P., Bleloch, A. L., Palmer, R. E., & Lauhon, L. J. (2008). High-resolution detection of Au catalyst atoms in Si nanowires. *Nature Nanotechnology, 3*, 168–173.

Allen, L. J., Findlay, S. D., Oxley, M. P., & Rossouw, C. J. (2003). Lattice-resolution contrast from a focus coherent electron probe. Part I. *Ultramicroscopy, 96*, 47–63.

Allen, L. J., McBride, W., O'Leary, N. L., & Oxley, M. P. (2004a). Exit wave reconstruction at atomic resolution. *Ultramicroscopy, 100*, 91–104.

Allen, L. J., McBride, W., O'Leary, N. L., & Oxley, M. P. (2004b). Investigation of the effects of partial coherence on exit wave reconstruction. *Journal of Microscopy, 216*, 70–75.

Allen, L. J., & Oxley, M. P. (2001). Phase retrieval from series of images obtained by defocus variation. *Optics Communications, 199*, 65–75.

Alshammari, A., Köckritz, A., Kalevaru, V. N., Bagabas, A., & Martin, A. (2012). Influence of single use and combination of reductants on the size, morphology and growth steps of gold nanoparticles in colloidal mixture. *Open Journal of Physical Chemistry, 2*(4), 252–261.

Altantzis, T., Coutino-Gonzalez, E., Baekelant, W., Martinez, G. T., Abakumov, A. M., Van Tendeloo, G., Roeffaers, M. B. J., Bals, S., & Hofkens, J. (2016). Direct observation of luminescent silver clusters confined in faujasite zeolites. *ACS Nano, 10*, 7604.

Altantzis, T., Lobato, I., De Backer, A., Béché, A., Zhang, Y., Basak, S., Porcu, M., Xu, Q., Sánchez-Iglesias, A., Liz-Marzán, L. M., Van Tendeloo, G., Van Aert, S., & Bals, S. (2019). Three-dimensional quantification of the facet evolution of Pt nanoparticles in a variable gaseous environment. *Nano Letters, 19*, 477–481.

Anguiano, E., & Aguilar, M. (1999). A cross-measurement procedure (CMP) for near noise-free imaging in scanning microscopes. *Ultramicroscopy, 76*, 39–47.

Armstrong, N., & Hibbert, D. B. (2009). An introduction to Bayesian methods for analyzing chemistry data. Part 1: An introduction to Bayesian theory and methods. *Chemometrics and Intelligent Laboratory Systems, 97*, 194–210.

Aveyard, R., Ferrando, R., Johnston, R. L., & Yuan, J. (2014). Modeling nanoscale inhomogeneities for quantitative HAADF STEM imaging. *Physical Review Letters*, *113*, Article 075501.

Balandin, A. A., Ghosh, S., Bao, W., Calizo, I., Teweldebrhan, D., Miao, F., & Lau, C. N. (2008). Superior thermal conductivity of single-layer graphene. *Nano Letters*, *8*, 902–907.

Bals, S., Batenburg, K. J., Verbeeck, J., Sijbers, J., & Van Tendeloo, G. (2007). Quantitative three-dimensional reconstruction of catalyst particles for bamboo-like carbon nanotubes. *Nano Letters*, 7(12), 3669–3674.

Bals, S., Casavola, M., van Huis, M. A., Van Aert, S., Batenburg, K. J., Van Tendeloo, G., & Vanmaekelbergh, D. (2011). Three-dimensional atomic imaging of colloidal core-shell nanocrystals. *Nano Letters*, *11*(8), 3420–3424.

Bals, S., Van Aert, S., Romero, C. P., Lauwaet, K., Van Bael, M. J., Schoeters, B., Partoens, B., Yücelen, E., Lievens, P., & Van Tendeloo, G. (2012). Atomic scale dynamics of ultrasmall germanium clusters. *Nature Communications*, *3*, 897.

Bals, S., Van Aert, S., Van Tendeloo, G., & Ávila-Brande, D. (2006). Statistical estimation of atomic positions from exit wave reconstruction with a precision in the picometer range. *Physical Review Letters*, *96*, Article 096106.

Banfield, J. D., & Raftery, A. E. (1993). Model-based Gaussian and non-Gaussian clustering. *Biometrics*, *49*, 803.

Bartesaghi, A., Merk, A., Banerjee, S., Matthies, D., Wu, X., Milne, J. L., & Subramaniam, S. (2015). 2.2 Å resolution cryo-EM structure of β-galactosidase in complex with a cell-permeant inhibitor. *Science*, *348*, 1147–1151.

Batenburg, K. J. (2007). A network flow algorithm for reconstructing binary images from discrete X-rays. *Journal of Mathematical Imaging and Vision*, *27*, 175–191.

Batson, P. E., Dellby, N., & Krivanek, O. L. (2002). Sub-ångstrom resolution using aberration corrected electron optics. *Nature*, *418*, 617–620.

Béché, A., Goris, B., Freitag, B., & Verbeeck, J. (2016). Development of a fast electromagnetic beam blanker for compressed sensing in scanning transmission electron microscopy. *Applied Physics Letters*, *108*(9), Article 093103.

Becker, J., Zins, I., Jakab, A., Khalavka, Y., Schubert, O., & Sönnichsen, C. (2008). Plasmonic focusing reduces ensemble linewidth of silver-coated gold nanorods. *Nano Letters*, *8*(6), 1719–1723.

Belz, J., Beyer, A., & Volz, K. (2018). Atomic-scale 3D reconstruction of antiphase boundaries in gap on (001) silicon by STEM. *Micron*, *114*, 32–41.

Bergström, J. (2012). Bayesian evidence for non-zero θ_{13} and CP-violation in neutrino oscillations. *The Journal of High Energy Physics*, *8*, 1–20.

Berkels, B., Binev, P., Blom, D. A., Dahmen, W., Sharpley, R. C., & Vogt, T. (2014). Optimized imaging using non-rigid registration. *Ultramicroscopy*, *138*, 46–56.

Bermejo, F. J., Howells, W. S., Jiménez-Ruiz, M., González, M. A., Price, D. L., Saboungi, M. L., & Cabrillo, C. (2004). Origin of the complex dielectric relaxation spectra of molecular glass formers. *Physical Review B*, *69*, Article 174201.

Bettens, E., Van Dyck, D., den Dekker, A. J., Sijbers, J., & van den Bos, A. (1999). Model-based two-object resolution from observations having counting statistics. *Ultramicroscopy*, 77, 37–48.

Biermans, E., Molina, L., Batenburg, K. J., Bals, S., & Van Tendeloo, G. (2010). Measuring porosity at the nanoscale by quantitative electron tomography. *Nano Letters*, 7, 5014–5019.

Biernacki, C., Celeux, G., & Govaert, G. (1998). *Assessing a mixture model for clustering with the integrated classification likelihood* (Technical Report 3521). Rhônes-Alpes: INRIA.

Biernacki, C., Celeux, G., & Govaert, G. (1999). An improvement of the NEC criterion for assessing the number of clusters in a mixture model. *Pattern Recognition Letters*, *20*(3), 267–272.

Biernacki, C., & Govaert, G. (1997). Using the classification likelihood to choose the number of clusters. *Computing Science and Statistics*, *29*, 451–457.

Binev, P., Dahmen, W., DeVore, R., Lamby, P., Savu, D., & Sharpley, R. (2012). Compressed sensing and electron microscopy. In *Modeling nanoscale imaging in electron microscopy* (pp. 73–126). Springer.

Binns, C. (2001). Nanoclusters deposited on surfaces. *Surface Science Reports*, *44*, 1–49.

Botti, S., Schindlmayr, A., Del Sole, R., & Reining, L. (2007). Time-dependent density-functional theory for extended systems. *Reports on Progress in Physics*, *70*, 357–407.

Broeckx, J., Op de Beeck, M., & Van Dyck, D. (1995). A useful approximation of the exit wave function in coherent STEM. *Ultramicroscopy*, *60*, 71–80.

Broersen, P. M. T., & Wensink, H. E. (1993). On finite sample theory for autoregressive model order selection. *IEEE Transactions on Signal Processing*, *41*(1), 194–204.

Broersen, P. M. T., & Wensink, H. E. (1996). On the penalty factor for autoregressive order selection in finite samples. *IEEE Transactions on Signal Processing*, *44*, 748–752.

Brown, H. G., D'Alfonso, A. J., Chen, Z., Morgan, A. J., Weyland, M., Zheng, C., Fuhrer, M. S., Findlay, S. D., & Allen, L. J. (2016). Structure retrieval with fast electrons using segmented detectors. *Physical Review B*, *93*, Article 134116.

Brown, H. G., Ishikawa, R., Sánchez-Santolino, G., Lugg, N. R., Ikuhara, Y., Allen, L. J., & Shibata, N. (2017). A new method to detect and correct sample-tilt in scanning transmission electron microscopy bright-field imaging. *Ultramicroscopy*, *173*, 76–83.

Brown, M. W., Kohlmeyer, A., Plimpton, S. J., & Tharrington, A. N. (2012). Implementing molecular dynamics on hybrid high performance computers - particle-particle particle-mesh. *Computer Physics Communications*, *183*, 449–459.

Brown, M. W., Wang, P., Plimpton, S. J., & Tharrington, A. N. (2011). Implementing molecular dynamics on hybrid high performance computers - short range forces. *Computer Physics Communications*, *182*, 898–911.

Brown, M. W., & Yamada, M. (2013). Implementing molecular dynamics on hybrid high performance computers - three-body potentials. *Computer Physics Communications*, *184*, 2785–2793.

Brunelli, R. (2009). *Template matching techniques in computer vision: Theory and practice*. Wiley Press.

Buban, J. P., Ramasse, Q., Gipson, B., Browning, N. D., & Stahlberg, H. (2010). High-resolution low-dose scanning transmission electron microscopy. *Journal of Electron Microscopy*, *59*, 103–112.

Buffat, P.-A. (1977). Size effect modifications of the Debye-Waller factor in small gold particles. *Solid State Communications*, *23*, 547–550.

Burnham, K. P., & Anderson, D. R. (2002). *Model selection and multimodal inference*. Springer-Verlag.

Callaway, E. (2015). The revolution will not be crystallized: A new method sweeps through structural biology. *Nature*, *525*, 172–176.

Campbell, M. G., Veesler, D., Cheng, A., Potter, C. S., & Carragher, B. (2015). 2.8 Å resolution reconstruction of the thermoplasma acidophilum 20S proteasome using cryo-electron microscopy. *eLife*, *4*, Article e06380.

Cardinal, M. F., Rodriguez-González, B., Alvarez-Puebla, R. A., Pérez-Juste, J., & Liz-Marzán, L. M. (2010). Modulation of localized surface plasmons and SERS response in gold dumbbells through silver coating. *Journal of Physical Chemistry C*, *114*, 10417–10423.

Celeux, G., & Soromenho, G. (1996). An entropy criterion for assessing the number of clusters in a mixture model. *Journal of Classification*, *13*(2), 195–212.

Chang, L. Y., Barnard, A. S., Gontard, L. C., & Dunin-Borkowski, R. E. (2010). Resolving the structure of active sites on platinum catalytic nanoparticles. *Nano Letters*, *10*, 3073–3076.

Chen, C.-C., Chun, Z., White, E. R., Chiu, C.-Y., Scott, M. C., Regan, B. C., Marks, L. D., Huang, Y., & Miao, J. (2013). Three-dimensional imaging of dislocations in a nanoparticle at atomic resolution. *Nature*, *496*, 74–77.

Chen, J., Wiley, B., Li, Z.-Y., Campbell, D., Saeki, F., Cang, H., Au, L., Lee, J., Li, X., & Xia, Y. (2005). Gold nanocages: Engineering their structure for biomedical applications. *Advanced Materials*, *17*(18), 2255–2261.

Cheng, Y. (2018). Single-particle cryo-EM: How did it get here and where will it go. *Science*, *361*, 876–880.

Chiu, W., Downing, K. H., Dubochet, J., Glaeser, R. M., Heide, H. G., Knapek, E., Kopf, D. A., Lamvik, M. K., Lepault, J., Robertson, J. D., Zeitler, J. D., & Zemlin, F. (1987). Cryoprotection in electron microscopy. *Journal of Microscopy*, *141*, 385–391.

Claeskens, G. (2016). Statistical model choice. *Annual Review of Statistics and Its Application*, *3*, 10.1–10.24.

Claeskens, G., & Hjort, N. L. (2003). The focused information criterion. *Journal of the American Statistical Association*, *98*, 879–899.

Claeskens, G., & Hjort, N. L. (2008). *Model selection and model averaging*. Cambridge University Press.

Claridge, S. A., Castleman, A. W., Khanna, S. N., Murray, C. B., Sen, A., & Weiss, P. S. (2009). Cluster-assembled materials. *ACS Nano*, *3*, 244–255.

Clark, B. C., Herman, R., & Wallis, R. F. (1965). Theoretical mean-square displacements for surface atoms in face-centered cubic lattices with applications to nickel. *Physical Review*, *139*(3A), A860–A867.

Coene, W., & Janssen, G. (1992). Phase retrieval through focus variation for ultra-resolution in field-emission transmission electron microscopy. *Physical Review Letters*, *69*(26), 3743–3746.

Coene, W. M. J., Thust, A., Op de Beeck, M., & Van Dyck, D. (1996). Maximum-likelihood method for focus-variation image reconstruction in high resolution transmission electron microscopy. *Ultramicroscopy*, *64*, 109–135.

Corsaro, E., & De Ridder, J. (2014). Diamonds: A new Bayesian nested sampling tool - application to peak bagging of solar-like oscillations. *Astronomy & Astrophysics*, *571*, Article A71.

Cortie, M. B., & McDonagh, A. M. (2011). Synthesis and optical properties of hybrid and alloy plasmonic nanoparticles. *Chemical Reviews*, *111*, 3713–3735.

Cox, R. T. (1946). Probability, frequency and reasonable expectation. *American Journal of Physics*, *14*(1), 1–13.

Csencsits, R., & Gronsky, R. (1987). Damage of zeolite Y in the TEM and its effects on TEM images. *Ultramicroscopy*, *23*, 421–431.

D'Agostini, G. (2003). Bayesian inference in processing experimental data: Principles and basic applications. *Reports on Progress in Physics*, *66*, 1383–1419.

Daniel, M.-C., & Astruc, D. (2004). Gold nanoparticles: Assembly, supramolecular chemistry, quantum-size-related properties, and applications toward biology, catalysis, and nanotechnology. *Chemical Reviews*, *104*(1), 293–346.

Day, N. E. (1969). Estimating the components of a mixture of normal distributions. *Biometrika*, *56*(3), 463–474.

De Backer, A., De wael, A., Gonnissen, J., & Van Aert, S. (2015). Optimal experimental design for nano-particle atom-counting from high-resolution STEM images. *Ultramicroscopy*, *151*, 46–55.

De Backer, A., Jones, L., Lobato, I., Altantzis, T., Goris, B., Nellist, P. D., Bals, S., & Van Aert, S. (2017). Three-dimensional atomic models from a single projection using Z-contrast imaging: Verification by electron tomography and opportunities. *Nanoscale*, *9*, 8791–8798.

De Backer, A., Martinez, G. T., MacArthur, K. E., Jones, L., Béché, A., Nellist, P. D., & Van Aert, S. (2015). Dose limited reliability of quantitative annular dark field scanning transmission electron microscopy for nano-particle atom-counting. *Ultramicroscopy*, *151*, 56–61.

De Backer, A., Martinez, G. T., Rosenauer, A., & Van Aert, S. (2013). Atom counting in HAADF STEM using a statistical model-based approach: Methodology, possibilities, and inherent limitations. *Ultramicroscopy*, *134*, 23–33.

De Backer, A., Van Aert, S., & Van Dyck, D. (2011). High precision measurements of atom column positions using model-based exit wave reconstruction. *Ultramicroscopy*, *111*, 1475–1482.

De Backer, A., van den Bos, K. H. W., Van den Broek, W., Sijbers, J., & Van Aert, S. (2016). StatSTEM: An efficient approach for accurate and precise model-based quantification of atomic resolution electron microscopy images. *Ultramicroscopy*, *171*, 104–116.

De Cruz, L., Ireland, D. G., Vancraeyveld, P., & Ryckebusch, J. (2011). Bayesian model selection for electromagnetic kaon production on the nucleon. *International Journal of Modern Physics A*, *26*(03n04), 642–644.

De Graef, M. (2003). *Introduction to conventional transmission electron microscopy*. Cambridge University Press.

de Jong, A. F., & Van Dyck, D. (1993). Ultimate resolution and information in electron microscopy II. The information limit of transmission electron microscopes. *Ultramicroscopy*, *49*, 66–80.

De wael, A., De Backer, A., Jones, L., Nellist, P. D., & Van Aert, S. (2017). Hybrid statistics-simulations based method for atom-counting from ADF STEM images. *Ultramicroscopy*, *177*, 69.

De wael, A., De Backer, A., Jones, L., Varambhia, A., Nellist, P. D., & Van Aert, S. (2020). Measuring dynamic structural changes of nanoparticles at the atomic scale using scanning transmission electron microscopy. *Physical Review Letters*, *124*, Article 106105.

Debono, I. (2014). Bayesian model selection for dark energy using weak lensing forecasts. *Monthly Notices of the Royal Astronomical Society. Letters*, *437*(1), 887–897.

Dellisanti, C. (2015). A barrier-breaking resolution. *Nature Structural & Molecular Biology*, *22*, 361.

Dempster, A. P., Laird, N. M., & Rubin, D. B. (1977). Maximum likelihood from incomplete data via the EM algorithm (with discussion). *Journal of the Royal Statistical Society, Series B*, *39*, 1–38.

den Dekker, A. J., Gonnissen, J., De Backer, A., Sijbers, J., & Van Aert, S. (2013). Estimation of unknown structure parameters from high-resolution (S)TEM images: What are the limits? *Ultramicroscopy*, *134*, 34–43.

den Dekker, A. J., Sijbers, J., & Van Dyck, D. (1999). How to optimize the design of a quantitative HREM experiment so as to attain the highest precision. *Journal of Microscopy*, *194*, 95–104.

den Dekker, A. J., Van Aert, S., van den Bos, A., & Van Dyck, D. (2005). Maximum likelihood estimation of structure parameters from high resolution electron microscopy images. Part I: A theoretical framework. *Ultramicroscopy*, *104*, 83–106.

den Dekker, A. J., Van Aert, S., Van Dyck, D., van den Bos, A., & Geuens, P. (2001). Does a monochromator improve the precision in quantitative HRTEM? *Ultramicroscopy*, *89*, 275–290.

Dodge, Y. (2003). *The Oxford dictionary of statistical terms*. Oxford University Press.

Donoho, D. L. (2006). For most large undetermined systems of linear equations, the minimal l^1-norm solution is also the sparsest solution. *Communications on Pure and Applied Analysis*, *59*, 797–829.

Downing, K. H., & Li, H. (2001). Accurate recording and measurement of electron diffraction data in structural and difference Fourier studies of proteins. *Microscopy and Microanalysis*, *7*, 407–417.

Du Pasquier, A., Plitz, I., Menocal, S., & Amatucci, G. (2003). A comparative study of Li-ion battery, supercapacitor and nonaqueous asymmetric hybrid devices for automotive applications. *Journal of Power Sources*, *115*, 171–178.

E, H., MacArthur, K. E., Pennycook, T. J., Okunishi, E., D'Alfonso, A. J., Lugg, N. R., Allen, L. J., & Nellist, P. D. (2013). Probe integrated scattering cross sections in the analysis of atomic resolution HAADF STEM images. *Ultramicroscopy*, *133*, 109–119.

Egerton, R. F., Li, P., & Malac, M. (2004). Radiation damage in the TEM and SEM. *Micron*, *35*, 399–409.

Erni, R., Heinrich, H., & Kostorz, G. (2003). Quantitative characterisation of chemical inhomogeneities in Al-Ag using high-resolution Z-contrast STEM. *Ultramicroscopy*, *94*, 125–133.

Erni, R., Rossell, M. D., Kisielowski, C., & Dahmen, U. (2009). Atomic-resolution imaging with a sub-50-pm electron probe. *Physical Review Letters*, *102*, Article 096101.

Erni, R., Rossell, M. D., & Nakashima, P. N. H. (2010). Optimization of exit-plane waves restored from HRTEM through-focal series. *Ultramicroscopy*, *110*, 151–161.

Everitt, B. S., & Skrondal, A. (2010). *Cambridge dictionary of statistics*. Cambridge University Press.

Faruqi, A. R., & McMullan, G. (2018). Direct imaging detector for electron microscopy. *Nuclear Instruments & Methods in Physics Research. Section A*, *878*, Article 180.

Fatermans, J., den Dekker, A. J., Müller-Caspary, K., Gauquelin, N., Verbeeck, J., & Van Aert, S. (2020). Atom column detection from simultaneously acquired ABF and ADF STEM images. *Ultramicroscopy*, *219*, Article 113046.

Fatermans, J., den Dekker, A. J., Müller-Caspary, K., Lobato, I., O'Leary, C. M., Nellist, P. D., & Van Aert, S. (2018). Single atom detection from low contrast-to-noise ratio electron microscopy images. *Physical Review Letters*, *121*, Article 056101.

Fatermans, J., Van Aert, S., & den Dekker, A. J. (2019). The maximum a posteriori probability rule for atom column detection from HAADF STEM images. *Ultramicroscopy*, *201*, 81–91.

Feroz, F., Balan, S. T., & Hobson, M. P. (2011). Bayesian evidence for two companions orbiting HIP 5158. *Monthly Notices of the Royal Astronomical Society. Letters*, *416*(1), L104–L108.

Fertig, J., & Rose, H. (1981). Resolution and contrast of crystalline objects in high-resolution scanning transmission electron microscopy. *Optik*, *59*(5), 407–429.

Findlay, S. D., Kohno, Y., Cardamone, L. A., Ikuhara, Y., & Shibata, N. (2014). Enhanced light element imaging in atomic resolution scanning transmission electron microscopy. *Ultramicroscopy*, *136*, 31–41.

Findlay, S. D., & LeBeau, J. M. (2013). Detector non-uniformity in scanning transmission electron microscopy. *Ultramicroscopy*, *124*, 52–60.

Findlay, S. D., Shibata, N., Sawada, H., Okunishi, E., Kondo, Y., & Ikuhara, Y. (2010). Dynamics of annular bright field imaging in scanning transmission electron microscopy. *Ultramicroscopy*, *110*(7), 903–923.

Findlay, S. D., Shibata, N., Sawada, H., Okunishi, E., Kondo, Y., Yamamoto, T., & Ikuhara, Y. (2009). Robust atomic resolution imaging of light elements using scanning transmission electron microscopy. *Applied Physics Letters*, *95*(19), Article 191913.

Foiles, S. M., Baskes, M. I., & Daw, M. S. (1986). Embedded-atom-method function for the fcc metals Cu, Ag, Au, Ni, Pd, Pt, and their alloys. *Physical Review B*, *33*, 7983–7991.

Frank, J. (Ed.). (1992). *Electron tomography: Three-dimensional imaging with the transmission electron microscope*. New York and London: Plenum Press.

Frieden, B. R. (Ed.). (1998). *Physics from Fisher information - a unification*. Cambridge: Cambridge University Press.

Gao, H. X., & Peng, L.-M. (1999). Parameterization of the temperature dependence of the Debye-Waller factors. *Acta Crystallographica. Section A*, *55*, 926–932.

Gao, P., Britson, J., Nelson, C. T., Jokisaari, J. R., Duan, C., Trassin, M., Baek, S.-H., Guo, H., Li, L., Wang, Y., Chu, Y.-H., Minor, A. M., Eom, C.-B., Ramesh, R., Chen, L.-Q., & Pan, X. (2014). Ferroelastic domain switching dynamics under electrical and mechanical excitations. *Nature Communications*, *5*, Article 3801.

Gao, P., Kumamoto, A., Lugg, N., Shibata, N., & Ikuhara, Y. (2018). Picometer-scale atom position analysis in annular bright-field STEM imaging. *Ultramicroscopy*, *184*, 177–187.

García, R. A. (2013). Stellar pulsations: Impact of new instrumentation and new insights. In J. C. Suárez, R. Garrido, L. A. Balona, & J. Christensen-Dalsgaard (Eds.), *Advances in solid state physics: Vol. 31*. Springer.

Gauch, H. G. (2003). *Scientific method in practice*. Cambridge University Press.

Geman, S., Bienenstock, E., & Doursat, R. (1992). Neural networks and the bias/variance dilemma. *Neural Computation*, *4*, 1–58.

Geuchies, J. J., van Overbeek, C., Evers, W. H., Goris, B., De Backer, A., Gantapara, G. P., Rabouw, F. T., Hilhorst, J., Peters, J. L., Konovalov, O., Petukhov, A. V., Dijkstra, M., Siebbeles, L. D. A., Van Aert, S., Bals, S., & Vanmaekelbergh, D. (2016). In situ study of the formation mechanism of two-dimensional superlattices from PbSe nanocrystals. *Nature Materials*, *15*, 1248–1254.

Geuens, P., & Van Dyck, D. (2002). The S-state model: A work horse for HRTEM. *Ultramicroscopy*, *93*, 179–198.

Geuens, P., & Van Dyck, D. (2005). The S-state model for electron channeling in high-resolution electron microscopy. *Advances in Electronics and Electron Physics*, *136*, 111–226.

Girit, C. O., Meyer, J. C., Erni, R., Rosell, M. D., Kisielowski, C., Yang, L., Park, C.-H., Crommie, M. F., Cohen, M. L., Louie, S. G., & Zettl, A. (2009). Graphene at the edge: Stability and dynamics. *Science*, *323*, 1705–1708.

Gómez-Graña, S., Pérez-Juste, J., Alvarez-Puebla, R. A., Guerrero-Martínez, A., & Liz-Marzán, L. M. (2013). Self-assembly of Au@Ag nanorods mediated by gemini surfactants for highly efficient SERS-active supercrystals. *Advanced Optical Materials*, *1*(7), 477–481.

Gonnissen, J., De Backer, A., den Dekker, A. J., Martinez, G. T., Rosenauer, A., Sijbers, J., & Van Aert, S. (2014). Optimal experimental design for the detection of light atoms from high-resolution scanning transmission electron microscopy images. *Applied Physics Letters*, *105*, Article 063116.

Gonnissen, J., De Backer, A., den Dekker, A. J., Sijbers, J., & Van Aert, S. (2016). Detecting and locating light atoms from high-resolution stem images: The quest for a single optimal design. *Ultramicroscopy*, *170*, 128–138.

Goris, B., Bals, S., Van den Broek, W., Carbó-Argibay, E., Gómez-Graña, S., Liz-Marzán, L. M., & Van Tendeloo, G. (2012). Atomic-scale determination of surface facets in gold nanorods. *Nature Materials*, *11*, 930–935.

Goris, B., De Backer, A., Van Aert, S., Gómez-Graña, S., Liz-Marzán, L. M., Van Tendeloo, G., & Bals, S. (2013). Three-dimensional elemental mapping at the atomic scale in bimetallic nanocrystals. *Nano Letters*, *13*, 4236–4241.

Goris, B., De Beenhouwer, J., De Backer, A., Zanaga, D., Batenburg, K. J., Sánchez-Iglesias, A., Liz-Marzán, L. M., Van Aert, S., Bals, S., Sijbers, J., & Van Tendeloo, G. (2015). Measuring lattice strain in three dimensions through electron microscopy. *Nano Letters*, *15*(10), 6996–7001.

Goris, B., Van den Broek, W., Batenburg, K. J., Heidari, H., & Bals, S. (2012). Electron tomography based on a total variation minimization reconstruction technique. *Ultramicroscopy*, *113*, 120–130.

Goshtasby, A. A. (2005). *2-D and 3-D image registration: For medical, remote sensing, and industrial applications*. Wiley Press.

Gregory, P. C. (2005). *Bayesian logical data analysis for the physical sciences*. Cambridge University Press.

Grieb, T., Müller, K., Fritz, R., Schowalter, M., Neugebohrn, N., Knaub, N., Volz, K., & Rosenauer, A. (2012). Determination of the chemical composition of GaNAs using STEM HAADF imaging and STEM strain state analysis. *Ultramicroscopy, 117*, 15–23.

Grzelczak, M., Pérez-Juste, J., Mulvaney, P., & Liz-Marzán, L. M. (2008). Shape control in gold nanoparticle synthesis. *Chemical Society Reviews, 37*, 1783–1791.

Gu, L., Zhu, C., Li, H., Yu, Y., Li, C., Tsukimoto, S., Maier, J., & Ikuhara, Y. (2011). Direct observation of lithium staging in partially delithiated $LiFePO_4$ at atomic resolution. *Journal of the American Chemical Society, 133*, 4661–4663.

Gull, S., & Daniell, G. J. (1978). Image reconstruction from incomplete and noisy data. *Nature, 272*, 686–690.

Gull, S. F., & Skilling, J. (1984). Maximum entropy method in image processing. *IEE Proceedings, 131F*, 646–659.

Habeck, M. (2011). Statistical mechanics analysis of sparse data. *Journal of Structural Biology, 173*, 541–548.

Hadermann, J., Abakumov, A., Van Rompaey, S., Perkisas, T., Filinchuk, Y., & Van Tendeloo, G. (2012). Crystal structure of a lightweight borohydride from submicrometer crystallites by precession electron diffraction. *Chemistry of Materials, 24*, 3401.

Hadermann, J., Abakumov, A. M., Turner, S., Hafideddine, S., Khasanova, N. R., Antipov, E. V., & Van Tendeloo, G. (2011). Solving the structure of Li ion battery materials with precession electron diffraction: Application to Li_2CoPO_4F. *Chemistry of Materials, 23*, 3540.

Haider, M., Hartel, P., Müller, H., Uhlemann, S., & Zach, J. (2010). Information transfer in a TEM corrected for spherical and chromatic aberration. *Microscopy and Microanalysis, 16*, 393–408.

Haider, M., Rose, H., Uhlemann, S., Kabius, B., & Urban, K. (1998). Towards 0.1 nm resolution with the first spherical corrected transmission electron microscope. *Journal of Electron Microscopy, 47*(5), 395–405.

Haider, M., Uhlemann, S., Schwan, E., Rose, H., Kabius, B., & Urban, K. (1998). Electron microscopy image enhanced. *Nature, 392*, 768–769.

Haigh, S., Sawada, H., & Kirkland, A. I. (2009). Optimal tilt magnitude determination for aberration-corrected super resolution exit wave reconstruction. *Philosophical Transactions of the Royal Society of London. A, 367*, 3755–3771.

Haight, F. (1967). *Handbook of the Poisson distribution*. Wiley.

Hammel, M., & Rose, H. (1995). Optimum rotationally symmetric detector configurations for phase-contrast imaging in scanning transmission electron microscopy. *Ultramicroscopy, 58*, 403–415.

Hannan, E. J., & Quinn, B. G. (1979). The determination of the order of an autoregression. *Journal of the Royal Statistical Society, 41*, 190–195.

Hao, X., Zhai, J., Kong, L. B., & Xu, Z. (2014). A comprehensive review on the progress of lead zirconate-based antiferroelectric materials. *Progress in Materials Science, 63*, 1–57.

Hartel, P., Rose, D., & Dinges, C. (1996). Conditions and reasons for incoherent imaging in STEM. *Ultramicroscopy, 63*, 63–114.

Hashimoto, A., Shimojo, M., Mitsuishi, K., & Takeguchi, M. (2009). Three-dimensional imaging of carbon nanostructures by scanning confocal electron microscopy. *Journal of Applied Physics, 106*, Article 086101.

Hashimoto, A., Shimojo, M., Mitsuishi, K., & Takeguchi, M. (2010). Three-dimensional optical sectioning by scanning confocal electron microscopy with a stage-scanning system. *Microscopy and Microanalysis, 16*, 233–238.

Hasselblad, V. (1969). Estimation of finite mixture of distributions from the exponential family. *Journal of the American Statistical Association, 64*, 1459–1471.

Hawkes, P. W. (1992). The electron microscope as a structure projector. In *Electron tomography: Three-dimensional imaging with the transmission electron microscope*. Plenum Press.

Hawkes, P. W. (2015). The correction of electron lens aberrations. *Ultramicroscopy, 156*, A1–A64.

He, X., Gu, L., Zhu, C., Yu, Y., Li, C., Hu, Y.-S., Tsukimoto, S., Maier, J., Ikuhara, Y., & Duan, X. (2011). Direct imaging of lithium ions using aberration corrected annular-bright-field scanning transmission electron microscopy and associated contrast mechanisms. *Materials Express, 1*, 43–50.

Henderson, R. (1995). The potential and limitations of neutrons, electrons and X-rays for atomic resolution microscopy of unstained biological molecules. *Quarterly Reviews of Biophysics, 28*, 171–193.

Henglein, A. (2000). Preparation and optical absorption spectra of Au core-Pt shell and Pt core-Au shell colloidal nanoparticles in aqueous solution. *Journal of Physical Chemistry B, 104*, 2201–2203.

Herman, G. T., & Kuba, A. (2007). *Advances in discrete tomography and its applications*. Birkhäuser Boston.

Heron, J. T., Schlom, D. G., & Ramesh, R. (2014). Electric field control of magnetism using $BiFeO_3$-based heterostructures. *Applied Physics Reviews, 1*, Article 021303.

Herrmann, K.-H. (1997). Image recording in microscopy. In *Handbook of microscopy - applications in materials science, solid-state physics and chemistry, methods II* (pp. 885–921). Weinheim VCH.

Higuchi, T., Tsukamoto, T., Sata, N., Ishigame, M., Tezuka, Y., & Shin, S. (1998). Electronic structure of p-type $SrTiO_3$ by photoemission spectroscopy. *Physical Review B, 57*, 6978.

Hjort, N. L., & Claeskens, G. (2006). Focused information criterion and model averaging for the Cox hazard regression model. *Journal of the American Statistical Association, 98*, 900–916.

Hodak, J. H., Henglein, A., & Hartland, G. V. (2000). Coherent excitation of acoustic breathing modes in bimetallic core-shell nanoparticles. *Journal of Physical Chemistry B, 104*, 5053–5055.

Hoffman, R., Minkin, V. I., & Carpenter, B. K. (1997). Ockham's razor and chemistry. *International Journal for Philosophy of Chemistry, 3*, 3–28.

Holmes, A. T. (2014). Simple Bayesian method for improved analysis of quasi-two-dimensional scattering data. *Physical Review B, 90*, Article 024514.

Houben, L., Thust, A., & Urban, K. (2006). Atomic-precision determination of the reconstruction of a 90° tilt boundary in $YBa_2Cu_3O_{7-\delta}$ by aberration corrected HRTEM. *Ultramicroscopy, 106*, 200–214.

Hovden, R., & Muller, D. A. (2012). Efficient elastic imaging of single atoms on ultrathin supports in a scanning transmission electron microscope. *Ultramicroscopy, 109*, 59–65.

Hsieh, W.-K., Chen, F.-R., Kai, J.-J., & Kirkland, A. I. (2004). Resolution extension and exit wave reconstruction in complex HREM. *Ultramicroscopy, 98*, 99–114.

Huang, R., Hitosugi, T., Findlay, S. D., Fisher, C. A. J., Ikuhara, Y. H., Moriwake, H., Oki, H., & Ikuhara, Y. (2011). Real-time direct observation of Li in $LiCoO_2$ cathode material. *Applied Physics Letters, 98*, Article 051913.

Huang, T., & Xu, X.-H. N. (2010). Synthesis and characterization of tunable rainbow colored colloidal silver nanoparticles using single-nanoparticle plasmonic microscopy and spectroscopy. *Journal of Materials Chemistry, 20*, 9867–9876.

Huijben, M., Rijnders, G., Blank, D. H. A., Bals, S., Van Aert, S., Verbeeck, J., Van Tendeloo, G., Brinkman, A., & Hilgenskamp, H. (2006). Electronically coupled complementary interfaces between perovskite band insulators. *Nature Materials, 5*, 556–560.

Hwang, J., Zhang, J. Y., D'Alfonso, A. J., Allen, L. J., & Stemmer, S. (2013). Three-dimensional imaging of individual dopant atoms in $SrTiO_3$. *Physical Review Letters, 111*, Article 266101.

Hÿtch, M. J., Putaux, J.-L., & Pénisson, J.-M. (2003). Measurement of the displacement field of dislocations to 0.03 Å by electron microscopy. *Nature, 423*, 270–273.

Hÿtch, M. J., Snoeck, E., & Kilaas, R. (1998). Quantitative measurement of displacement and strain fields from hrem micrographs. *Ultramicroscopy, 74*, 131–146.

Isaacson, M., Kopf, D., Ohtsuki, M., & Utlaut, M. (1979). Atomic imaging using the dark-field annular detector in the STEM. *Ultramicroscopy, 4*, 101–104.

Ishikawa, R., Lupini, A. R., Findlay, S. D., & Pennycook, S. J. (2014). Quantitative annular dark field electron microscopy using single electron signals. *Microscopy and Microanalysis, 20*, 99.

Ishikawa, R., Lupini, A. R., Hinuma, Y., & Pennycook, S. J. (2015). Large-angle illumination STEM: Toward three-dimensional atom-by-atom imaging. *Ultramicroscopy, 151*, 122–129.

Ishikawa, R., Mishra, R., Lupini, A. R., Findlay, S. D., Taniguchi, T., Pantelides, S. T., & Pennycook, S. J. (2014). Direct observation of dopant atom diffusion in a bulk semiconductor crystal enhanced by a large size mismatch. *Physical Review Letters, 113*, Article 155501.

Ishizuka, K. (2002). A practical approach for STEM image simulation based on the FFT multislice method. *Ultramicroscopy, 90*, 71–83.

Jain, P. K., El-Sayed, I. H., & El-Sayed, M. A. (2007). Au nanoparticles target cancer. *Nano Today, 2*, 18–29.

Jaynes, E. (1957). Information theory and statistical mechanics. *Physical Review, 106*, 620–630.

Jaynes, E. T. (2003). *Probability theory: The logic of science*. Cambridge University Press.

Jesson, D. E., & Pennycook, S. J. (1995). Incoherent imaging of crystals using thermally scattered electrons. *Proceedings of the Royal Society A, 449*, 273–293.

Jia, C. L., Lentzen, M., & Urban, K. (2004). High-resolution transmission electron microscopy using negative spherical aberration. *Microscopy and Microanalysis, 10*, 174–184.

Jia, C. L., Mi, S. B., Barthel, J., Wang, D. W., Dunin-Borkowski, R. E., Urban, K. W., & Thust, A. (2014). Determination of the 3D shape of a nanoscale crystal with atomic resolution from a single image. *Nature Materials, 13*, 1044–1049.

Jia, C. L., Mi, S. B., Faley, J., Poppe, U., Schubert, J., & Urban, K. (2009). Oxygen octahedron reconstruction in the $SrTiO_3/LaAlO_3$ heterointerfaces investigated using aberration-corrected ultrahigh-resolution transmission electron microscopy. *Physical Review B, 79*, Article 081405(R).

Jia, C.-L., Mi, S.-B., Urban, K., Vrejoiu, I., Alexe, M., & Hesse, D. (2008). Atomic-scale study of electric dipoles near charged and uncharged domain walls in ferroelectric films. *Nature Materials, 7*, 57–61.

Jia, C. L., Mi, S. B., Urban, K., Vrejoiu, I., Alexe, M., & Hesse, D. (2009). Effect of a single dislocation in a heterostructure layer on the local polarization of a ferroelectric layer. *Physical Review Letters, 102*, Article 117601.

Jinschek, J. R., Batenburg, K. J., Calderon, H. A., Kilaas, R., Radmilovic, V., & Kisielowski, C. (2008). 3-D reconstruction of the atomic positions in a simulated gold nanocrystal based on discrete tomography: Prospects of atomic resolution electron tomography. *Ultramicroscopy, 108*, 589–604.

Johnson, C. L., Snoeck, E., Ezcurdia, M., Rodriguez-González, B., Pastoriza-Santos, I., Liz-Marzán, L. M., & Hÿtch, M. J. (2008). Effects of elastic anisotropy on strain distributions in decahedral gold nanoparticles. *Nature Materials, 7*, 120–124.

Jones, L., MacArthur, K. E., Fauske, V. T., van Helvoort, A. T. J., & Nellist, P. D. (2014). Rapid estimation of catalyst nanoparticle morphology and atomic-coordination by high-resolution Z-contrast electron microscopy. *Nano Letters, 14*(11), 6336–6341.

Jones, L., & Nellist, P. D. (2013). Identifying and correcting scan noise and drift in the scanning transmission electron microscope. *Microscopy and Microanalysis, 19*, 1050–1060.

Jones, L., Yang, H., Pennycook, T. J., Marshall, M. S. J., Van Aert, S., Browning, N. D., Castell, M. R., & Nellist, P. D. (2015). Smart align - a new tool for robust non-rigid registration of scanning microscope data. *Advanced Structural and Chemical Imaging, 1*, Article 8.

Karakulina, O. M., Demortiere, A., Dachraoui, W., Abakumov, A. M., & Hadermann, J. (2018). In situ electron diffraction tomography using a liquid electrochemical transmission electron microscopy cell for crystal structure determination of cathode materials for Li-ion batteries. *Nano Letters, 18*, Article 6286.

Katz-Boon, H., Rossouw, C. J., Weyland, M., Funston, A. M., Mulvaney, P., & Etheridge, J. (2011). Three-dimensional morphology and crystallography of gold nanorods. *Nano Letters, 11*, 273–278.

Katz-Boon, H., Walsh, M., Dwyer, C., Mulvaney, P., Funston, A. M., & Etheridge, J. (2015). Stability of crystal facets in gold nanorods. *Nano Letters, 15*, 1635–1641.

Kay, S. M. (2009). *Fundamentals of statistical signal processing, volume II: Detection theory*. New Jersey: Prentice-Hall, Inc.

Ke, X., Bals, S., Cott, D., Hantschel, T., Bender, H., & Van Tendeloo, G. (2010). Three-dimensional analysis of carbon nanotube networks in interconnects by electron tomography without missing wedge artefacts. *Microscopy and Microanalysis, 16*, 210–217.

Khoshouei, M., Radjainia, M., Baumeister, W., & Danev, R. (2017). Cryo-EM structure of haemoglobin at 3.2 Å determined with the Volta phase plate. *Nature Communications, 8*, Article 16099.

Kim, Y.-M., He, J., Biegalski, M. D., Ambaye, H., Lauter, V., Christen, H. M., Pantelides, S. T., Pennycook, S. J., Kalinin, S. V., & Borisevich, A. Y. (2012). Probing oxygen vacancy concentration and homogeneity in solid-oxide fuel-cell cathode materials on the subunit-cell level. *Nature Materials, 11*, 888–894.

Kim, Y.-M., Morozovska, A., Eliseev, E., Oxley, M. P., Mishra, R., Selbach, S. V., Grande, T., Pantelides, S. T., Kalinin, S. V., & Borisevich, A. Y. (2014). Direct observation of ferroelectric field effect and vacancy-controlled screening at the $BiFeO_3/La_xSr_{1-x}MnO_3$ interface. *Nature Materials, 13*, 1019–1025.

Kimoto, K., Asaka, T., Yu, X., Nagai, T., Matsui, Y., & Ishizuka, K. (2010). Local crystal structure analysis with several picometer precision using scanning transmission electron microscopy. *Ultramicroscopy, 110*, 778–782.

Kirkland, A. I., & Meyer, R. R. (2004). "Indirect" high-resolution transmission electron microscopy: Aberration measurement and wavefunction reconstruction. *Microscopy and Microanalysis, 10*, 401–413.

Kirkland, A. I., Saxton, W. O., Chau, K.-L., Tsuno, K., & Kawasaki, M. (1995). Super-resolution by aperture synthesis: Tilt series reconstruction in CTEM. *Ultramicroscopy, 57*, 355–374.

Kisielowski, C., Freitag, B., Bischoff, M., van Lin, H., Lazar, S., Knippels, G., Tiemeijer, P., van der Stam, M., von Harrach, S., Stekelenburg, M., Haider, M., Uhlemann, S., Müller, H., Hartel, P., Kabius, B., Miller, D., Petrov, I., Olson, E. A., Donchev, T., . . . Dahmen, U. (2008). Detection of single atoms and buried defects in three dimension by aberration-corrected electron microscope with 0.5-Å information limit. *Microscopy and Microanalysis, 14*, 469–477.

Kisielowski, C., Hetherington, C. J. D., Wang, Y. C., Kilaas, R., O'Keefe, M. A., & Thust, A. (2001). Imaging columns of light elements carbon, nitrogen and oxygen with sub ångstrom resolution. *Ultramicroscopy, 89*, 243–263.

Kisielowski, C., Principe, E., Freitag, B., & Hubert, D. (2001). Benefits of microscopy with super resolution. *Physica B, 308*(310), 1090–1096.

Klabunde, K. J. (Ed.). (2002). *Nanoscale materials in chemistry*. John Wiley and Sons, Inc.

Klenov, D. O., Findlay, S. D., Allen, L. J., & Stemmer, S. (2007). Influence of orientation on the contrast of high-angle annular dark-field images of silicon. *Physical Review B, 76*(1), Article 014111.

Klingstedt, M., Sundberg, M., Eriksson, L., Haigh, S., Kirkland, A., Grüner, D., De Backer, A., Van Aert, S., & Terasaki, O. (2012). Exit wave reconstruction from focal series of HRTEM images, single crystal XRD and total energy studies on Sb_xWO_{3+y} (x ~ 0.11). *Zeitschrift für Kristallographie, 227*, 341–349.

Knapek, E., & Dubochet, J. (1980). Beam damage to organic material is considerably reduced in cryo-electron microscopy. *Journal of Molecular Biology, 141*, 147–161.

Knuth, K. H., Habeck, M., Malakar, N. K., Mubeen, A. M., & Placek, B. (2015). Bayesian evidence and model selection. *Digital Signal Processing, 47*, 50–67.

Knuth, K. H., Placek, B., & Richards, Z. (2012). Detection and characterization of non-transiting extra-solar planets in Kepler data using reflected light variations. In *Proceedings of the conference on intelligent data understanding 2012* (pp. 31–38).

Kolodiazhnyi, T., & Petric, A. (2005). The applicability of Sr-deficient n-type $SrTiO_3$ for SOFC anodes. *Journal of Electroceramics, 15*, 5–11.

Koneti, S., Roiban, L., Dalmas, F., Langlois, C., Gay, A.-S., Cabiac, A., Grenier, T., Banjak, H., Maxim, V., & Epicier, T. (2019). Fast electron tomography: Applications to beam sensitive samples and in situ tem or operando environmental TEM studies. *Materials Characterization, 151*, 480–495.

Konishi, S., & Kitagawa, G. (2008). *Information criteria and statistical modeling*. Springer.

Kook, S., Zhang, R., Chan, Q. N., Aizawa, T., Kondo, K., Pickett, L. M., Cenker, E., Bruneaux, G., Andersson, O., Pagels, J., & Nordin, E. Z. (2016). Automated detection of primary particles from transmission electron microscope (TEM) images of soot aggregates in diesel engine environments. *SAE International Journal of Engines, 9*, 279–296.

Koster, A. J., van den Bos, A., & van der Mast, K. D. (1987). An autofocus method for a TEM. *Ultramicroscopy, 21*, 209–222.

Kramberger, C., & Meyer, J. C. (2016). Progress in structure recovery from low dose exposures: Mixed molecular adsorption, exploitation of symmetry and reconstruction from the minimum signal level. *Ultramicroscopy, 170*, 60–68.

Kramberger, C., Mittelberger, A., Hofer, C., & Meyer, J. C. (2017). Analysis of point defects in graphene using low dose scanning transmission electron microscopy imaging and maximum likelihood reconstruction. *Physica Status Solidi B, 254*, Article 1700176.

Krause, F. F., Schowalter, M., Grieb, T., Müller-Caspary, K., Mehrtens, T., & Rosenauer, A. (2016). Effects of instrument imperfections on quantitative scanning transmission electron microscopy. *Ultramicroscopy, 161*, 146–160.

Krivanek, O. L., Chisholm, M. F., Nicolosi, V., Pennycook, T. J., Corbin, G. J., Dellby, N., Murfitt, M. F., Own, C. S., Szilagyi, Z. S., Oxley, M. P., Pantelides, S. T., & Pennycook, S. J. (2010). Atom-by-atom structural and chemical analysis by annular dark-field electron microscopy. *Nature, 464*, 571–574.

Krivanek, O. L., Dellby, N., & Murfitt, M. F. (2009). In *Handbook of charged particle optics* (pp. 601–641). CRC Press.

Krivanek, O. L., Lovejoy, T. C., & Dellby, N. (2015). Aberration-corrected stem for atomic-resolution imaging and analysis. *Journal of Microscopy, 259*, 165–172.

Kullback, S. (1959). *Information theory and statistics. Springer texts in statistics*. New York: John Wiley and Sons.

Kullback, S., & Leibler, R. (1951). On information and sufficiency. *The Annals of Mathematical Statistics, 22*, 79–86.

Kundu, P., Turner, S., Van Aert, S., Ravishankar, N., & Van Tendeloo, G. (2014). Atomic structure of quantum gold nanowires: Quantification of the lattice strain. *ACS Nano, 8*, 599–606.

Lagarias, J. C., Reeds, J. A., Wright, M. H., & Wright, P. E. (1998). Convergence properties of the Nelder-Mead simplex method in low dimension. *SIAM Journal on Optimization, 9*(1), 112–147.

Lawton, W. H., & Sylvestre, E. A. (1971). Elimination of linear parameters in nonlinear regression. *Technometrics, 13*(3), 461–467.

LeBeau, J. M., Findlay, S. D., Allen, L. J., & Stemmer, S. (2008). Quantitative atomic resolution scanning transmission electron microscopy. *Physical Review Letters*, *100*, Article 206101.

LeBeau, J. M., Findlay, S. D., Allen, L. J., & Stemmer, S. (2010). Standardless atom counting in scanning transmission electron microscopy. *Nano Letters*, *10*, 4405–4408.

LeBeau, J. M., & Stemmer, S. (2008). Experimental quantification of annular dark-field images in scanning transmission electron microscopy. *Ultramicroscopy*, *108*, 1653–1658.

Lee, C., Wei, X., Kysar, J. W., & Hone, J. (2008). Measurement of the elastic properties and intrinsic strength of monolayer graphene. *Science*, *321*, 385–388.

Lee, S., Oshima, Y., Sawada, H., Hosokawa, F., Okunishi, E., Kaneyama, T., Kondo, Y., Nikita, S., Takagi, H., Tanishiro, Y., & Takayanagi, K. (2011). Counting lithium ions in the diffusion channel of a LiV_2O_4. *Journal of Applied Physics*, *109*, Article 113530.

Lever, J., Krzywinski, M., & Altman, N. (2016). Model selection and overfitting. *Nature Methods*, *13*, 703–704.

Levi, M. D., Salitra, G., Markovsky, B., Teller, H., Aurbach, D., Heider, U., & Heider, L. (1999). Solid-state electrochemical kinetics of Li-ion intercalation into $Li_{1-x}CoO_2$: Simultaneous application of electroanalytical techniques SSCV, PITT, and EIS. *Journal of the Electrochemical Society*, *146*(4), Article 1279.

Levin, B. D., Zachmann, M. J., Werner, J. G., Sahore, R., Nguyen, K. X., Han, Y., Xie, B., Ma, L., Archer, L. A., Giannelis, E. P., Wiesner, U., Kourkoutis, L. F., & Muller, D. A. (2017). Characterization of sulfur and nanostructured sulfur battery cathodes in electron microscopy without sublimation artefacts. *Microscopy and Microanalysis*, *23*, 155–162.

Li, H., Wang, Z., & Chen, L. (2009). Research on advanced materials for Li-ion batteries. *Advanced Materials*, *21*, 4593–4607.

Li, L., Gao, P., Nelson, C. T., Jokisaari, J. R., Zhang, Y., Kim, S.-J., Melvill, A., Adamo, C., Schlom, D. G., & Pan, X. (2013). Atomic scale structure changes induced by charged domain walls in ferroelectric materials. *Nano Letters*, *13*(11), 5218–5223.

Li, Z., Baker, M. L., Jiang, W., Estes, M. K., & Prasad, B. V. (2009). Rotavirus architecture at subnanometer resolution. *Journal of Virology*, *83*(4), 1754–1766.

Li, Z. Y., Young, N. P., Di Vece, M., Palomba, S., Palmer, R. E., Bleloch, A. L., Curley, B. C., Johnston, R. L., Jiang, J., & Yuan, J. (2008). Three-dimensional atomic-scale structure of size-selected gold nanoclusters. *Nature*, *451*, 46–48.

Lichte, H. (1986). Electron holography approaching atomic resolution. *Ultramicroscopy*, *20*, 293–304.

Liu, M. Z., & Guyot-Sionnest, P. (2004). Synthesis and optical characterization of Au/Ag core/shell nanorods. *Journal of Physical Chemistry B*, *108*, 5882–5888.

Liz-Marzàn, L. M., & Kamat, P. V. (Eds.). (2003). *Nanoscale materials*. Springer US.

Lobato, I., Van Aert, S., & Verbeeck, J. (2016). Progress and new advances in simulating electron microscopy datasets using MULTEM. *Ultramicroscopy*, *168*, 17–27.

Lobato, I., & Van Dyck, D. (2015). MULTEM: A new multislice program to perform accurate and fast electron diffraction and imaging simulation using graphics processing units with CUDA. *Ultramicroscopy*, *156*, 9–17.

Lubk, A., Rossell, M. D., Seidel, J., Chu, Y. H., Ramesh, R., Hytch, M. J., & Snoeck, E. (2013). Electromechanical coupling among edge dislocations, domain walls, and nanodomains in $BiFeO_3$ revealed by unit-cell-wise strain and polarization maps. *Nano Letters*, *13*, 1410–1415.

Luo, W., Duan, W., Louie, S. G., & Cohen, M. L. (2004). Structural and electronic properties of n-doped and p-doped $SrTiO_3$. *Physical Review B*, *70*, Article 214109.

Ma, Y. Y., Li, W. Y., Cho, E. C., Li, Z. Y., Yu, T. K., Zeng, J., Xie, Z. X., & Xia, Y. N. (2010). Au/Ag core-shell nanocubes with finely tuned and well-controlled sizes, shell thicknesses, and optical properties. *ACS Nano*, *4*, 6725–6734.

MacArthur, K. E., Jones, L. B., & Nellist, P. D. (2014). How flat is your detector? Non-uniform annular detector sensitivity in stem quantification. *Journal of Physics. Conference Series, 522*(1), Article 012018.

Maccagnano-Zacher, S. E., Mkhoyan, K. A., Kirkland, E. J., & Silcox, J. (2008). Effects of tilt on high-resolution ADF STEM imaging. *Ultramicroscopy, 108*, 718–726.

Mackay, A. L. (1962). A dense non-crystallographic packing of equal spheres. *Acta Crystallographica, 15*, 916.

MacLaren, I., & Ramasse, Q. M. (2014). Aberration-corrected scanning transmission electron microscopy for atomic-resolution studies of functional oxides. *International Materials Reviews, 59*(3), 115–131.

Madsen, J., Liu, P., Kling, J., Birkedal Wagner, J., Hansen, T. W., Winther, O., & Schiotz, J. (2018). A deep learning approach to identify local structures in atomic-resolution transmission electron microscopy images. *Advanced Theory and Simulations, 1*, Article 1800037.

Malakar, N. K., Gladkov, D., & Knuth, K. H. (2013). Modeling a sensor to improve its efficacy. *Journal of Sensors, 2013*, Article 481054.

Mana, G., Giuliano Alba, P. A., & Lago, S. (2014). Bayesian estimate of the degree of a polynomial given a noisy data sample. *Measurement, 55*, 564–570.

Martin, L. W., Chu, Y.-H., & Ramesh, R. (2010). Advances in the growth and characterization of magnetic, ferroelectric, and multiferroic oxide thin films. *Materials Science & Engineering. R, Reports, 68*, 89–133.

Martinez, G. T., De Backer, A., Rosenauer, A., Verbeeck, J., & Van Aert, S. (2014). The effect of probe inaccuracies on the quantitative model-based analysis of high angle annular dark field scanning transmission electron microscopy images. *Micron, 63*, 57–63.

Martinez, G. T., Jones, L., De Backer, A., Béché, A., Verbeeck, J., Van Aert, S., & Nellist, P. D. (2015). Quantitative STEM normalisation: The importance of the electron flux. *Ultramicroscopy, 159*, 46–58.

Martinez, G. T., Rosenauer, A., De Backer, A., Verbeeck, J., & Van Aert, S. (2014). Quantitative composition determination at the atomic level using model-based high-angle annular dark field scanning transmission electron microscopy. *Ultramicroscopy, 137*, 12–19.

Matheron, G. (1963). Principles of geostatistics. *Economic Geology, 58*, 1246–1266.

McGibbon, A. J., Pennycook, S. J., & Jesson, D. E. (1999). Crystal structure retrieval by maximum entropy analysis of atomic resolution incoherent images. *Journal of Microscopy, 195*(1), 44–57.

McLachlan, G., & Peel, D. (2000). *Finite mixture models. Wiley series in probability and statistics.* John Wiley and Sons, Inc.

McQuarrie, A. D., & Tsai, C.-L. (1998). *Regression and time series model selection.* World Scientific.

Mehrtens, T., Müller, K., Schowalter, M., Hu, D., Schaadt, D. M., & Rosenauer, A. (2013). Measurement of indium concentration profiles and segregation efficiencies from high-angle annular dark field-scanning transmission electron microscopy images. *Ultramicroscopy, 131*, 1–9.

Mehrtens, T., Schowalter, M., Tytko, D., Choi, P., Raabe, D., Hoffmann, L., Jonen, H., Rossow, U., Hangleiter, A., & Rosenauer, A. (2013). Measurement of the indium concentration in high indium content InGaN layers by scanning transmission electron microscopy and atom probe tomography. *Applied Physics Letters, 102*, Article 132112.

Meng, Y., Zhang, Z., Yin, H., & Ma, T. (2018). Automatic detection of particle size distribution by image analysis based on local adaptive canny edge detection and modified circular hough transform. *Micron, 106*, 34–41.

Merk, A., Bartesaghi, A., Banerjee, S., Falconieri, V., Rao, P., Davis, M. I., Pragani, R., Boxer, M. B., Earl, L. A., Milne, J. L., & Subramaniam, S. (2016). Breaking cryo-EM resolution barriers to facilitate drug discovery. *Cell, 165*, 1698–1707.

Meyer, J. C., Kotakoski, J., & Mangler, C. (2014). Atomic structure from large-area, low-dose exposures of materials: A new route to circumvent radiation damage. *Ultramicroscopy, 145*, 13–21.

Meyer, R. R., Kirkland, A. I., & Saxton, W. O. (2002). A new method for the determination of the wave aberration function for high resolution TEM 1. Measurement of the symmetric aberrations. *Ultramicroscopy, 92*, 89–109.

Midgley, P. A., & Dunin-Borkowski, R. E. (2009). Electron tomography and holography in materials science. *Nature Materials, 8*, 271–280.

Midgley, P. A., & Weyland, M. (2003). 3D electron microscopy in the physical sciences: The development of Z-contrast and EFTEM tomography. *Ultramicroscopy, 96*, 413–431.

Miedema, M. A. O., van den Bos, A., & Buist, A. H. (1994). Experimental design of exit wave reconstruction from a transmission electron microscope defocus series. *IEEE Transactions on Instrumentation and Measurement, 43*, 181–186.

Migunov, V., Ryll, H., Zhuge, X., Simson, M., Struder, L., Batenburg, K. J., Houben, L., & Dunin-Borkowski, R. E. (2015). Rapid low dose electron tomography using a direct electron detection camera. *Scientific Reports, 5*, Article 14516.

Mir, J. A., Plackett, R., Shipsey, I., & dos Santos, J. M. F. (2017). Using the medipix3 detector for direct electron imaging in the range 60kev-200kev in electron microscopy. *Journal of Instrumentation, 12*, Article C11015.

Mitchell, R. H. (Ed.). (2003). *Perovskites - modern and ancient*. Almaz.

Mkhoyan, K. A., Maccagnano-Zacher, S. E., Kirkland, E. J., & Silcox, J. (2008). Effect of amorphous layers on ADF-STEM imaging. *Ultramicroscopy, 108*, 791–803.

Monserrat, D., Vispa, A., Pardo, L. C., Tolchenov, R., Mukhopadhyay, S., & Fernandez-Alonso, F. (2015). Fabada goes mantid to answer an old question: How many lines are there? *Journal of Physics. Conference Series, 663*, Article 012009.

Mood, A. M., Graybill, F. A., & Boes, D. C. (1974). *Introduction to the theory of statistics* (third edition). McGraw-Hill Book Company.

Morcrette, M., Rozier, P., Dupont, L., Mugnier, E., Sannier, L., Galy, J., & Tarascon, J.-M. (2003). A reversible copper extrusion-insertion electrode for rechargeable Li batteries. *Nature Materials, 2*, 755–761.

Mory, C., Tence, M., & Colliex, C. (1985). Theoretical-study of the characteristics of the probe for a stem with a field-emission gun. *Journal de Microscopie et de Spectroscopie Électroniques, 10*(5), 381–387.

Muller, D. A. (1999). Why changes in bond lengths and cohesion lead to core-level shifts in metals, and consequences for the spatial difference method. *Ultramicroscopy, 78*, 163–174.

Muller, D. A., Kirkland, E. J., Thomas, M. G., Grazul, J. L., Fitting, L., & Weyland, M. (2006). Room design for high-performance electron microscopy. *Ultramicroscopy, 106*, 1033–1040.

Myung, I. J. (2000). The importance of complexity in model selection. *Journal of Mathematical Psychology, 44*, 190–204.

Myung, I. J. (2003). Tutorial on maximum likelihood estimation. *Journal of Mathematical Psychology, 47*, 90–100.

Myung, I. J., & Pitt, M. A. (1997). Applying Occam's razor in modeling cognition: A Bayesian approach. *Psychonomic Bulletin & Review, 4*, 79–95.

Nakanishi, N., Kotaka, Y., & Yamazaki, T. (2006). An expanded approach to noise reduction from high-resolution STEM images based on the maximum entropy method. *Ultramicroscopy, 106*, 233–239.

Nakanishi, N., Yamazaki, T., Recnik, A., Ceh, M., Kawasaki, M., Watanabe, K., & Shiojiri, M. (2002). Retrieval process of high-resolution HAADF-STEM images. *Journal of Electron Microscopy, 51*(6), 383–390.

Nebel, C. E. (2003). Semiconductor materials: From gemstone to semiconductor. *Nature Materials, 2*, 431–432.

Nellist, P. D. (2007). Scanning transmission electron microscopy. In P. W. Hawkes, & J. H. C. Spence (Eds.), *Science of microscopy: Vol. 1, chapter 2* (pp. 65–132). Springer New York.

Nellist, P. D., Behan, G., Kirkland, A. I., & Hetherington, C. J. D. (2006). Confocal operation of a transmission electron microscope with two spherical aberration correctors. *Applied Physics Letters, 89*, Article 124105.

Nellist, P. D., Chisholm, M. F., Dellby, N., Krivanek, O. L., Murfitt, M. F., Szilagyi, Z. S., Lupini, A. R., Borisevich, A. Y., Sides Jr., W. H., & Pennycook, S. J. (2004). Direct sub-angstrom imaging of a crystal lattice. *Science, 305*, 1741.

Nellist, P. D., & Pennycook, S. J. (1998). Accurate structure determination from image reconstruction in ADF STEM. *Journal of Microscopy, 190*(1/2), 159–170.

Nellist, P. D., & Pennycook, S. J. (1999). Incoherent imaging using dynamically scattered coherent electrons. *Ultramicroscopy, 78*, 111–124.

Nellist, P. D., & Pennycook, S. J. (2000). The principles and interpretation of annular dark-field Z-contrast imaging. *Advances in Imaging and Electron Physics, 113*, 147–203.

Nellist, P. D., & Rodenburg, J. M. (1994). Beyond the conventional information limit: The relevant coherence function. *Ultramicroscopy, 54*(1), 61–74.

Newmark, P. (1982). Cryo-transmission microscopy: Fading hopes. *Nature, 299*, 386–387.

Nishimura, S., Kobayashi, G., Ohoyama, K., Kanno, R., Yashima, M., & Yamada, A. (2008). Experimental visualization of lithium diffusion in Li_xFePO_4. *Nature Materials, 7*, 707–711.

Nitta, N., Wu, F., & Lee, J. T. (2015). Li-ion battery materials: Present and future. *Materials Today, 18*, 252–264.

Novoselov, K. S., Geim, A. K., Morozov, S. V., Jiang, D., Zhang, Y., Dubonos, S. V., Grigorieva, I. V., & Firsov, A. A. (2004). Electric field effect in atomically thin carbon films. *Science, 306*, 666–669.

Ohtomo, A., & Hwang, H. Y. (2004). A high-mobility electron gas at the $LaAlO_3/SrTiO_3$ heterointerface. *Nature, 427*, 423–426.

Ohtomo, A., Muller, D. A., Grazul, J. L., & Hwang, H. Y. (2002). Artificial charge-modulation in atomic-scale perovskite titanate superlattices. *Nature, 419*, 378–380.

Okunishi, E., Ishikawa, I., Sawada, H., Hosokawa, F., Hori, M., & Kondo, Y. (2009). Visualization of light elements at ultrahigh resolution by STEM annular bright field microscopy. *Microscopy and Microanalysis, 15*(S2), 164–165.

Oliveira, M. M., Bowen, B., McKenna, R., & Chang, Y.-S. (2001). Fast digital image inpainting. In *Proceedings of the international conference on visualization, imaging and image processing (VIIP 2001)*.

Olson, G. B. (1997). Computational design of hierarchically structure materials. *Science, 277*, 1237–1242.

Olson, G. B. (2000). Designing a new material world. *Science, 288*, 993–998.

Oltalan, V., Uzun, A., Gates, B. C., & Browning, N. D. (2010). Towards full-structure determination of bi metallic nanoparticles with an aberration-corrected electron microscope. *Nature Nanotechnology, 5*, 843–847.

Op de Beeck, M., Van Dyck, D., & Coene, W. (1996). Wave function reconstruction in HRTEM: The parabola method. *Ultramicroscopy, 64*, 167–183.

Ophus, C., Ciston, J., & Nelson, C. T. (2016). Correcting nonlinear drift distortion of scanning probe and scanning transmission electron microscopes from image pairs with orthogonal scan directions. *Ultramicroscopy, 162*, 1–9.

Oshima, Y., Sawada, H., Hosokawa, F., Okunishi, E., Kaneyama, T., Kondo, Y., Niitaka, S., Takagi, H., Tanishiro, Y., & Takayanagi, K. (2010). Direct imaging of lithium atoms in LiV_2O_4 by spherical aberration-corrected electron microscopy. *Journal of Electron Microscopy, 59*, 457–461.

Papoulis, A., & Pillai, S. (2002). *Probability, random variable, and stochastic processes*. McGraw-Hill.

Park, K., Drummy, L. F., & Vaia, R. A. (2011). Ag shell morphology on Au nanorod core: Role of Ag precursor complex. *Journal of Materials Chemistry, 21*, 15608–15618.

Pastoriza-Santos, I., Sánchez-Iglesias, A., García de Abajo, F. J., & Liz-Marzán, L. (2007). Environmental optical sensitivity of gold nanodecahedra. *Advanced Functional Materials, 17*, 1443–1450.

Pecharromán, C., Pérez-Juste, J., Mata-Osoro, G., Liz-Marzán, L. M., & Mulvaney, P. (2008). Redshift of surface plasmon modes of small gold rods due to their atomic roughness and end-cap geometry. *Physical Review B, 77*, Article 035418.

Peng, L.-M. (1997). Anisotropic thermal vibrations and dynamical electron diffraction by crystals. *Acta Crystallographica. Section A, 53*, 663–672.

Pennycook, S. J., & Boatner, L. A. (1988). Chemically sensitive structure-imaging with a scanning transmission electron microscope. *Nature, 336*, 565–567.

Pennycook, S. J., & Nellist, P. D. (Eds.). (2011). *Scanning transmission electron microscopy*. Springer.

Pérez-Juste, J., Pastoriza-Santos, I., Liz-Marzán, L., & Mulvaney, P. (2005). Gold nanorods: Synthesis, characterization and applications. *Coordination Chemistry Reviews, 249*, 1870–1901.

Placek, B., Knuth, K. H., & Angerhausen, D. (2014). Exonest: Bayesian model selection applied to the detection and characterization of exoplanets via photometric variations. *The Astrophysical Journal, 795*, 112.

Pounds, K., & Vaughan, S. (2006). X-ray reflection in the nenear seyfert 2 galaxy NGC 1068. *Monthly Notices of the Royal Astronomical Society, 368*(2), 707–714.

Ralph, T. R., & Hogarth, M. P. (2002). Catalysis for low temperature fuel cells. Part II: The anode challenges. *Platinum Metals Review, 46*(3), 117–135.

Rayleigh, L. (1896). On the theory of optical images, with special reference to the microscope. *Philosophical Magazine, 42*, 167–195.

Recnik, A., Möbus, G., & Sturm, S. (2005). Image-warp: A real-space restoration method for high-resolution STEM images using quantitative HRTEM analysis. *Ultramicroscopy, 103*, 285–301.

Rehn, L. (1992). Ion-beams in high-temperature superconductivity research. *Nuclear Instruments and Methods, 64*, 161–168.

Retsky, M. (1974). Observed single atom elastic cross sections in a scanning electron microscope. *Optik, 41*, 127.

Ricolleau, C., Le Bouar, Y., Amara, H., Landon-Cardinal, O., & Alloyeau, D. (2013). Random vs realistic amorphous carbon models for high resolution microscopy and electron diffraction. *Journal of Applied Physics, 114*, Article 213504.

Ricolleau, C., Nelayah, J., Oikawa, T., Kohno, Y., Braidy, N., Wang, G., Hue, F., & Alloyeau, D. (2012). High resolution imaging and spectroscopy using Cs-corrected TEM with cold FEG JEM-ARM200F. *JEOL News, 47*(1), 2–8.

Rose, H. (1974). Phase contrast in scanning transmission electron microscopy. *Optik, 39*, 416–436.

Rose, H. H. (2009). Historical aspects of aberration correction. *Journal of Electron Microscopy, 58*(3), 77–85.

Rosenauer, A., Gries, K., Müller, K., Pretorius, A., Schowalter, M., Avramescu, A., Engl, K., & Lutgen, S. (2009). Measurement of specimen thickness and composition in $Al_xGa_{1-x}N/GaN$ using high-angle annular dark field images. *Ultramicroscopy, 109*, 1171–1182.

Rosenauer, A., Mehrtens, T., Müller, K., Gries, K., Schowalter, M., Satyam, P. V., Bley, S., Tessarek, C., Hommel, D., Sebald, K., Seyfried, M., Gutowski, J., Avramescu, A., Engl, K., & Lutgen, S. (2011). Composition mapping in InGaN by scanning transmission electron microscopy. *Ultramicroscopy, 111*, 1316–1327.

Rosenauer, A., & Schowalter, M. (2008). Stemsim - a new software tool for simulation of STEM HAADF Z-contrast imaging. In A. Cullis, & P. Midgley (Eds.), *Microscopy of semiconducting materials 2007. Springer proceedings in physics: Vol. 120* (pp. 170–172). Springer Netherlands.

Ryll, H., Simson, M., Holl, P., Huth, M., Ihle, S., Kondo, Y., Kotula, P., Liebel, A., Müller-Caspary, K., Rosenauer, A., Sagawa, R., Schmidt, J., Soltau, H., & Strüder, L. (2016). A pnccd-based, fast direct single electron imaging camera for TEM and STEM. *Journal of Instrumentation, 11*, Article P04006.

Sagawa, R., Yang, H., Jones, L., Simson, M., Huth, M., Soltau, H., Nellist, P. D., & Kondo, Y. (2017). Development of fast pixelated STEM detector and its applications using 4-dimensional dataset. *Microscopy and Microanalysis, 23*(Suppl 1), 52–53.

Saghi, Z., Benning, M., Leary, R., Macias-Montero, M., Borras, A., & Midgley, P. A. (2015). Reduced-dose and high-speed acquisition strategies for multi-dimensional electron microscopy. *Advanced Structural and Chemical Imaging, 1*(1), 1–10.

Salmons, B. S., Katz, D. R., & Trawick, M. L. (2010). Correction of distortion due to thermal drift in scanning probe microscopy. *Ultramicroscopy, 110*, 339–349.

Sammut, C., & Webb, G. I. (2011). Bias-variance decomposition. In *Encyclopedia of machine learning* (pp. 100–101). Springer.

Sánchez-Iglesias, A., Carbó-Argibay, E., Glaria, A., Rodriguez-González, B., Pérez-Juste, J., Pastoriza-Santos, I., & Liz-Marzán, L. M. (2010). Rapid epitaxial growth of Ag on Au nanoparticles: From Au nanorods to core-shell Au/Ag octahedrons. *Chemistry, 16*, 5558–5563.

Sánchez-Iglesias, A., Pastoriza-Santos, I., Pérez-Juste, J., Rodríguez-González, B., García de Abajo, F. J., & Liz-Marzán, L. M. (2006). Synthesis and optical properties of gold nanodecahedra with size control. *Advanced Materials, 18*, 2529–2534.

Sang, X., Du, K., Zhuo, M., & Ye, H. (2009). On the accuracy of maximum entropy reconstruction of high-resolution Z-contrast STEM images. *Micron, 40*, 247–254.

Sang, X., & LeBeau, J. M. (2013). Revolving scanning transmission electron microscopy: Correcting sample drift distortion without prior knowledge. *Ultramicroscopy, 138*, 28–35.

Sang, X., & LeBeau, J. M. (2016). Characterizing the response of a scintillator-based detector to single electrons. *Ultramicroscopy, 161*, 3–9.

Sawada, H., Tanishori, Y., Ohashi, N., Tomita, T., Hosokawa, F., Kaneyama, T., Kondo, Y., & Takayanagi, K. (2009). STEM imaging of 47-pm-separated atomic columns by a spherical aberration-corrected electron microscope with a 300-kv cold field emission gun. *Journal of Electron Microscopy, 58*, 357–361.

Schaffer, J. (2015). What not to multiply without necessity. *Australasian Journal of Philosophy, 93*, 644–664.

Schlapbach, L., & Zuttel, A. (2001). Hydrogen-storage materials for mobile applications. *Nature, 414*, 353–358.

Schwarz, G. (1978). Estimating the dimension of a model. *The Annals of Statistics, 6*(2), 461–464.

Scott, M. C., Chen, C.-C., Mecklenburg, M., Zhu, C., Xu, R., Ercius, P., Dahmen, U., Regan, B. C., & Miao, J. (2012). Electron tomography at 2.4-angstrom resolution. *Nature, 483*, 444–448.

Seidel, J., Martin, L. W., He, Q., Zhan, Q., Chu, H.-Y., Rother, A., Hawkridge, M. E., Maksymovych, P., Yu, P., Gajek, M., Balke, N., Kalinin, S. V., Gemming, S., Wang, F., Catalan, G., Scott, J. F., Spaldin, N. A., Orenstein, J., & Ramesh, R. (2009). Conduction at domain walls in oxide multiferroics. *Nature Materials, 8*, 229–234.

Seki, T., Ikuhara, Y., & Shibata, N. (2018). Theoretical framework of statistical noise in scanning transmission electron microscopy. *Ultramicroscopy, 193*, 118–125.

Shah, A. B., Ramasse, Q. M., May, S. J., Kavich, J., Wen, J. G., Zhai, X., Eckstein, J. N., Freeland, J., Bhattacharya, A., & Zuo, J. M. (2010). Presence and spatial distribution of interfacial electronic states in $LaMLa_3$-$SrMnO_3$ superlattices. *Physical Review B, 82*, Article 115112.

Shashilov, V. A., Sikirzhytski, V., Popova, L. A., & Lednev, I. K. (2010). Quantitative methods for structural characterization of proteins based on deep UV resonance Raman spectroscopy. *Methods, 52*(1), 23–37.

Sheehy, D. E., & Schmalian, J. (2009). Optical transparency of graphene as determined by the fine-structure constant. *Physical Review B, 80*, Article 193411.

Shibata, N., Kohno, Y., Findlay, S. D., Sawada, H., Kondo, Y., & Ikuhara, Y. (2010). New area detector for atomic-resolution scanning transmission electron microscopy. *Journal of Electron Microscopy, 59*(6), 473–479.

Singhal, A., Yang, J. C., & Gibson, J. M. (1997). STEM-based mass spectroscopy of supported Re clusters. *Ultramicroscopy, 67*, 191–206.

Sirohi, D., Chen, Z., Sun, L., Klose, T., Pierson, T. C., Rossmann, M. G., & Kuhn, R. J. (2016). The 3.8 Å resolution cryo-EM structure of Zika virus. *Science, 352*, 467–470.

Sivia, D. S., & Carlile, C. J. (1992). Molecular spectroscopy and Bayesian spectral analysis - how many lines are there? *Journal of Chemical Physics, 96*(1), Article 170.

Sivia, D. S., Carlile, C. J., & Howells, W. S. (1992). Bayesian analysis of quasielastic neutron scattering data. *Physica, 182*, 341–348.

Sivia, D. S., David, W. I. F., & Knight, K. S. (1993). An introduction to Bayesian model selection. *Physica D, 66*, 234–242.

Sivia, D. S., & Skilling, J. (2006). *A Bayesian tutorial.* Oxford University Press.

Skilling, J., & Bryan, R. K. (1984). Maximum entropy reconstruction: General algorithm. *Monthly Notices of the Royal Astronomical Society, 211*, 111–124.

So, Y.-G., & Kimoto, K. (2012). Effect of specimen misalignment on local structure analysis using annular dark-field imaging. *Journal of Electron Microscopy, 61*, 207–215.

Song, F., Wang, X., Powles, R. C., He, L., Marks, N. A., Zhao, S., Wan, J., Liu, Z., Zhou, J., Ringer, S. P., Han, M., & Wang, G. (2010). Calibrating the atomic balance by carbon nanoclusters. *Applied Physics Letters, 96*, Article 033103.

Springborg, M. (2000). *Methods of electronic-structure calculations: From molecules to solids.* Chichester, UK: John Wiley and Sons.

Stoica, P., & Selen, Y. (2004). Model-order selection: A review of information criterion rules. *IEEE Signal Processing Magazine, 21*, 36–47.

Stoica, P., Selén, Y., & Li, J. (2004). Multi-modal approach to model-selection. *Digital Signal Processing, 14*, 399–412.

Stuart, A., & Ord, K. (1994). *Kendall's advanced theory of statistics: Vol. I - distribution theory.* Edward Arnold.

Tach, R. E., & Tach, S. S. (1971). Damage to biological samples caused by the electron beam during electron microscopy. *Biophysical Journal, 11*, 204–210.

Talapatra, S., Ganesan, P. G., Kim, T., Vajtai, R., Huang, M., Shima, M., Ramanath, G., Srivastava, D., Deevi, S. C., & Ajayan, P. M. (2005). Irradiation-induced magnetism in carbon nanostructures. *Physical Review Letters, 95*, Article 097201.

Tarascon, J.-M., & Armand, M. (2001). Issues and challenges facing rechargeable lithium batteries. *Nature, 414*, 359–367.

Tate, M. W., Purohit, P., Chamberlain, D., Nguyen, K. X., Hovden, R., Chang, C. S., Deb, P., Turgut, E., Heron, J. T., Schlom, D. G., Ralph, D. C., Fuchs, G. D., Shanks, K. S., Philipp, H. T., Muller, D. A., & Gruner, S. M. (2016). High dynamic range pixel array detector for scanning transmission electron microscopy. *Microscopy and Microanalysis, 22*, 237–249.

Tedsree, K., Li, T., Jones, S., Chan, C. W. A., Yu, K. M. K., Bogot, P. A. J., Marquies, E. A., Smith, G. D. W., & Tsang, S. C. E. (2011). Hydrogen production from formic acid decomposition at room temperature using a Ag-Pd core-shell nanocatalyst. *Nature Nanotechnology*, *6*, 302–307.

Thust, A., Coene, W. M. J., Op de Beeck, M., & Van Dyck, D. (1996). Focal-series reconstruction in HRTEM: Simulation studies on non-periodic objects. *Ultramicroscopy*, *64*, 211–230.

Trotta, R. (2008). Bayes in the sky: Bayesian inference and model selection in cosmology. *Contemporary Physics*, *49*(2), 71–104.

Ugurlu, O., Haus, J., Gunawan, A. A., Thomas, M. G., Maheshwari, S., Tsapatsis, M., & Mkhoyan, K. A. (2011). Radiolysis to knock-on damage transition in zeolites under electron beam irradiation. *Physical Review B*, *83*, Article 113408.

Unwin, P. N., & Henderson, R. (1975). Molecular structure determination by electron microscopy of unstained crystalline specimens. *Journal of Molecular Biology*, *94*, 425–440.

Urban, K. W. (2008). Studying atomic structures by aberration-corrected transmission electron microscopy. *Science*, *321*, 506–510.

Van Aert, S., Batenburg, K. J., Rossell, M. D., Erni, R., & Van Tendeloo, G. (2011). Three-dimensional atomic imaging of crystalline nanoparticles. *Nature*, *470*, 374–377.

Van Aert, S., Chang, L. Y., Bals, S., Kirkland, A. I., & Van Tendeloo, G. (2009). Effect of amorphous layers on the interpretation of restored exit waves. *Ultramicroscopy*, *109*, 237–246.

Van Aert, S., De backer, A., Jones, L., Martinez, G. T., Bèchè, A., & Nellist, P. D. (2019). Control of knock-on damage for 3D atomic scale quantification of nanostructures: Making every electron count in scanning transmission electron microscopy. *Physical Review Letters*, *122*, Article 066101.

Van Aert, S., De Backer, A., Martinez, G. T., Goris, B., Bals, S., & Van Tendeloo, G. (2013). Procedure to count atoms with trustworthy single-atom sensitivity. *Physical Review B*, *87*, Article 064107.

Van Aert, S., den Dekker, A., & Van Dyck, D. (2004). How to optimize the experimental design of quantitative atomic resolution TEM experiments? *Micron*, *35*, 425–429.

Van Aert, S., den Dekker, A. J., van den Bos, A., & Van Dyck, D. (2002). High-resolution electron microscopy: From imaging toward measuring. *IEEE Transactions on Instrumentation and Measurement*, *51*(4), 611–615.

Van Aert, S., den Dekker, A. J., van den Bos, A., & Van Dyck, D. (2004). Statistical experimental design for quantitative atomic resolution transmission electron microscopy. *Advances in Imaging and Electron Physics*, *130*, 1–164.

Van Aert, S., den Dekker, A. J., van den Bos, A., Van Dyck, D., & Chen, J. H. (2005). Maximum likelihood estimation of structure parameters from high resolution electron microscopy images. Part II: A practical example. *Ultramicroscopy*, *104*, 107–125.

Van Aert, S., den Dekker, A. J., Van Dyck, D., & van den Bos, A. (2002a). High-resolution electron microscopy and electron tomography: Resolution versus precision. *Journal of Structural Biology*, *138*, 21–33.

Van Aert, S., den Dekker, A. J., Van Dyck, D., & van den Bos, A. (2002b). Optimal experimental design of STEM measurement of atom column positions. *Ultramicroscopy*, *90*, 273–289.

Van Aert, S., Geuens, P., Van Dyck, D., Kisielowski, C., & Jinscheck, J. R. (2007). Electron channelling based crystallography. *Ultramicroscopy*, *107*, 551–558.

Van Aert, S., Van den Broek, W., Goos, P., & Van Dyck, D. (2012). Model-based electron microscopy: From images toward precise numbers for unknown structure parameters. *Micron*, *43*(4), 509–515.

Van Aert, S., & Van Dyck, D. (2001). Do smaller probes in a STEM result in more precise measurement of the distances between atom columns? *Philosophical Magazine. B*, *81*, 1833–1846.

Van Aert, S., Van Dyck, D., & den Dekker, A. J. (2006). Resolution of coherent and incoherent imaging systems reconsidered - classical criteria and a statistical alternative. *Optics Express*, *9*(14), 3830–3839.

Van Aert, S., Verbeeck, J., Erni, R., Bals, S., Luysberg, M., Van Dyck, D., & Van Tendeloo, G. (2009). Quantitative atomic resolution mapping using high-angle annular dark field scanning transmission electron microscopy. *Ultramicroscopy*, *109*, 1236–1244.

van Benthem, K., Lupini, A. R., Kim, M., Baik, H. S., Doh, S., Lee, J.-H., Oxley, M. P., Findlay, S. D., Allen, L. J., Luck, J. T., & Pennycook, S. J. (2005). Three-dimensional imaging of individual hafnium atoms inside a semiconductor device. *Applied Physics Letters*, *87*, Article 034104.

van Benthem, K., Lupini, A. R., Oxley, M. P., Findlay, S. D., Allen, L. J., & Pennycook, S. J. (2006). Three-dimensional ADF imaging of individual atoms by through-focal series scanning transmission electron microscopy. *Ultramicroscopy*, *106*, 1062–1068.

van den Bos, A. (1982). Parameter estimation. In *Handbook of measurement science: Vol. 1* (pp. 331–377). Chichester: Wiley.

van den Bos, A. (2007). *Parameter estimation for scientists and engineers*. Wiley.

van den Bos, A., & den Dekker, A. J. (2001). *Resolution reconsidered - conventional approaches and an alternative*. *Advances in imaging and electron physics: Vol. 117*. San Diego: Academic Press (pp. 241–360).

van den Bos, K. H. W., De Backer, A., Martinez, G. T., Winckelmans, N., Bals, S., Nellist, P. D., & Van Aert, S. (2016). Unscrambling mixed elements using high angle annular dark field scanning transmission electron microscopy. *Physical Review Letters*, *116*(24), Article 246101.

van den Bos, K. H. W., Janssens, L., De Backer, A., Nellist, P. D., & Van Aert, S. (2019). The atomic lensing model: New opportunities for atom-by-atom metrology of heterogeneous nanomaterials. *Ultramicroscopy*, *203*, 155–162.

Van den Broek, W., & Koch, C. T. (2012). Method for retrieval of the three-dimensional object potential by inversion of dynamical electron scattering. *Physical Review Letters*, *109*, Article 245502.

Van den Broek, W., & Koch, C. T. (2013). General framework for quantitative three-dimensional reconstruction from arbitrary detection geometries in tem. *Physical Review B*, *87*, Article 184108.

Van den Broek, W., Van Aert, S., Goos, P., & Van Dyck, D. (2011). Throughput maximization of particle radius measurements through balancing size versus current of the electron probe. *Ultramicroscopy*, *111*, 940–947.

Van Dyck, D. (2002). High-resolution electron microscopy. *Advances in Imaging and Electron Physics*, *123*, 105–171.

Van Dyck, D., Bettens, E., Sijbers, J., Op de Beeck, M., van den Bos, A., & den Dekker, A. J. (1997). From high resolution image to atomic structure: How far are we? *Scanning Microscopy*, *11*, 467–478.

Van Dyck, D., & Chen, J. H. (1999a). A simple theory for dynamical electron diffraction in crystals. *Solid State Communications*, *109*(8), 501–505.

Van Dyck, D., & Chen, J. H. (1999b). Towards an exit wave in closed analytical form. *Acta Crystallographica. Section A*, *55*, 212–215.

Van Dyck, D., Jinscheck, J. R., & Chen, F.-R. (2012). 'Big Bang' tomography as a new route to atomic-resolution electron tomography. *Nature*, *486*, 243–246.

Van Dyck, D., & Op de Beeck, M. (1990). New direct methods for phase and structure retrieval in HREM. In *Proceedings of the XIIth international congress for electron microscopy* (pp. 26–27). Seattle: San Fransisco Press.

Van Dyck, D., & Op de Beeck, M. (1996). A simple intuitive theory for electron diffraction. *Ultramicroscopy*, *64*, 99–107.

Van Dyck, D., Op de Beeck, M., & Coene, W. (1993). A new approach to object wave function reconstruction in electron microscopy. *Optik*, *93*(3), 103–107.

Van Dyck, D., Van Aert, S., den Dekker, A. J., & van den Bos, A. (2003). Is atomic resolution transmission electron microscopy able to resolve and refine amorphous structures? *Ultramicroscopy, 98*, 27–42.

Vanrompay, H., Béché, A., Verbeeck, J., & Bals, S. (2019). Experimental evaluation of undersampling schemes for electron tomography of nanoparticles. *Particle & Particle Systems Characterization, 36*(7), Article 1900096.

Velazco, A., Nord, M., Béché, A., & Verbeeck, J. (2020). Evaluation of different rectangular scan strategies for stem imaging. *Ultramicroscopy, 215*, Article 113021.

Verbeeck, J., Béché, A., & Van den Broek, W. (2012). A holographic method to measure source size broadening in STEM. *Ultramicroscopy, 120*, 35–40.

Vieira, M. A. C., Bakic, P. R., & Maidment, A. D. A. (2013). Effect of denoising on the quality of reconstructed images in digital breast tomosynthesis. *Proceedings - SPIE, 8668*, Article 86680C.

Vispa, A., Busch, S., Ll, T. J., Unruh, T., Fernandez-Alonso, F., & Pardo, L. C. (2016). A robust comparison of dynamical scenarios in a glass forming liquid. *Physical Chemistry, 18*, Article 3975.

Vispa, A., Monserrat, D., Cuello, G. J., Fernandez-Alonso, F., Mukhopadhyay, S., Demmel, F., Tamarit, J. Ll., & Pardo, L. C. (2017). On the microscopic mechanism behind the purely orientation disorder-disorder transition in the plastic phase of 1-chloroadamantane. *Physical Chemistry Chemical Physics, 19*(30), 20259–20266.

von Colbe, J. B., Ares, J. R., Barale, J., Baricco, M., Buckley, C., Capurso, G., Gallandat, N., Grant, D. M., Guzik, M. N., & Jacob, I. (2019). Applications of hydrides in hydrogen storage and compression: Achievements, outlook and perspectives. *International Journal of Hydrogen Energy, 44*, 7780–7808.

von Harrach, H. (1995). Instrumental factors in high-resolution FEG STEM. *Ultramicroscopy, 58*, 1–5.

Vonck, J., & Mills, D. J. (2017). Advances in high-resolution cryo-EM of oligomeric enzymes. *Current Opinion in Structural Biology, 46*, 48–54.

Voyles, P. M., Muller, D. A., Grazul, J. L., Citrin, P. H., & Gossmann, H.-J. L. (2002). Atomic-scale imaging of individual dopant atoms and clusters in highly n-type bulk Si. *Nature, 416*, 826–829.

Walther, T., & Ross, I. M. (2013). Aberration corrected high-resolution transmission and scanning transmission electron microscopy of thin perovskite layers. *Physics Procedia, 40*, 49–55.

Wang, A., Chen, F. R., Van Aert, S., & Van Dyck, D. (2010). Direct structure inversion from exit waves. Part I: Theory and simulations. *Ultramicroscopy, 110*, 527–534.

Wang, A., Chen, F. R., Van Aert, S., & Van Dyck, D. (2012). Direct structure inversion from exit waves. Part II: A practical example. *Ultramicroscopy, 116*, 77–85.

Wang, A., Van Aert, S., Goos, P., & Van Dyck, D. (2012). Precision of three-dimensional atomic scale measurements from HRTEM images: What are the limits? *Ultramicroscopy, 114*, 20–30.

Wang, P., Behan, G., Kirkland, A. I., Nellist, P. D., Cosgriff, E. C., D'Alfonso, A. J., Morgan, A. J., Allen, L. J., Hashimoto, A., Takeguchi, M., Mitsuishi, K., & Shimojo, M. (2011). Bright-field scanning confocal electron microscopy using a double aberration-corrected transmission electron microscope. *Ultramicroscopy, 111*, 877–886.

Wang, P., Behan, G., Takeguchi, M., Hashimoto, A., Mitsuishi, K., Shimojo, M., Kirkland, A. I., & Nellist, P. D. (2010). Nanoscale energy-filtered scanning confocal electron microscopy using a double-aberration-corrected transmission electron microscope. *Physical Review Letters, 104*, Article 200801.

Watanabe, K., Kotaka, Y., Nakanishi, N., Yamazaki, T., Hashimoto, I., & Shiojiri, M. (2002). Deconvolution processing of HAADF STEM images. *Ultramicroscopy, 92*, 191–199.

Welvaert, M., & Rosseel, Y. (2013). On the definition of signal-to-noise ratio and contrast-to-noise ratio for fMRI data. *PLoS ONE, 8*, Article e77089.

Westin, C.-F., Knutsson, H., & Kikinis, R. (2000). Adaptive image filtering. In *Handbook of medical imaging*. Academic Press.

Wiesendanger, R. (Ed.). (1994). *Scanning probe microscopy and spectroscopy - method and applications*. Cambridge University Press.

Willets, K. A., & Van Duyne, R. P. (2007). Localized surface plasmon resonance spectroscopy and sensing. *Annual Review of Physical Chemistry, 58*, 267–297.

Williams, D. B., & Carter, C. B. (2009). *Transmission electron microscopy: A textbook for materials science* (second edition). Springer.

Wilson, E. B. (1927). Probable inference, the law of succession, and statistical inference. *Journal of the American Statistical Association, 22*, 209–212.

Wolf, S. G., Houben, L., & Elbaum, M. (2014). Cryo-scanning transmission electron tolography of vitrified cells. *Nature Methods, 11*, 423–428.

Wolfe, J. H. (1970). Pattern clustering by multivariate mixture analysis. *Multivariate Behavioral Research, 5*, 329–350.

Xiang, Y. U., Wu, X. C., Liu, D. F., Li, Z. Y., Chu, W. G., Feng, L. L., Zhang, K., Zhou, W. Y., & Xie, S. S. (2008). Gold nanorod-seeded growth of silver nanostructures: From homogeneous coating to anisotropic coating. *Langmuir, 24*, 3465–3470.

Xu, R., Chen, C.-C., Wu, L., Scott, M. C., Theis, W., Ophus, C., Bartels, M., Yang, Y., Ramezani-Dakhel, H., Sawaya, M. R., Heinz, H., Marks, L. D., Ercius, P., & Miao, J. (2015). Three-dimensional coordinates of individual atoms in materials revealed by electron tomography. *Nature Materials, 14*, 1099–1103.

Xu, X., Beckman, S. P., Specht, P., Weber, E. R., Chrzan, D. C., Erni, R. P., Arslan, I., Browning, N., Bleloch, A., & Kisielowski, C. (2005). Distortion and segragation in a dislocation core region at atomic resolution. *Physical Review Letters, 95*, Article 145501.

Yang, H., Pennycook, T. J., & Nellist, P. D. (2015). Efficient phase contrast imaging in STEM using a pixelated detector. Part II: Optimisation of imaging conditions. *Ultramicroscopy, 151*, 232–239.

Yang, P., & Tarascon, J.-M. (2012). Towards systems materials engineering. *Nature Materials, 11*, 560–563.

Yang, Z. S., & Chang, H. T. (2006). Anisotropic syntheses of boat-shaped core-shell Au-Ag nanocrystals and nanowires. *Nanotechnology, 17*, 2304–2310.

Yankovich, A. B., Berkels, B., Dahmen, W., Binev, P., Sanchez, S. I., Bradley, S. A., Li, A., Szlufarska, I., & Voyles, P. (2014). Picometre-precision analysis of scanning transmission electron microscopy images of platinum nanocatalysts. *Nature Communications, 5*, 4155.

Young, N. P., Li, Z. Y., Chen, Y., Palomba, S., Di Vece, M., & Palmer, R. E. (2008). Weighing supported nanoparticles: Size-selected clusters as mass standards in nanometrology. *Physical Review Letters, 101*, Article 246103.

Yu, P., Lee, J.-S., Okamoto, S., Rossell, M. D., Huijben, M., Yang, C.-H., He, Q., Zhang, J. X., Yang, S. Y., Lee, M. J., Ramasse, Q. M., Erni, R., Chu, H.-Y., Arena, D., Kao, C. C., Martin, L. W., & Ramesh, R. (2010). Interface ferromagnetism and orbital reconstruction in $BiFeO_3$-$La_{0.7}Sr_{0.3}MnO_3$ heterointerfaces. *Physical Review Letters, 105*, Article 027201.

Zanchet, D., Hall, B. D., & Ugarte, D. (2001). X-ray characterization of nanoparticles. In *Characterization of nanophase materials* (pp. 13–36). Wiley-VCH.

Zeches, R. J., Rossell, M. D., Zhang, J. X., Hatt, A. J., He, Q., Yang, C.-H., Kumar, A., Wang, C. H., Melvill, A., Adamo, C., Sheng, G., Chu, Y.-H., Ihlefeld, J. F., Erni, R., Ederer, C., Gopalan, V., Chen, L. Q., Schlom, D. G., Spaldin, N. A., ... Ramesh, R. (2009). A strain-driven morphotrophic phase boundary in $BiFeO_3$. *Science, 326*, 977–980.

Zhao, J., Zhang, J. X., Yonzon, C. R., Haes, A. J., & Van Duyne, R. P. (2006). Localized surface plasmon resonance biosensors. *Nanomedicine, 1*, 219–228.

Zhou, D., Müller-Caspary, K., Sigle, W., Krause, F. F., Rosenauer, A., & van Aken, P. A. (2016). Sample tilt effects on atom column position determination in ABF-STEM imaging. *Ultramicroscopy*, *160*, 110–117.

Zhou, J., Yang, Y., Ercius, P., & Miao, J. (2020). Atomic electron tomography in three and four dimensions. *MRS Bulletin*, *45*, 290–297.

Zhu, S.-E., Yuan, S., & Janssen, G. C. A. M. (2014). Optical transmittance of multilayer graphene. *Europhysics Letters*, *108*, Article 17007.

Zucchini, W. (2000). An introduction to model selection. *Journal of Mathematical Psychology*, *44*, 41–61.

Index

FSC
www.fsc.org
MIX
Papier | Fördert gute Waldnutzung
FSC® C083411

Kolibri 360 GmbH
Ferdinand-Jühlke-Straße 7
99095 Erfurt, Deutschland
produktsicherheit@kolibri360.de